Methods in Molecular Biology

Series Editor
John M. Walker
School of Life and Medical Sciences
University of Hertfordshire
Hatfield, Hertfordshire, AL10 9AB, UK

For further volumes:
http://www.springer.com/series/7651

Plant Genomics

Methods and Protocols

Edited by

Wolfgang Busch

Gregor Mendel Institute (GMI), Austrian Academy of Sciences,
Vienna Biocenter (VBC), Vienna, Austria

Humana Press

Editor
Wolfgang Busch
Gregor Mendel Institute (GMI)
Austrian Academy of Sciences
Vienna Biocenter (VBC)
Vienna, Austria

ISSN 1064-3745 ISSN 1940-6029 (electronic)
Methods in Molecular Biology
ISBN 978-1-4939-7001-8 ISBN 978-1-4939-7003-2 (eBook)
DOI 10.1007/978-1-4939-7003-2

Library of Congress Control Number: 2017937865

Printed on acid-free paper

This Humana Press imprint is published by Springer Nature
The registered company is Springer Science+Business Media LLC
The registered company address is: 233 Spring Street, New York, NY 10013, U.S.A.

Preface

One of the central questions of biology is how the genome of an organism encodes all the information necessary for its operation. Finding comprehensive answers to this is a monumental task. While efforts to answer this question are still in their infancy and it is not yet clear how to best approach this, there is no doubt that the problem of decoding the genome requires knowledge of the genome sequences (information), phenotypes (the final output), and the molecular processes linking the two. The term genomics is being used to classify a broad spectrum of methods and approaches currently in use to answer these questions. It is also frequently used to distinguish studies that involve multiple genes from those that are focused on a single gene.

The last few years have seen tremendous advances in multiple technical areas that have enabled unprecedented progress in genomics. There are three areas that I consider outstanding. The most obvious one is the development of the so-called next-generation sequencing. This has enabled the sequencing of whole genomes at reasonable cost and has not only allowed for sequencing the genomes of many plant species but has also allowed for the accurate determination of genotypes of large mutant collections and natural strains across multiple plants species. Moreover, these sequencing methods are being very successfully used for the sequencing-based elucidation of chromatin features and transcriptomes at a genome-wide scale as well as for a diverse set of large-scale molecular assays whose outputs are DNA sequences. The second outstanding area is related to the efficient assessment of phenotypes at a very large scale. This has been driven by an increase in throughput and accuracy in quantifying molecular phenotypes such as transcriptomes, proteins, metabolites, as well as phenotypes that relate to growth and morphology. The latter was possible through advances in high-throughput image acquisition and computer-vision-based image processing. Importantly, combined with the ever-increasing numbers of genomes available, these advances in the quantification of phenotypes have enabled the genome-wide mapping of phenotypes onto the genome, such as through genome-wide association mapping. The third area that I'd like to mention relates to methods of molecular biology. Enabled by lab automation and robotics, new highly efficient methods for molecular cloning, and the availability of cheap next-generation sequencing, genome-scale datasets of molecular interactions can now be produced. This area also includes the rapid evolution of genome-editing methods with TALENs or CRISPR/Cas9. With these tools, it has now become possible to test genetic hypotheses beyond just a few genes and even at the genome scale. In the same vein, recent progress in microscopy has allowed for the investigation of highly resolved molecular interactions in vivo, thereby significantly extending our view beyond the single gene/protein to a network based one. Overall, it is an exhilarating time to be studying biology; for the first time, we truly have the means to generate and test hypotheses at a genome-wide scale.

In this book, I have assembled protocols that revolve around these three pillars of progress, spanning genotypes, phenotypes, and the molecular processes in between. Importantly, they are not restricted to the predominant model species *Arabidopsis thaliana*, and I hope that this will encourage and facilitate other researchers to expand their research to other species. These protocols were written by leading scientists in their fields and are very much at the forefront of what is currently state of the art in plant genomics. I hope that this book will serve as an inspiration for further studies in plant genomics and will enable a widespread use of these methods.

Vienna, Austria *Wolfgang Busch*

Contents

Contributors

CHRISTOPHER ADAMCHEK • *Department of Molecular, Cellular, and Developmental Biology, Yale University, New Haven, CT, USA*
MARKUS ALBERT • *Zentrum für Molekularbiologie der Pflanzen, University Tübingen, Tübingen, Germany*
STIG U. ANDERSEN • *Department of Molecular Biology and Genetics, Aarhus University, Aarhus C, Denmark*
ELIN AXELSSON • *Gregor Mendel Institute (GMI), Austrian Academy of Sciences, Vienna Biocenter (VBC), Vienna, Austria*
ANNE-MAARIT BÅGMAN • *Department of Plant and Genome Center, Davis, CA, USA*
CLAUDE BECKER • *Department of Molecular Biology, Max Planck Institute for Developmental Biology, Tübingen, Germany; Gregor Mendel Institute, Austrian Academy of Sciences, Vienna Biocenter (VBC), Vienna, Austria*
PHILIP N. BENFEY • *Department of Biology and Howard Hughes Medical Institute, Duke University, Durham, NC, USA*
FREDERIC BERGER • *Gregor Mendel Institute (GMI), Austrian Academy of Sciences, Vienna Biocenter (VBC), Vienna, Austria*
SIOBHAN M. BRADY • *Department of Plant and Genome Center, Davis, CA, USA*
WOLFGANG BUSCH • *Gregor Mendel Institute (GMI), Austrian Academy of Sciences, Vienna Biocenter (VBC), Vienna, Austria*
MELINKA A. BUTENKO • *Department of Biosciences, Section for Genetics and Evolutionary Biology, University of Oslo, Oslo, Norway*
XIAOFEI CHENG • *Division of Plant Biology, The Samuel Roberts Noble Foundation, Ardmore, OK, USA*
ANDREA CHINI • *Department of Plant Molecular Genetics, National Centre for Biotechnology (CNB-CSIC), Madrid, Spain*
NATALIE M. CLARK • *Department of Plant and Microbial Biology, North Carolina State University, Raleigh, NC, USA; Biomathematics Graduate Program, North Carolina State University,Raleigh, NC, USA*
BRUCE CORLISS • *Department of Biomedical Engineering, University of Virginia, Charlottesville, VA, USA*
GIULIA DEVESCOVI • *Bacteriology Group, International Centre for Genetic Engineering and Biotechnology (ICGEB), Trieste, Italy*
FRIEDRICH FAUSER • *Botanical Institute II, Karlsruhe Institute of Technology, Karlsruhe, Germany*
ANN FEKE • *Department of Molecular, Cellular, and Developmental Biology, Yale University, New Haven, CT, USA*
DANIELE L. FILIAULT • *Gregor Mendel Institute (GMI), Austrian Academy of Sciences, Vienna Biocenter (VBC), Vienna, Austria*
ALLISON GAUDINIER • *Department of Plant and Genome Center, Davis, CA, USA*
JOSHUA M. GENDRON • *Department of Molecular, Cellular, and Developmental Biology, Yale University, New Haven, CT, USA*

MARCO GIOVANNETTI • *Gregor Mendel Institute (GMI), Austrian Academy of Sciences, Vienna Biocenter (VBC), Vienna, Austria*
CHRISTIAN GÖSCHL • *Gregor Mendel Institute (GMI), Austrian Academy of Sciences, Vienna Biocenter (VBC), Vienna, Austria*
STEFAN GROB • *Institute of Human Genetics, UMR9002 CNRS-UM, Montpellier, France*
JÖRG HAGMANN • *Computomics GmbH, Tübingen, Germany*
TOMOKAZU KAWASHIMA • *Gregor Mendel Institute (GMI), Austrian Academy of Sciences, Vienna Biocenter (VBC), Vienna, Austria Department of Plant and Soil Sciences, University of Kentucky, Lexington, KY, USA*
NICK KROM • *Department of Computing Services, The Samuel Roberts Noble Foundation, Ardmore, OK, USA*
CHIN-MEI LEE • *Department of Molecular, Cellular, and Developmental Biology, Yale University, New Haven, CT, USA*
JESSICA LOPEZ • *Department of Biology, University of Alabama at Birmingham, Birmingham, AL, USA*
ANNA MAŁOLEPSZY • *Department of Molecular Biology and Genetics, Aarhus University, Aarhus C, Denmark*
JULIN N. MALOOF • *Department of Plant Biology, University of California, Davis, Davis, CA, USA*
RAVI MARUTHACHALAM • *School of Biology, Indian Institute of Science Education and Research (IISER), Thiruvananthapuram, Kerala, India*
MAXWELL R. MCREYNOLDS • *Department of Plant Pathology and Microbiology, Iowa State University, Ames, IA, USA*
M. SHAHID MUKHTAR • *University of Alabama at Birmingham, Birmingham, AL, USA; Nutrition Obesity Research Center, University of Alabama at Birmingham, Birmingham, AL, USA*
TERRY MUN • *Department of Molecular Biology and Genetics, Aarhus University, Aarhus C, Denmark*
KIRANKUMAR S. MYSORE • *Division of Plant Biology, The Samuel Roberts Noble Foundation, Ardmore, OK, USA*
DMITRI A. NUSINOW • *Donald Danforth Plant Science Center, St. Louis, MO, USA*
AKIHISA OSAKABE • *Gregor Mendel Institute (GMI), Austrian Academy of Sciences, Vienna Biocenter (VBC), Vienna, Austria*
ANNALISA POLVERARI • *Laboratory of Phytopathology, Department of Biotechnology, University of Verona, Verona, Italy*
HOLGER PUCHTA • *Botanical Institute II, Karlsruhe Institute of Technology, Karlsruhe, Germany*
MARIA RITA PUTTILLI • *Laboratory of Phytopathology, Department of Biotechnology, University of Verona, Verona, Italy*
NIELS SANDAL • *Department of Molecular Biology and Genetics, Aarhus University, Aarhus C, Denmark*
SANTOSH B. SATBHAI • *Gregor Mendel Institute, Austrian Academy of Sciences, Vienna Biocenter (VBC), Vienna, Austria*
SIMON SCHIML • *Botanical Institute II, Karlsruhe Institute of Technology, Karlsruhe, Germany*
DANELLE K. SEYMOUR • *Department of Ecology and Evolutionary Biology, University of California, Irvine, Irvine, CA, USA*

RÜDIGER SIMON • *Institute for Developmental Genetics, Cluster of Excellence on Plant Sciences (CEPLAS), and Center for Advanced imaging (CAi), Heinrich Heine University, Düsseldorf, Germany*

MARC SOMSSICH • *Institute for Developmental Genetics, Heinrich Heine University, Düsseldorf, Germany; School of Biosciences, University of Melbourne, Melbourne, VIC, Australia*

GAOYUAN SONG • *Department of Plant Pathology and Microbiology, Iowa State University, Ames, IA, USA*

ROSANGELA SOZZANI • *Department of Plant and Microbial Biology, North Carolina State University, Raleigh, NC, USA; Biomathematics Graduate Program, North Carolina State University, Raleigh, NC, USA*

ERIN E. SPARKS • *Department of Biology and Howard Hughes Medical Institute, Duke University, Durham, NC, USA*

DELYANA STOEVA • *Gregor Mendel Institute, Austrian Academy of Sciences, Vienna Biocenter (VBC), Vienna, Austria*

JENS STOUGAARD • *Department of Molecular Biology and Genetics, Aarhus University, Aarhus C, Denmark*

MICHELLE TANG • *Department of Plant and Genome Center, Davis, CA, USA*

MICHAEL UDVARDI • *Division of Plant Biology, The Samuel Roberts Noble Foundation, Ardmore, OK, USA*

ELODIE VANDELLE • *Laboratory of Phytopathology, Department of Biotechnology, University of Verona, Verona, Italy*

VITTORIO VENTURI • *Bacteriology Group, International Centre for Genetic Engineering and Biotechnology (ICGEB), Trieste, Italy*

JUSTIN W. WALLEY • *Department of Plant Pathology and Microbiology, Iowa State University, Ames, IA, USA*

JIANGQI WEN • *Division of Plant Biology, The Samuel Roberts Noble Foundation, Ardmore, OK, USA*

MARI WILDHAGEN • *Department of Biosciences, Section for Genetics and Evolutionary Biology, University of Oslo, Oslo, Norway*

RAMESH YELAGANDULA • *Gregor Mendel Institute (GMI), Austrian Academy of Sciences, Vienna Biocenter (VBC), Vienna, Austria; Institute of Molecular Biotechnology of the Austrian Academy of Sciences (IMBA), Vienna Biocenter (VBC), Vienna, Austria*

SHULAN ZHANG • *Division of Plant Biology, The Samuel Roberts Noble Foundation, Ardmore, OK, USA*

Part I

Genotypes

Chapter 1

CRISPR/Cas-Mediated *In Planta* Gene Targeting

Simon Schiml, Friedrich Fauser, and Holger Puchta

Abstract

The recent emergence of the CRISPR/Cas system has boosted the possibilities for precise genome engineering approaches throughout all kingdoms of life. The most common application for plants is targeted mutagenesis, whereby a Cas9-mediated DNA double-strand break (DSB) is repaired by mutagenic nonhomologous end joining (NHEJ). However, the site-specific alteration of a genomic sequence or integration of a transgene relies on the precise repair by homologous recombination (HR) using a suitable donor sequence: this poses a particular challenge in plants, as NHEJ is the preferred repair mechanism for DSBs in somatic tissue. Here, we describe our recently developed *in planta* gene targeting (ipGT) system, which works via the induction of DSBs by Cas9 to activate the target and the targeting vector at the same time, making it independent of high transformation efficiencies.

Key words Gene technology, Genome engineering, Double-strand break repair, Engineered nucleases, Cas9

1 Introduction

Modern methods for genome engineering in plants, but also in other eukaryotes, rely on the targeted induction of a DSB into the DNA. Thus, natural DSB repair mechanisms can be stimulated and exploited to achieve a desired outcome. Basically, DSBs in somatic plant tissues can be repaired via two distinctive pathways [1]. The major pathway is marked by NHEJ, involving processing of the DSB ends followed by a ligation reaction. Due to this processing, NHEJ generally incorporates small insertions or deletions into the genomic sequence, thus potentially generating a frameshift in an open reading frame. This approach is therefore referred to as targeted mutagenesis. The second pathway is homologous recombination, where a homologous donor sequence can be utilized as repair template for an error-free repair [2]. If an ectopic sequence is offered, usually termed donor sequence, that is, homologous or a transgene flanked by homologies, the respective sequence can be inserted into the repaired site, hence changing its information in a predefined manner.

Wolfgang Busch (ed.), *Plant Genomics: Methods and Protocols*, Methods in Molecular Biology, vol. 1610,
DOI 10.1007/978-1-4939-7003-2_1, © Springer Science+Business Media LLC 2017

This experimental approach is termed gene targeting (GT) and can be used to perform amino acid exchanges or to guide a transgene to a desired position within the genome [3].

Widespread use of both targeted mutagenesis and gene targeting has become possible through the development of programmable nucleases which enable the induction of a precise DSB at a desired position in the genome [4]. The most recent yet most versatile class of programmable nucleases is formed by the bacterial clustered regularly interspaced short palindromic repeats (CRISPR)/CRISPR-associated (Cas) system. Originating from being a bacterial adaptive immune system, CRISPR/Cas was adapted as a programmable nuclease, which is composed of two components [5]. The small so-called sgRNA binds directly to its complementary sequence on the target DNA, next to the protospacer adjacent motif (PAM), usually "NGG." The endonuclease Cas9 then cleaves the DNA within the bound segment, 3 bp away from the PAM. Owing to this simple, yet highly efficient architecture together with its applicability to a vast range of organisms, RNA-guided Cas9 has rapidly become the most important tool for targeted genome engineering [6–9].

Achieving GT in plants is to date still challenging, as it relies on the rarely occurring repair via HR and therefore depends on highly efficient DSB induction together with the constant availability of a donor. Nevertheless, gene targeting efficiencies in plants are only in the percent range. As this requires a bigger number of transformation events, GT is hardly achievable for plant species with low transformation efficiencies. Within recent years, we were able to establish the efficient ipGT system in the model plant *Arabidopsis thaliana* that is independent of transformation efficiencies or the use of a mutant background to enhance GT [10, 11]. In this approach, a T-DNA is stably integrated that contains an expression cassette for a nuclease as well as the donor sequence, consisting of the desired transgene flanked by homologies to the desired target locus. Upon expression of the nuclease inside the plant cells, the donor sequence is excised via two DSBs and a third DSB is induced in the target locus, thus activating it for HR (Fig. 1). By using the flanking homologies, the donor sequence can then be integrated into the repaired locus. Since in plants the germ line is developed out of somatic tissue, the GT event can become heritable, thus creating offspring that stably carries the GT event.

Here, we describe the procedure to perform the *Streptococcus pyogenes* Cas9-mediated ipGT approach in *A. thaliana*, enabling stably heritable targeting of a transgene to a desired locus, for example, to specifically tag an endogenous gene, or the predefined alteration of a genomic sequence, to achieve amino acid substitutions.

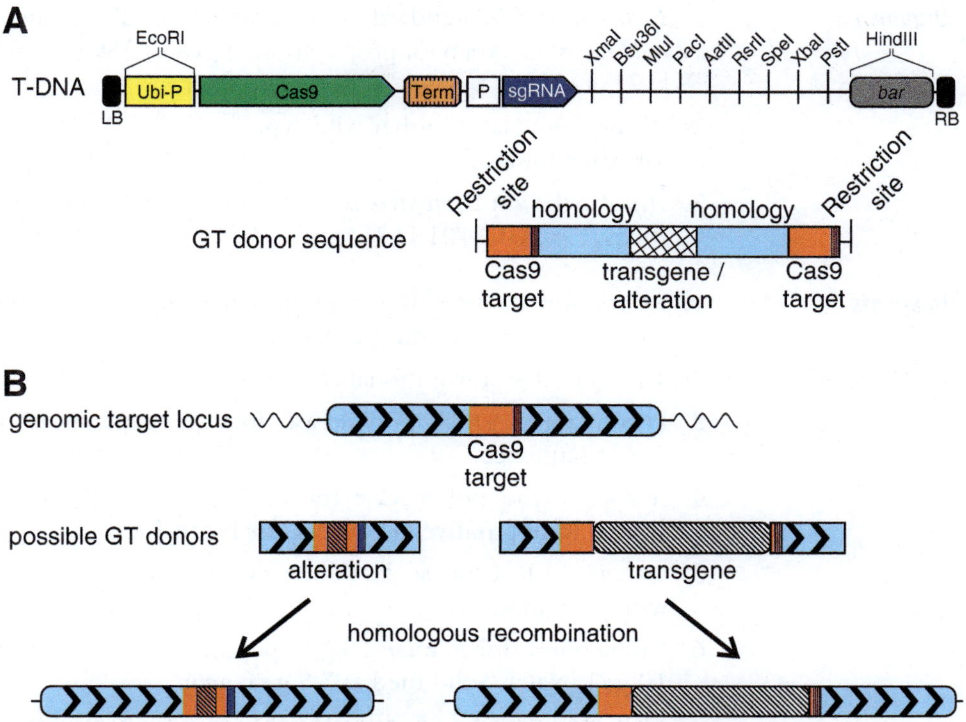

Fig. 1 Outline of the ipGT approach. (**a**) T-DNA construct. The Cas9 sequence is controlled by a constitutive ubiquitin promoter, which is exchangeable with EcoRI. The GT donor sequence can be cloned into the depicted restriction sites in the MCS. The general structure of the donor consists of the desired sequence flanked by homologies to the target locus, the Cas9 target sites, and the required restriction sites. The PPT resistance (*bar*) can be exchanged using HindIII. (**b**) Upon expression of the nuclease, a DSB is induced in the target locus and two DSBs in the T-DNA release the donor sequence. The latter should contain specific sequence alterations – silent mutation in case single amino acid changes will be targeted in an ORF to avoid Cas9 cutting. Alternatively, the cutting site may be replaced by foreign sequence flanked by the required homologies to ensure integration into the genomic site

2 Materials

2.1 Plasmids

All plasmids are available directly from the authors or through the Arabidopsis Biological Resource Center (ABRC). Full sequence information is deposited at www.botanik.kit.edu/crispr.

1. pDe-CAS9 (ABRC CD3-1928)

   Binary vector for stable transformation into plants via *A. tumefaciens*. Contains a constitutive Cas9 expression system, along with a Gateway destination sequence with a *ccdB* gene to take up the sgRNA expression sequence. Confers plant resistance to phosphinothricin (PPT). This plasmid also serves as the vector for the desired GT donor sequence.

2. pEn-Chimera (ABRC CD3-1931)

   Ampicillin-resistant Gateway entry vector, containing the sgRNA expression cassette, flanked by attL1/2 sites.

2.2 Organisms

1. *Escherichia coli*, standard cloning strain for all cloning steps; *ccdB*-resistant strain for propagation of pDe-CAS9 (e.g., DB3.1 [12], *see* **Note 1**).
2. *A. thaliana* plants, either wild-type (e.g., Col-0) or any transformable mutant.
3. *Agrobacterium tumefaciens*, any conventional transformation strain, e.g., GV3101 [13].

2.3 Reagents

1. Restriction enzyme BbsI, additional restriction enzymes as required (*see* Subheading 3, Fig. 1 and **Note 2**).
2. T4 Ligase for conventional cloning steps.
3. Proofreading DNA polymerase for the generation of the donor sequence.
4. A robust Taq polymerase for *E. coli* colony PCRs and for screening of putative GT plants (*see* **Note 3**).
5. Gateway LR Clonase II (ThermoFisher Scientific, supplied with proteinase K).
6. LB medium (for *E. coli)*: 10 g/L peptone, 5 g/L yeast extract, 10 g/L NaCl. Solid media: 7.5 g/L agar.
7. YEB medium (for *A. tumefaciens)*: 5 g/L beef extract and 5 g/L peptone. 1 g/L yeast extract, 5 g/L sucrose, 439 mg/L $MgSO_4$, and 7.5 g/L agar for solid media.
8. Germination medium (GM): 4.9 g/L Murashige & Skoog, 10 g/L sucrose, pH 5.7, and 8 g/L agar. For selection media, add ampicillin (100 mg/L), spectinomycin (100 mg/L), or PPT (6 mg/L).
9. TE buffer: 10 mM Tris–HCl and 1 mM EDTA at pH 8.

3 Methods

3.1 Experimental Design: Site-Specific Integration of a Foreign Sequence

1. After selecting the desired target locus for your transgene sequence, identify potential Cas9 target sites (*see* **Note 4**).
2. Add at least 400 bp upstream and downstream from the Cas9 cutting site to the desired transgene, therefore defining the exact position and orientation (*see* **Notes 5** and **6**).
3. For correct excision of the donor sequence, add the Cas9 target site including the PAM to the proximal and distal end of the donor sequence (*see* **Note 7**).

3.2 Specific Sequence Alteration

1. Select a Cas9 target site close to the site to be changed.
2. Choose the flanking homologies as described above; a total size of 0.8–1.5 kb is recommended (0.4–0.8 kb on either side).

3. Additionally to the desired base exchange(s), it is crucial to introduce silent mutations to abolish any Cas9 activity in your donor sequence (*see* **Note 8**). This should also enable the detection of the GT event by PCR, if a primer is used that can bind to the altered sequence but not to the genomic site.
4. As described above (*see* Subheading 3.1, **step 3**), flank the donor sequence by the correct Cas9 target sites (including the PAM) to enable its excision from the T-DNA (*see* **Note 7**).
5. Finally, add restriction sites to the donor sequence to enable cloning into the T-DNA construct. Fig. 1 depicts the T-DNA construct and the available restriction sites in the MCS (*see* **Note 2**).
6. The construct itself can be assembled by overlap extension PCR, Gibson assembly (generate fragments with a potent proofreading polymerase), or via gene synthesis (*see* **Note 9**).

3.3 Cloning of the T-DNA Construct for ipGT

1. Order oligonucleotides for your Cas9 target sequence. For an NGG PAM, the fw oligo should contain the 20 nt upstream of the PAM with ATTG added to the 5′ end. The second oligo should contain the reverse complement of the target sequence with AAAC added 5′.
2. Dilute and mix your oligos in ddH_2O to a final concentration of 2 pmol/μL for each oligo in a total volume of 50 μL. Incubate for 5 min at 95 °C and put at room temperature for an additional 10 min for annealing.
3. Digest pEn-Chimera with BbsI as recommended by the supplier for at least 2 h. Purify the reaction and dilute the final concentration to 5 ng/μL.
4. Perform a ligation reaction with 2 μL digested vector, 3 μL prepared oligos, 1 μL T4 ligase buffer, 1 μL T4 ligase, and 3 μL ddH_2O, and incubate as recommended by the vendor. Transform into *E. coli* and select for colonies on ampicillin-containing LB plates.
5. Set up a colony PCR as recommended by the vendor of the Taq polymerase to identify positive colonies, using your fw oligo and M13 rev as primers, which generate a band at approx. 370 bp.
6. Purify plasmids from a small number (1–4) of correct clones. Validate by sequencing with primer SS42.
7. Transfer the correct sgRNA expression sequence to pDe-CAS9 by Gateway cloning. Set up a reaction with 100 ng entry vector, 300 ng destination vector (pDe-CAS9), 4 μL TE buffer, and 1 μL LR Clonase II in a total volume of 10 μL, and incubate for 2–3 h at room temperature.
8. Stop the reaction by adding 1 μL Proteinase K for 10 min at 37 °C (crucial step).

9. Transform the whole reaction into *E. coli* (*see* **Note 1**), and select on spectinomycin-containing LB plates.
10. Check for correct clones by colony PCR with primers SS42/SS43, producing a 1 kb band.
11. Isolate correct plasmids. A control restriction digestion is possible with AflII and NheI, producing bands at approx. 5.9, 5, and 3.8 kb.
12. Add your GT donor sequence by conventional cloning. If one restriction enzyme is used, make sure to dephosphorylate the vector backbone prior to ligation (*see* **Note 10**).
13. Transform into *E. coli* and grow on spectinomycin-containing LB plates. Identify and verify correct clones by a suitable colony PCR, restriction digestion, and sequencing.
14. Transform your final plasmid into *A. tumefaciens* and further into *A. thaliana* (e.g., by floral dipping [14]).

3.4 Identification of GT Plants

1. Select primary transformant plants by sowing seeds from transformed plants on germination medium containing PPT.
2. Pick at least 40 plants for further cultivation in soil (*see* **Note 11**).
3. Qualitative control of the nuclease activity at this stage is optional. This can be done by T7 endonuclease assay, restriction digestion assay, or high-resolution melting analysis.
4. Isolate DNA from single leaves of the primary transformants with a fast extraction protocol [15]. Set up a PCR as depicted in Fig. 2, using one primer outside of the homologous region and a corresponding primer within the transgene or specific for the defined sequence alteration. That way, only the correct GT event should produce a band (*see* **Note 12**).
5. Cultivate plants in soil for progeny seeds.
6. Test the T2 lines for single-locus integration of the T-DNA by sowing a small amount (~50–100) of progeny seeds from each T1 plant on PPT selection medium, and verify a correct Mendelian segregation pattern after 10–14 d (75% germination, representing homozygous and heterozygous plants with respect to the T-DNA).
7. For ten or more correctly segregating T2 lines, sow at least 100 seeds without applying a selection marker (*see* **Note 13**). Isolate DNA from each plant and check again for the GT event with PCR. Cultivate candidate plants in soil individually to obtain progeny seeds.
8. In T3, check for absence of the T-DNA by sowing a small amount of seeds on selection medium. Confirm the presence of the GT event to assure stable inheritance (*see* **Note 14**).

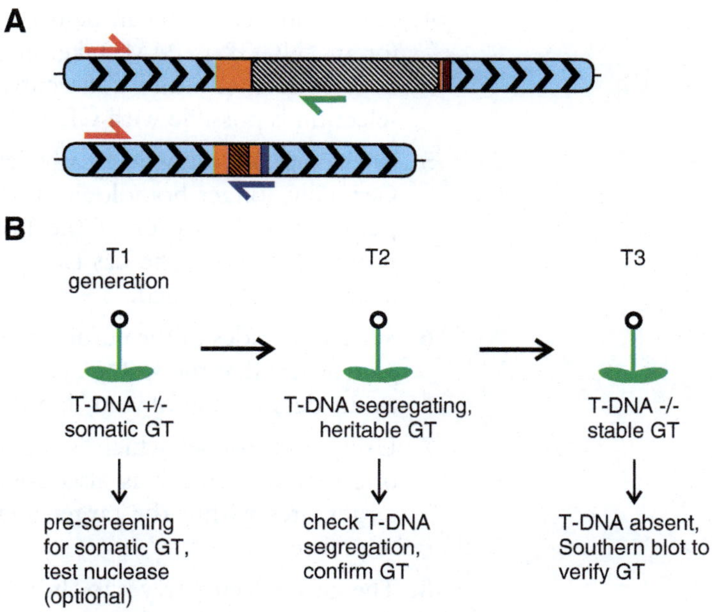

Fig. 2 Identification of GT plants. (**a**) PCR-based identification. Primers should be placed outside the homologous donor flanks and inside the transgene or the altered sequence segment (*see* **Note 12**). (**b**) Simplified procedure for stable GT plants. In T1, confirm functionality of the GT approach by PCR-based prescreening as depicted above. Correct segregation of the T-DNA has to be confirmed in T2, along with further PCRs for the GT event. In absence of the T-DNA in T3 (plants not germinating on selection marker), any positive GT PCR indicates stable inheritance. The procedures in T2/T3 may be repeated in later generations until the stable event is obtained

9. To assure the correct, two-sided GT (i.e., both flanks were correctly integrated without the occurrence of NHEJ) and physical linkage, verification with a Southern blot is highly recommended, exploiting the introduction (or destruction) of novel restriction sites along with the GT event.

4 Notes

1. All cloning steps require a conventional *E. coli* strain, e.g., DH5α. The *ccdB*-resistant strain is only used to amplify pDe-CAS9 in advance to any cloning steps described here.
2. Choose restriction enzymes for the cloning according to Fig. 1, avoiding restriction sites that are already present in the donor construct design.
3. We recommend DreamTaq polymerase (ThermoFisher Scientific) for colony and plant screening PCRs.

4. Target site selection can be done manually simply by looking for an “NGG” as PAM; take the 20 bp upstream by avoiding to have five or more consecutive T in it. Software-assisted selection is possible with CRISPR-P or CCTop [16, 17].
5. The actual size can vary greatly depending on your experiment. Generally, longer homologies (0.8–1 kb) should improve HR frequencies. However, if the homology contains parts of a promoter region, one has to be aware of potential expression from the T-DNA itself.
6. Since both sides of the cutting site are included in the homologies, assure that the sgRNA cannot bind there, as it would lead to the degradation of your donor sequence.
7. Using a vector set which is capable of expressing more than one sgRNA [10], it is also possible to have different Cas9 target sites within the target locus and to release the donor sequence.
8. The most effective way is to alter the PAM and the seed region of the target site [18]. Note that for *S. pyogenes* Cas9, NAG is also reported to be a functional PAM sequence [19]. Furthermore, if you plan to confirm your GT experiment with a Southern blot, consider also degrading a restriction site in the donor sequence that is present in the genomic sequence or vice versa. CRISPR-P illustrates restriction sites in a chosen target site [17].
9. Having the donor sequence synthesized is the easiest, yet most expensive method. However, overlap PCR or Gibson assembly may be challenging due to reoccurring sequence elements flanking the construct.
10. Since the orientation of the donor construct within the T-DNA is arbitrary, cloning with one restriction enzyme is generally sufficient. With two enzymes, however, no dephosphorylation is required.
11. All numbers given in this paragraph are based on experience and may be scaled up to actually detect the GT event.
12. Assure the functionality of the PCR: primers may be tested individually with genomic DNA and the T-DNA construct with suitable corresponding primers. Also, consider performing a similar PCR for the downstream flank of your GT event, to exclude false positives such as one-sided GT events. Note that the GT may be a rare event, so a high number of PCR cycles (>40) is required.
13. Preferably choose lines that were already tested positive for the GT in T1, as these indicate a functional integration site of the T-DNA.
14. The described processes can be repeated in following generations if necessary.

Acknowledgment

We thank Amy Whitbread for the critical reading of the manuscript. Our work on Cas9-mediated genome engineering and GT was funded by the European Research Council (Advanced Grant "COMREC") as well as the Federal Ministry of Education and Research (PLANT 2030, Pflanzenbiotechnologie fur die Zukunft – TAMOCRO, Grant 0315948).

References

1. Puchta H (2005) The repair of double-strand breaks in plants: mechanisms and consequences for genome evolution. J Exp Bot 56(409):1–14. doi:10.1093/jxb/eri025
2. Puchta H, Dujon B, Hohn B (1996) Two different but related mechanisms are used in plants for the repair of genomic double-strand breaks by homologous recombination. Proc Natl Acad Sci U S A 93(10):5055–5060
3. Steinert J, Schiml S, Puchta H (2016) Homology-based double-strand break-induced genome engineering in plants. Plant Cell Rep 35(7):1429–1438. doi:10.1007/s00299-016-1981-3
4. Puchta H, Fauser F (2014) Synthetic nucleases for genome engineering in plants: prospects for a bright future. Plant J 78(5):727–741. doi:10.1111/tpj.12338
5. Jinek M, Chylinski K, Fonfara I et al (2012) A programmable dual-RNA-guided DNA endonuclease in adaptive bacterial immunity. Science 337(6096):816–821. doi:10.1126/science.1225829
6. Schiml S, Puchta H (2016) Revolutionizing plant biology: multiple ways of genome engineering by CRISPR/Cas. Plant Methods 12:8. doi:10.1186/s13007-016-0103-0
7. Hsu PD, Lander ES, Zhang F (2014) Development and applications of CRISPR-Cas9 for genome engineering. Cell 157(6):1262–1278. doi:10.1016/j.cell.2014.05.010
8. Mali P, Yang L, Esvelt KM et al (2013) RNA-guided human genome engineering via Cas9. Science 339(6121):823–826. doi:10.1126/science.1232033
9. Cong L, Ran FA, Cox D et al (2013) Multiplex genome engineering using CRISPR/Cas systems. Science 339(6121):819–823. doi:10.1126/science.1231143
10. Schiml S, Fauser F, Puchta H (2014) The CRISPR/Cas system can be used as nuclease for in planta gene targeting and as paired nickases for directed mutagenesis in Arabidopsis resulting in heritable progeny. Plant J 80(6):1139–1150. doi:10.1111/tpj.12704
11. Fauser F, Roth N, Pacher M et al (2012) In planta gene targeting. Proc Natl Acad Sci U S A 109(19):7535–7540. doi:10.1073/pnas.1202191109
12. Bernard P, Couturier M (1992) Cell killing by the F plasmid CcdB protein involves poisoning of DNA-topoisomerase II complexes. J Mol Biol 226(3):735–745
13. Koncz C, Kreuzaler F, Kalman Z et al (1984) A simple method to transfer, integrate and study expression of foreign genes, such as chicken ovalbumin and alpha-actin in plant tumors. EMBO J 3(5):1029–1037
14. Clough SJ, Bent AF (1998) Floral dip: a simplified method for Agrobacterium-mediated transformation of *Arabidopsis thaliana*. Plant J 16(6):735–743
15. Zhang J, Stewart JM (2000) Economical and rapid method for extracting cotton genomic DNA. J Cotton Sci 4:193–201
16. Stemmer M, Thumberger T, Del Sol Keyer M et al (2015) CCTop: an intuitive, flexible and reliable crispr/cas9 target prediction tool. PLoS One 10(4):e0124633. doi:10.1371/journal.pone.0124633
17. Lei Y, Lu L, Liu H et al (2014) CRISPR-P: a web tool for synthetic single-guide RNA design of CRISPR-system in plants. Mol Plant 7(9):1494–1496. doi:10.1093/mp/ssu044
18. Jiang W, Bikard D, Cox D et al (2013) RNA-guided editing of bacterial genomes using CRISPR-Cas systems. Nat Biotechnol 31(3):233–239. doi:10.1038/nbt.2508
19. Hsu PD, Scott DA, Weinstein JA et al (2013) DNA targeting specificity of RNA-guided Cas9 nucleases. Nat Biotechnol 31(9):827–832. doi:10.1038/nbt.2647

Chapter 2

User Guide for the *LORE1* Insertion Mutant Resource

Terry Mun, Anna Małolepszy, Niels Sandal, Jens Stougaard, and Stig U. Andersen

Abstract

Lotus japonicus is a model legume used in the study of plant-microbe interactions, especially in the field of biological nitrogen fixation due to its ability to enter into a symbiotic relationship with a soil bacterium, *Mesorhizobium loti*. The *LORE1* mutant population is a valuable resource for reverse genetics in *L. japonicus* due to its non-transgenic nature, high tagging efficiency, and low copy count. Here, we outline a workflow for identifying, ordering, and establishing homozygous *LORE1* mutant lines for a gene of interest, *LjFls2*, including protocols for growth and genotyping of a segregating *LORE1* population.

Key words *Lotus japonicus*, *LORE1*, Reverse genetics, Mutagenesis, Genotyping

1 Introduction

Lotus japonicus is a well-characterized model legume [1] that is widely used in the study of biological nitrogen fixation when entering a symbiotic relationship with its compatible symbiont *Mesorhizobium loti* [2]. The published genome sequence of *L. japonicus* [3], combined with the public release of *Lotus* Base, a central information portal for the model legume [4], enables researchers to tap into the wealth of genomics and expression data from *Lotus*. Additional proteomic data from *Lotus* are available separately [5].

Since the discovery of mobile genetic elements in maize [6], their mutagenic nature has been widely utilized for large-scale mutagenesis in various model plants, such as *Tnt1* in *Medicago truncatula* [7] and *Tos17* in rice [8]. The endogenous *Lotus* retrotransposon element 1 (*LORE1*) was first discovered in a nodulation mutant, *Nin* [9]. Its subsequent successful derepression in tissue culture [10] culminated in the establishment of large mutant populations, comprising more than 134,000 mutant lines and 640,000 annotated insertions [11–13]. The non-transgenic nature of the *LORE1* element, its low copy number, and its high tagging efficiency posit *LORE1* as a valuable resource in forward and reverse

Wolfgang Busch (ed.), *Plant Genomics: Methods and Protocols*, Methods in Molecular Biology, vol. 1610, DOI 10.1007/978-1-4939-7003-2_2, © Springer Science+Business Media LLC 2017

genetic studies in *L. japonicus* [12]. The *LORE1* resource has been used in various forward [11, 12] and reverse genetics studies [14–18], but applications in the former are considered beyond the scope of this chapter and will not be discussed further.

Here, we describe the complete workflow of a typical researcher aiming to generate homozygous *LORE1* mutants for the purpose of downstream characterization and genetic studies of a gene of interest. We have selected a gene encoding the putative *Lotus* ortholog of the flagellin receptor, *FLS2* from *Arabidopsis thaliana* (AT5G46330), as a candidate. In this workflow, a researcher will be guided through the procedure for identification of *Lotus* orthologs/homologs for genes of interest by (1) searching for exonic *LORE1* insertions in the candidate gene; (2) ordering the *LORE1* mutants of interest; (3) germinating, growing, and genotyping a segregating *LORE1* F0 population; and finally (4) selecting and setting up homozygous *LORE1* mutants for seed production.

2 Materials

Prepare all solutions using ultrapure water and analytical grade reagents. Prepare and store all reagents at room temperature and away from direct sunlight unless otherwise stated.

2.1 Seed Cleaning, Scarification, and Germination

1. Sandpaper.
2. Ceramic mortar.
3. 1% (v/v) hypochlorite solution.
4. Conserve (Dow Agroscience, Denmark).
5. Sterile forceps.
6. Greiner square petri dishes (120 × 120 × 10 mm; Sigma-Aldrich, Denmark).
7. Sterile aluminum foil (cut to 30 × 120 mm).
8. Parafilm M (Bemis Company Inc., USA).
9. Growth chamber or room, with the ability to regulate day/night cycle, light intensity, temperature, and humidity.
10. UV light.
11. Simple household blender.
12. 1 mm and 2 mm Metal gauzes.
13. Wax paper bag.

2.2 Plant Growth

1. Solution A: 291.4 g/L $CaCl_2.2H_2O$.
2. Solution B: 68.5 g/L KH_2PO_4, 113.4 g/L K_2HPO_4.
3. Solution C: 4.9 g/L ferric citrate.

4. Solution D: 123.3 g/L $MgSO_4 \cdot 7H_2O$, 87.0 g/L K_2SO_4, 0.338 g/L $MnSO_4 \cdot 7H_2O$, 0.247 g/L H_3BO_4, 0.288 g/L $ZnSO_4 \cdot 7H_2O$, 0.100 g/L $CuSO_4 \cdot 5H_2O$, 0.056 g/L $CoSO_4 \cdot 7H_2O$, 0.048 g/L $Na_2MoO_4 \cdot 2H_2O$.
5. 1/4 B&D medium: Per 1 L 1.4% (w/v) of Agar Noble, add 125 μL of solutions A, B, C, and D (in that order) [19]. Optional nitrate supplemented is achieved with 1 mM KNO_3 (*see* **Note 1**).

2.3 DNA Extraction Components

1. Tungsten carbide beads.
2. Tissuelyzer (QIAGEN, Denmark).
3. Isopropanol.
4. 70% ethanol.
5. Nanodrop (Thermo Fisher Scientific, USA).
6. Chloroform:isoamyl alcohol 24:1.
7. Rapid DNA extraction buffer: 200 mM Tris–HCl adjusted to pH 7.5, 250 mM NaCl, 25 mM 0.5 M EDTA adjusted to pH 8.0, and 0.5% (w/v) sodium dodecyl sulfate.
8. CTAB DNA extraction buffer: 2% (w/v) CTAB, 0.1 mM Tris–HCl adjusted to pH 8.0, 1.4 M NaCl, and 20 mM EDTA adjusted to pH 8.0. Add 0.5% (v/v) of β-mercaptoethanol immediately before use.
9. TE buffer: 10 mM Tris–HCl adjusted to pH 7.5 and 1 mM EDTA adjusted to pH 8.0.

2.4 Genotyping PCR

1. λ DNA (Fermentas, Germany).
2. PstI restriction enzyme (Fermentas, Germany).
3. Gel visualization equipment.
4. Genotyping PCR master mix (per reaction):
   2 μL of each forward and reverse primers (2.5 μM).
   0.1 μL of 20 μM dNTP.
   2.0 μL of manufacturer-supplied 10× reaction buffer.
   0.1 μL of Taq polymerase.

The master mix should be topped up to a total of 15 μL with ultrapure water. Each PCR reaction comprises 15 μL of master mix and 5 μL of extracted DNA.

2.5 Gel Electrophoresis Components

1. 5× loading buffer: 25% (v/v) glycerol, 0.8% (w/v) bromophenol blue, and 0.8% (w/v) xylene cyanol.
2. TAE buffer: 4.84 g/L Tris, 10% (v/v) 0.5 M EDTA adjusted to pH 8.0, and 5.71% (v/v) glacial acetic acid.
3. DNA ladder: Add 333 μL of 0.3 mg/mL λ DNA to 40 μL of 10× PstI enzyme buffer. Add 5 μL of 40 unit μL^{-1} of PstI. Top up the final mixture to a total volume of 400 μL, and incubate mixture overnight at 37 °C. Add 100 μL of 5× loading buffer before storing in −20 °C.

3 Methods

All wet lab procedures are performed at room temperature unless otherwise stated. In this section, we outline the workflow of a researcher interested in generating homozygous *LORE1* mutants of a gene of interest, in this case a putative *Lotus* ortholog of the *AtFLS2* gene.

3.1 Identification and BLAST Search for the Lotus Ortholog(s) of AtFLS2

1. Retrieve the amino acid sequence of *AtFLS2* (AT5g46330) from Araport [20]. The sequence is available from https://apps.araport.org/thalemine/report.do?id=1097852.
2. Search the retrieved sequence against the *L. japonicus* MG20 v3.0 protein database on *Lotus* BLAST (https://lotus.au.dk/blast/).
3. Retrieve the amino sequence of the top candidate (Lj4g3v0281040.1, *LjFls2*) from the SeqRet tool on *Lotus* Base (https://lotus.au.dk/tools/seqret), and BLAST against *Arabidopsis* TAIR protein database to validate the orthologous relationship.

3.2 Search for LORE1 Lines with Exonic LORE1 Insertions in LjFls2

1. Search for all *LORE1* mutant lines that contain genic insertions in *LjFls2* either by (1) using the TREX tool on *Lotus* Base (https://lotus.au.dk/tools/trex), selecting version 3.0 as the genome to be searched against, and then selecting "*LORE1* lines" in the drop-down contextual menu when hovering over the gene name on the results page, or (2) using the *LORE1* search page (https://lotus.au.dk/lore1/search), selecting version 3.0 as the reference genome and using "Lj4g3v0281040" as the gene ID in the filtering option. You should be presented with 45 mutant lines.
2. Select the *LORE1* lines of interest containing exonic insertions in *LjFls2* for further study (*see* **Note 2**).
3. Download the results by exporting a CSV file from the "download options" at the top of the page. You may download the entire search or check specific rows on the results page. The CSV file will contain other useful metadata for each insertion, such as the forward and reverse primer sequences used for genotyping (*see* Subheading 3.7).

3.3 Order LORE1 Lines of Interest

1. If required, apply for the necessary phytosanitary certificate(s) for the destination country by contacting the *Lotus* Base team (https://lotus.au.dk/meta/contact). The person placing the order shall bear the cost of said certificate(s). Should a phytosanitary certificate be required, the *LORE1* seeds shipment will only be dispatched when the relevant authorities have issued the certificate.

2. Place an order for *LORE1* lines of interest at the *LORE1* order page (https://lotus.au.dk/lore1/order). Currently, *LORE1* seeds are being shipped without the need to sign any material transfer agreements. The shipping time will usually be between 2 and 6 weeks depending on the geographical location of the recipient.

3.4 Seed Scarification and Germination

1. Scarify 16–20 *LORE1* seeds by abrasion of seed coats with sandpaper in a mortar until superficial layers of the seed coat have been removed and the seeds turn a lighter shade of brown.
2. Shake scarified seeds for 10 min at room temperature on a rotary shaker at 200 rpm in the presence of 1% (v/v) hypochlorite solution. Wash seeds four times with sterile water.
3. Stratify seeds by overnight incubation at 4 °C in 0.075% (v/v) Conserve (Dow Agroscience, Denmark, *see* **Note 3**) in the dark (*see* **Note 4**).
4. Germinate seeds on sterile filter paper wetted with 0.075% (v/v) Conserve under a 16 h/8 h day/night regime at 21 °C, with a light intensity of 200 μmol m^{-2} s^{-1} at the plant level, for at least three days [1].

3.5 Lotus Growth

1. Remove seed coat with a pair of sterile forceps (*see* **Note 5**).
2. Transfer seedlings onto square petri dishes (measuring 120 × 120 × 10 mm) poured with 50 mL of plant growth medium covered by a layer of wet sterile filter paper. The medium should be poured slanted and allowed to set as such, by propping up one end of the plate by around 8 mm. Shield roots from light by (1) placing a sterile wedge, made from a 30 × 120 mm aluminum foil folded into half lengthwise, onto the roots, and (2) wrapping upright petri dishes with opaque materials—black cupboard paper or aluminum foil—up till the height of the sterile aluminum wedge. Close the square petri dishes with parafilm tape (*see* **Note 6**). Incubate upright petri dishes at 21 °C under a 16 h/8 h day/night regime with a light intensity of 200 μmol m^{-2} s^{-1} in a growth chamber for the entire growth period. The simplified stepwise setup of plates for plant growth is summarized in Fig. 1.

3.6 DNA Extraction from Lotus Plants

Depending on the required quality of the DNA preparations, either Subheading 3.6.1 or 3.6.2 is performed.

3.6.1 Rapid DNA Extraction from Lotus Plants

Perform all centrifugation steps at 4 °C. This section describes a quick and dirty protocol used for DNA extraction. If a purer sample is desired, i.e., for sequencing, the CTAB protocol is recommended (*see* Subheading 3.6.2).

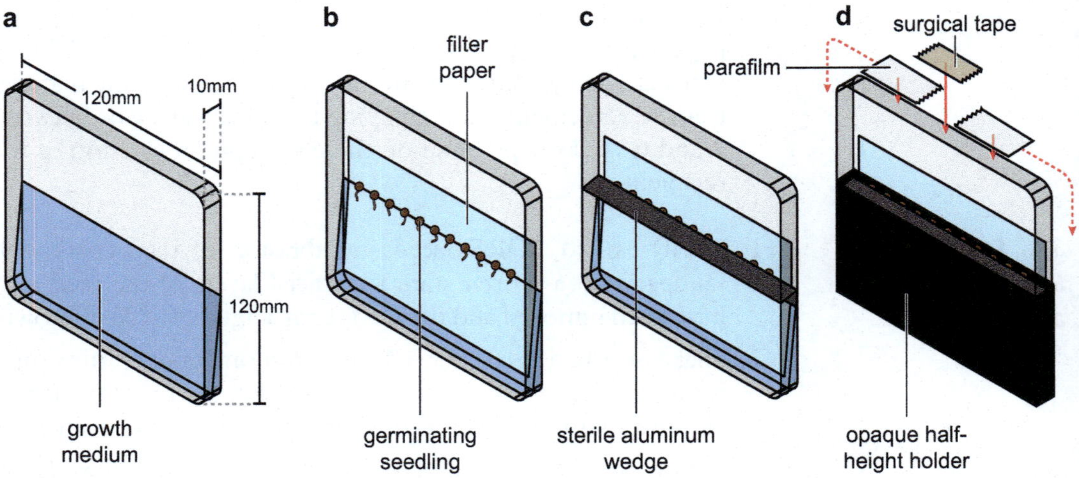

Fig. 1 The stepwise setup of the slanted plate system for growing Lotus plants: (**a**) 50 mL plant growth media is poured slated onto a square petri dish and allowed to set; (**b**) a maximum of ten germinating seedlings are transferred onto the medium and are equally spaced apart; (**c**) a sterile aluminum wedge is placed on top of the seedlings to shield roots from light above; and (**d**) the plates are wrapped in opaque paper or aluminum foil to cover roots from lateral light sources. Objects in the diagram are not drawn to scale. Broken red lines indicate stretching of parafilm around perimeter of plate to achieve a seal. Solid red lines indicate adhesion to plate perimeter

1. Remove a single trefoil from a plant, and place it in an Eppendorf tube (EPT) or collection microtube (CMT), depending on the number of samples that have to be processed (*see* **Note 7**). Harvesting of plant material from *Lotus* plants can be performed as soon as the first or second trefoil appears.
2. Add 400 μL DNA extraction buffer and a single tungsten carbide bead to each sample, and grind samples for 2 × 3 min at 30 Hz in a Tissuelyzer.
3. Centrifuge lysed samples for 10 min at 13,200 × *g* (EPT) or 40 min at 2700 × *g* (CMT).
4. Transfer 300 μL of the supernatant into new tubes, and add an equivalent volume of ice-cold isopropanol.
5. Incubate mixture at room temperature for at least 2 min with occasional mixing by inverting the tubes.
6. Precipitate DNA by centrifuging samples for 10 min at 13,200 × g (EPT) or 40 min at 2700 × *g* (CMT).
7. Wash pellet using 300 μL of 70% ethanol, and then centrifuge samples again for 5 min at 13,200 × *g* (EPT) or 40 min at 2700 × *g* (CMT).
8. Discard supernatant and allow pellet to dry at 65 °C for 15–30 min.
9. Resuspend pellets in 75 μL of TE buffer by heating at 65 °C for 15 min.

10. Check the quality and quantity of extracted DNA using a Nanodrop (Thermo Fisher Scientific, USA), following manufacturer's protocol.
11. Use 5 μL of DNA for the genotyping PCR amplification.

3.6.2 High-Quality DNA Extraction from Lotus Plants

If a purer DNA sample is required, we recommend using the CTAB method of DNA extraction, previously described in Urbanski et al. [13] and adapted from Rogers and Bendich [21].

1. Add a single tungsten carbide bead per sample, and homogenize plant tissues using a Tissuelyzer for 2 × 45 s.
2. Add 600 μl of CTAB extraction buffer. Incubate samples for 20 min in a 65 °C hot water bath.
3. Add 600 μL of chloroform:isoamyl alcohol 24:1. Shake mixture vigorously on a rotary shaker for 15 min at room temperature.
4. Centrifuge mixture for 10 min at 12,000 × *g*.
5. Transfer supernatant to new sterile tubes, and add 600 μL of isopropanol. Invert mixture three times.
6. Centrifuge mixture for 1 min at 12,000 × *g*. Discard supernatant, and wash DNA pellet with 500 μL of 70% ethanol.
7. Centrifuge samples for 10 min at 12,000 × *g*.
8. Dry pellet for at least 15 min in a fume hood, and resuspend in 30 μL of ultrapure water.

3.7 Genotyping PCR and Gel Electrophoresis to Identify Homozygous LORE1 Mutants from a Segregating F0 Population

LORE1 mutants can be genotyped using a modified touchdown PCR protocol [22], outlined in Table 1 (*see* **Note 8**). Each *LORE1* mutant should be genotyped with two pairs of primers: one comprising the forward and reverse primers and the other comprising the forward primer and the *LORE1* P2 primer, 5′-CCATGGCGGTTCCGTGAATCTTAGG-3′ (*see* **Note 9**).

1. Add 5 μL of loading buffer to 20 μL of PCR sample, and load samples onto a 1.5% agarose gel with 0.5 mg/L ethidium bromide. Load 8 μL of PstI-digested λ DNA marker per row.
2. Resolve DNA bands for 30–90 min at 150 V in TAE buffer.
3. Visualize DNA bands under UV light. *LORE1* homozygous mutants are identified by successful amplification with the forward and P2 primer pair, but not by the forward and reverse primer pair (Fig. 2; also *see* **Note 10**).

3.8 Establishing Homozygous fls2 Mutants from the LORE1 Collection

1. Scarify, germinate, and grow homozygous *LORE1* mutants as described in Subheadings 3.4 and 3.5. Perform PCR genotyping as per Subheading 3.7 to select mutants homozygous for the *LORE1* insertion.

Table 1
The modified touchdown PCR program for *LORE1* genotyping

Step	Temperature/°C	Duration	Cycle(s)
Activation	95	3 min	1
Denaturation Elongation	95 72	30 s 1 m 15 s	5
Denaturation Annealing Elongation	95 72–62 (−1 °C per cycle) 72	30 s 30 s 45 s	10
Denaturation Annealing Elongation	95 62 72	30 s 30 s 45 s	20–30
Termination	72	10 m	1
Storage	10	∞	1

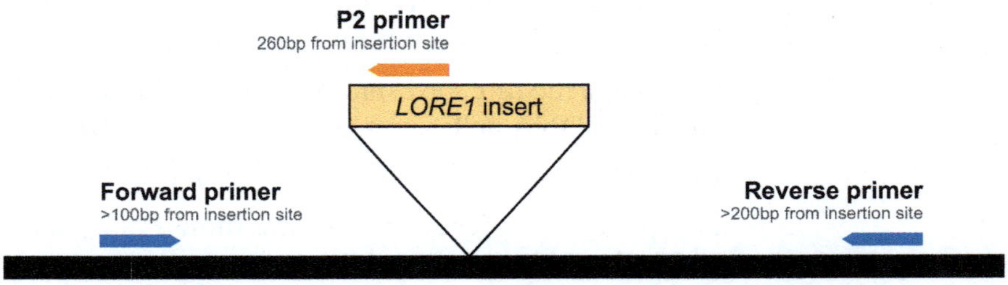

Fig. 2 The design of LORE1 genotyping primers. Note that the figure is not to scale. Forward and reverse primers are designed using Primer 3 [23] and are located at least 100 and 200 bp away from the LORE1 insertion site, respectively. The P2 primer binds to a region 264 bp downstream of the LORE1 5′ LTR

2. Sow homozygous mutants out for seed production in the greenhouse (14/10 h light/darkness and >70% relative humidity). If relative humidity drops below 65%, mature seed pods will burst, dispersing the seeds before they can be collected.
3. Transfer plants into larger pots when the root systems are well developed and extend throughout the pot volume. Two liter pots are well suited for growing plants for seed production. As *Lotus* have rather weak stems to support the relatively heavy and abundant foliage, tying the stems against vertical supports will help to control horizontal spread of plants.
4. Seed production in F0 homozygous plants will start around 1.5–2 months after sowing out in the greenhouse. Harvest seed pods, and allow them to dry for a week at room temperature and away from direct sunlight prior to processing.
5. Extract seeds by processing seed pods in a blender for 5–10 s.

6. Pass contents of the blender through a 2 mm metal gauze, which allows for *Lotus* seeds to pass through but not larger plant or pod debris.
7. Pass seeds through a 1 mm metal gauze, and shake to remove smaller particles. Seeds are typically larger than 1 mm, and do not pass through the second filter.
8. Store seeds in folded wax paper, at room temperature and away from direct sunlight and high levels of moisture. These seeds can be scarified, germinated, and grown (as described in Subheadings 3.4 and 3.5) for future studies.

3.9 Deposit Validated Homozygous *LORE1* Lines at Legume Base

To support research activities in the community and to ensure the continued availability of characterized *LORE1* mutants, we strongly encourage users of the LORE1 resource to deposit seeds of validated mutant lines at Legume Base, following the procedure outlined below.

1. Prepare a table in Microsoft Excel format with the following information for each homozygous *LORE1* line: name of mutant allele, ±1000 bp flanking sequence, original *LORE1* line ID, brief phenotype description, optional comments, and optional publication reference.
2. Send the table to the Legume Base curators at legume@brc.miyazaki-u.ac.jp.
3. Please enclose a phytosanitary certificate when shipping the seeds. If no phytosanitary certificate is available, please print "*Lotus japonicus* seeds enclosed" on the envelope.
4. Send the seeds to The National BioResource Project (*L. japonicus* and *Glycine max*) Office, Faculty of Agriculture, University of Miyazaki, Miyazaki 889–2192, Japan. Please mark the envelope "*LORE1* mutants."

4 Notes

1. Nitrate supplementation is only recommended for long-term growth of plants, and should not be used *in lieu* of inoculation with nodulating symbionts due to nitrate-based inhibition of nodulation [24, 25].
2. When establishing new *LORE1* homozygous mutant lines, we recommend selecting three or more alleles in order to confidently eliminate the effect of background mutants in future phenotyping or characterization experiments. For the same reasons, *LORE1* lines with a low number of exonic and total insertions are typically preferred.
3. Use of Conserve is recommended to eliminate any possible seed-borne thrips.

4. Stratification of scarified Lotus seeds is optional but strongly recommended to ensure uniform germination.
5. Removing the seed coat helps to prevent contamination of the sterile growth medium. Dried-out seed coats provide optimal growth conditions for fungal spores that might be harbored within.
6. Avoid sealing square petri dishes too tightly. To allow proper ventilation, either (1) cut a slit on the curved sides on the top of the petri dish or (2) leave a 3 cm area on the top edge free from parafilm, and cover instead with surgical tape, to allow venting of accumulated ethylene gas. The presence of ethylene, a pleiotropic and potent plant hormone, is known to inhibit root and shoot elongation, affect gravitropism, and inhibit nodulation in *Lotus*. We find that with this modification, it is still possible to retain sufficient moisture within the square petri dishes for up to six weeks, before the growth medium and filter paper start to dry out.
7. The choice between individual Eppendorf tubes (EPT) and arrays of collection microtubes (CMT) depends on the scale of the experiment. For a large number of samples to be collected and genotyped, the use of CMT is strongly recommended due to the ease of handling and processing. Individual EPTs can be overwhelmingly cumbersome and time-consuming for sample sizes exceeding 24.
8. The touchdown PCR protocol is used to ensure specificity of the PCR amplification process. However, if no temperature steps or ramping is available on the machine, a simple three-step PCR protocol may be used, with annealing temperature set to 62 °C.
9. The two genotyping PCR reactions performed per *LORE1* line should not be carried out together, i.e., mixing all three primers—forward, reverse, and P2—in the same reaction. Due to how the genotyping primers are designed, the PCR product of the forward + reverse and the forward + P2 primers may have similar sizes that cannot be resolved on the agarose gel.
10. Due to the reliance on the absence of DNA bands in the forward and reverse primer mix, *LORE1* homozygous mutants, when identified as such in the first PCR run, should be re-genotyped at least once more to confirm their genotype prior to sowing out for seed production.

Acknowledgments

This work is supported by the Danish National Research Foundation grant DNFR79.

References

1. Handberg K, Stougaard J (1992) *Lotus japonicus*, an autogamous, diploid legume species for classical and molecular-genetics. Plant J 2:487–496. doi:10.1111/j.1365-313X.1992.00487.x
2. Madsen LH et al (2010) The molecular network governing nodule organogenesis and infection in the model legume *Lotus japonicus*. Nat Commun 1:10. doi:10.1038/ncomms1009
3. Sato S et al (2008) Genome structure of the legume, *Lotus japonicus*. DNA Res 15:227–239. doi:10.1093/dnares/dsn008
4. Mun T, Bachmann A, Gupta V, Stougaard J, Andersen SU (2016) *Lotus* Base: an integrated information portal for the model legume *Lotus japonicus*. Sci Rep 6:39447. doi:10.1038/srep39447
5. Dam S et al (2009) The proteome of seed development in the model legume *Lotus japonicus*. Plant Physiol 149:1325–1340. doi:10.1104/pp.108.133405
6. Mc CB (1950) The origin and behavior of mutable loci in maize. Proc Natl Acad Sci U S A 36:344–355
7. Tadege M et al (2008) Large-scale insertional mutagenesis using the *Tnt1* retrotransposon in the model legume *Medicago truncatula*. Plant J 54:335–347. doi:10.1111/j.1365-313X.2008.03418.x
8. Hirochika H (2001) Contribution of the *Tos17* retrotransposon to rice functional genomics. Curr Opin Plant Biol 4:118–122
9. Madsen LH et al (2005) *LORE1*, an active low-copy-number TY3-*gypsy* retrotransposon family in the model legume *Lotus japonicus*. Plant J 44:372–381. doi:10.1111/j.1365-313X.2005.02534.x
10. Fukai E et al (2010) Derepression of the plant chromovirus *LORE1* induces germline transposition in regenerated plants. PLoS Genet 6:e1000868. doi:10.1371/journal.pgen.1000868
11. Fukai E et al (2012) Establishment of a *Lotus japonicus* gene tagging population using the exon-targeting endogenous retrotransposon *LORE1*. Plant J 69:720–730. doi:10.1111/j.1365-313X.2011.04826.x
12. Małolepszy A et al (2016) The *LORE1* insertion mutant resource. Plant J 88(2):306–317. doi:10.1111/tpj.13243
13. Urbanski DF, Małolepszy A, Stougaard J, Andersen SU (2012) Genome-wide *LORE1* retrotransposon mutagenesis and high-throughput insertion detection in *Lotus japonicus*. Plant J 69:731–741. doi:10.1111/j.1365-313X.2011.04827.x
14. Reid DE, Heckmann AB, Novak O, Kelly S, Stougaard J (2016) CYTOKININ OXIDASE/DEHYDROGENASE3 maintains cytokinin homeostasis during root and nodule development in *Lotus japonicus*. Plant Physiol 170:1060–1074. doi:10.1104/pp.15.00650
15. Małolepszy A et al (2015) The deubiquitinating enzyme AMSH1 is required for rhizobial infection and nodule organogenesis in *Lotus japonicus*. Plant J 83:719–731. doi:10.1111/tpj.12922
16. Rasmussen SR et al (2016) Intraradical colonization by arbuscular mycorrhizal fungi triggers induction of a lipochitooligosaccharide receptor. Sci Rep 6:29,733. doi:10.1038/srep29733
17. Wang C et al (2015) *Lotus japonicus* clathrin heavy Chain1 is associated with Rho-Like GTPase ROP6 and involved in nodule formation. Plant Physiol 167:1497–1510. doi:10.1104/pp.114.256107
18. Xue L et al (2015) Network of GRAS transcription factors involved in the control of arbuscule development in *Lotus japonicus*. Plant Physiol 167:854–871. doi:10.1104/pp.114.255430
19. Broughton WJ, Dilworth MJ (1971) Control of leghaemoglobin synthesis in snake beans. Biochem J 125:1075–1080
20. Krishnakumar V et al (2015) Araport: the Arabidopsis information portal. Nucleic Acids Res 43:D1003–D1009. doi:10.1093/nar/gku1200
21. Rogers SO, Bendich AJ (1985) Extraction of DNA from milligram amounts of fresh, herbarium and mummified plant tissues. Plant Mol Biol 5:69–76. doi:10.1007/BF00020088
22. Don RH, Cox PT, Wainwright BJ, Baker K, Mattick JS (1991) "Touchdown" PCR to circumvent spurious priming during gene amplification. Nucleic Acids Res 19:4008
23. Untergasser A et al (2012) Primer3—new capabilities and interfaces. Nucleic Acids Res 40:e115. doi:10.1093/nar/gks596
24. Gibson AH, Pagan JD (1977) Nitrate effects on the nodulation of legumes inoculated with nitrate-reductase-deficient mutants of *Rhizobium*. Planta 134:17–22. doi:10.1007/BF00390088
25. Streeter J, Nitrate G (1985) Inhibition of legume nodule growth and activity: II. Short term studies with high nitrate supply. Plant Physiol 77:325–328

Chapter 3

Enabling Reverse Genetics in *Medicago truncatula* Using High-Throughput Sequencing for *Tnt1* Flanking Sequence Recovery

Xiaofei Cheng*, Nick Krom*, Shulan Zhang, Kirankumar S. Mysore, Michael Udvardi, and Jiangqi Wen

Abstract

The genome sequence of *Medicago truncatula* was published and released in 2011. A *Tnt1* insertional mutant population with 21,700 independently regenerated lines was completed in 2012 at The Samuel Roberts Noble Foundation. With an estimated 25 insertions per line, the *Tnt1* mutant population harbors more than 500,000 insertions in the *M. truncatula* genome. Based on the genome size, average gene length, and random insertion of *Tnt1*into the genome, the mutant population affects about 90% of genes in the *M. truncatula* genome. Therefore, the mutant population enables functional characterization of most genes in the *M. truncatula* genome. From 2006 to 2011, we sequenced about 33,000 flanking sequence tags (FSTs) from 2600 *Tnt1* lines using TAIL-PCR followed by TA cloning, plasmid isolation, and traditional Sanger sequencing. To accelerate FST sequencing, we developed a two-dimensional DNA pooling strategy coupled with next-generation sequencing and produced about 380,000 FSTs from all 21,700 lines in a relatively short time. All FSTs are BLAST searchable in a web-based database. One can quickly search the database to find *M. truncatula* mutant lines with *Tnt1* insertions in most genes of interest.

Key words *Medicago truncatula*, Mutants, Next-generation sequencing, Reverse genetics, TAIL-PCR, *Tnt1*, Two-dimensional pooling

1 Introduction

Legumes play important roles in sustainable agriculture due to their unique symbiotic interactions with soil bacteria—rhizobia. *Medicago truncatula* is one of the model legume species for functional genomics studies of plant development, plant-microbe interactions, plant abiotic stresses, etc.

*These authors equally contributed to this work.

Wolfgang Busch (ed.), *Plant Genomics: Methods and Protocols*, Methods in Molecular Biology, vol. 1610, DOI 10.1007/978-1-4939-7003-2_3, © Springer Science+Business Media LLC 2017

With the completion of gene-rich region genome sequencing and annotation [1], functional characterization of thousands of genes in the *M. truncatula* genome became a top priority for researchers in the legume community. Mutant collections are irreplaceable resources for forward and reverse genetics to decipher gene function [2, 3]. From 2005 to 2012, we generated 21,700 retrotransposon *Tnt1*-tagged insertional lines in *M. truncatula* at The Samuel Roberts Noble Foundation. With an estimated 25 insertions per line [2], the *Tnt1* mutant population harbors more than 500,000 insertions in the *M. truncatula* genome, covering about 90% of *M. truncatula* genes. The high coverage of the mutant population has greatly enhanced researches in legume molecular genetics and functional genomics studies. By 2015, more than 80 papers resulting from the use of *Tnt1* mutants had been published.

Thermal asymmetric interlaced PCR (TAIL-PCR) [4] is an efficient and sensitive method to amplify unknown sequences adjacent to known inserted sequences [5]. Compared to other methods, like inverse PCR [6], splinkerette PCR [7], or adapter ligation PCR [8], TAIL-PCR has a number of advantages that facilitate and expedite the procedure of retrieving sequences that flank transposon insertion sites. For example, neither high-quality DNA templates nor DNA manipulations, such as restriction digestion or adapter ligation that may generate artifacts, are required prior to TAIL-PCR. Moreover, TAIL-PCR yields products of sufficient length and purity for direct sequencing. Unlike most other insertional mutants that have only one or two insertions in the genome, *Tnt1* mutants have an average of 25 insertions [2]. Thus, TAIL-PCR products from *Tnt1* mutants are typically many in numbers. In this case, the TAIL-PCR products cannot be sequenced directly using the Sanger sequencing method. Instead, the amplicons can be sequenced by next-generation sequencing or by traditional TA cloning and Sanger sequencing of individual colonies. During the process of TA cloning, transformation, and colony selection, some amplified TAIL-PCR amplicons may be lost at one of the steps due to ligation and/or transformation efficiency or due to the fact that not all colonies are sequenced for cost reasons. From 2006 to 2011, we sequenced about 33,000 flanking sequence tags (FSTs) from 2600 *Tnt1* lines using TAIL-PCR followed by TA cloning, plasmid isolation, and Sanger sequencing. Though FSTs recovered from Sanger sequencing are of high quality, the progress of sequencing the mutant population was slow and costly. To reduce the time and cost of FST sequencing, we developed a two-dimensional DNA pooling strategy and Illumina high-throughput sequencing to recover most FSTs in all 21,700 lines. In this chapter, we will describe the procedure of *Tnt1* FST recovery using the high-throughput sequencing approach in *M. truncatula*.

2 Materials

2.1 Plant Genomic DNA Isolation

1. Extraction buffer: 100 mM Tris–HCl, pH 8.0, 50 mM EDTA-Na_2, pH 8.0, 500 mM NaCl, 2-mercaptoethanol 350 μL/500 mL; before using, mix 9.35 mL of the above buffer with 0.625 mL of 20% SDS to make the working solution.
2. 3 M potassium acetate.
3. Other chemicals: chloroform, isopropanol, and 75% ethanol.

2.2 Oligonucleotides

1. Oligonucleotides at 200 nmol scale, desalted.
2. AD primers: *see* Table 1.
3. *Tnt1*-specific primers: For primary TAIL-PCR, primer *Tnt1*-Re (5′- CAGTGAACGAGCAGAACCTGTG-3′) is used in combination with different AD primers for all pooled DNA templates. For secondary TAIL-PCR, nested *Tnt1* primers with pool-specific bar codes are used (*see* Table 2).

2.3 PCR Reactions

1. Ex Taq™ (Takara Bio Inc.). All PCR reactions are performed in 40 μL total volume.
2. Agencourt AMPure XP magnetic beads (Beckman Coulter, Inc.).
3. Nanodrop Spectrophotometer.

2.4 Agarose Gel Electrophoresis

1. TAE buffer: 242 g Tris base, 57.1 mL acetic acid, 100 mL 0.5 M EDTA, distilled water (dH_2O) to 1 L, pH 8.5.
2. 6× DNA gel-loading buffer: for 100 mL, 200 mg bromophenol blue (Sigma), 33 mL 150 mM Tris–HCl, pH 7.6, 60 mL glycerol, dH_2O to 100 mL.
3. Agarose molecular biology grade.
4. Agarose gel electrophoresis apparatus.

Table 1
AD primer sequences

Primer name	Primer sequence (5′–3′)	Length	Degeneracy
AD1	NGTCGASWGANAWGAA	16 bp	128-fold
AD2	TGWGNAGSANCASAGA	16 bp	128-fold
AD3	AGWGNAGWANCAWAGG	16 bp	128-fold
AD5	SSTGGSTANATWATWCT	17 bp	64-fold
AD6	WGTGNAGWANCANAGA	16 bp	256-fold

W = A or T, S = G or C, N = A or T or G or C

Table 2
***Tnt1*-specific primers with bar codes for X pools and Y pools for the secondary TAIL-PCR**

SLX-*Tnt1-X1*	TGAACTTCGcaatgagatggacatcatca
SLX-*Tnt1*-X2	TCAAGCCTGcaatgagatggacatcatca
SLX-*Tnt1*-X3	GTACCAATCcaatgagatggacatcatca
SLX-*Tnt1*-X4	CTACTGCGTcaatgagatggacatcatca
SLX-*Tnt1*-X5	TGAGACGGTcaatgagatggacatcatca
SLX-*Tnt1*-X6	GCAGCAGGAcaatgagatggacatcatca
SLX-*Tnt1*-X7	CGAGTACCTcaatgagatggacatcatca
SLX-*Tnt1*-X8	CTATCGTTGcaatgagatggacatcatca
SLX-*Tnt1*-X9	TGCAACCAGcaatgagatggacatcatca
SLX-*Tnt1*-X10	GTCAGTAGAcaatgagatggacatcatca
SLX-*Tnt1*-X11	AGCTCTTAAcaatgagatggacatcatca
SLX-*Tnt1*-X12	GACTGAACTcaatgagatggacatcatca
SLX-*Tnt1*-X13	AGCTTCCGAcaatgagatggacatcatca
SLX-*Tnt1*-X14	GTGAATGGTcaatgagatggacatcatca
SLX-*Tnt1*-X15	CTGCATAGGcaatgagatggacatcatca
SLX-*Tnt1*-X16	CAGCGATTCcaatgagatggacatcatca
SLX-*Tnt1*-X17	CTGGACCTCcaatgagatggacatcatca
SLX-*Tnt1*-X18	CAGTTCCTGcaatgagatggacatcatca
SLX-*Tnt1*-X19	CTGCTACTTcaatgagatggacatcatca
SLX-*Tnt1*-X20	ACGTTGGAAcaatgagatggacatcatca
SLX-*Tnt1*-Y1	GTCATCTTGcaatgagatggacatcatca
SLX-*Tnt1*-Y2	GACCAGAGTcaatgagatggacatcatca
SLX-*Tnt1*-Y3	CTCGAAGAGcaatgagatggacatcatca
SLX-*Tnt1*-Y4	GTCTAGCTGcaatgagatggacatcatca
SLX-*Tnt1*-Y5	AGCTCATTCcaatgagatggacatcatca
SLX-*Tnt1*-Y6	TGTCTGACGcaatgagatggacatcatca
SLX-*Tnt1*-Y7	CGTCGTAGTcaatgagatggacatcatca
SLX-*Tnt1*-Y8	ACGTCGATAcaatgagatggacatcatca
SLX-*Tnt1*-Y9	TCGTACCATcaatgagatggacatcatca
SLX-*Tnt1*-Y10	CTGATCCGTcaatgagatggacatcatca
SLX-*Tnt1*-Y11	CGTCAAGTCcaatgagatggacatcatca

(continued)

Table 2 (continued)

SLX-*Tnt1*-Y12	AGTCTACTGcaatgagatggacatcatca
SLX-*Tnt1*-Y13	CATGCTCATcaatgagatggacatcatca
SLX-*Tnt1*-Y14	CACTAGGAGcaatgagatggacatcatca
SLX-*Tnt1*-Y15	GCTTGAATGcaatgagatggacatcatca
SLX-*Tnt1*-Y16	CGTACTGCGcaatgagatggacatcatca
SLX-*Tnt1*-Y17	CATGACGATcaatgagatggacatcatca
SLX-*Tnt1*-Y18	GCGACTCATcaatgagatggacatcatca
SLX-*Tnt1*-Y19	CACGGTGATcaatgagatggacatcatca
SLX-*Tnt1*-Y20	GTGCTGACTcaatgagatggacatcatca

Each primer contains a nine-nucleotide pool-specific bar code followed by 20 nucleotides from the reverse end of *Tnt1*

3 Methods

The *Tnt1* insertion mutant lines were generated by tissue culture and regeneration. First, *Tnt1* from *Nicotiana tabaccum* was introduced into *M. truncatula* (R108) through *Agrobacterium*-mediated transformation [9]. One of these transgenic lines, *tnk88-7-7*, was used as the starter line for large-scale mutant generation. Tissue culture and plant regeneration from the starter line were carried out as described previously [3, 9].

3.1 Preparation for TAIL-PCR

3.1.1 Genomic DNA Isolation

1. Collect approximately 0.3 g of fresh leaf tissue (typically two fully expanded leaflets) from each regenerated plant (R0) in 2 mL round-bottom tubes, freeze in liquid nitrogen, and grind to fine powder with glass beads using a Mini-Beadbeater (Biospec Products Inc.).
2. Add 0.5 mL of DNA extraction buffer (working solution) into each tube and mix well. Heat the samples in 65 °C water bath for 15 min. Mix 2–3 times by inverting tubes during incubation.
3. Add 200 μL of 3 M potassium acetate, invert to mix thoroughly, and set on ice for 10–15 min. Add 200 μL of chloroform and mix well by inverting tubes.
4. Spin at 17,000 × *g* for 10 min at room temperature.
5. Carefully transfer clear supernatant to a new tube containing 400 μL of isopropanol, and invert to mix well.
6. Place the tubes at −80 °C for 15–20 min or −20 °C overnight.
7. Spin at 17,000 × *g* for 15 min at 4 °C.
8. Pour off the liquid and add 1 mL of 75% ethanol to wash the pellet.

9. Air-dry the pellet for 20 min or in 37 °C incubator for 5 min.
10. Dissolve the pellet in 250 μL of autoclaved distilled deionized H_2O.

3.1.2 Two-Dimensional Genomic DNA Pooling

The goal of using high-throughput sequencing is to reduce cost and to save time. This goal can be achieved by reducing the total number of templates and therefore PCR reactions. By using a 20 × 20 two-dimensional DNA pooling strategy, the template number can be reduced from 400 samples to 40 in one Illumina MiSeq run. Figure 1 shows an example of DNA pooling table.

1. Prepare 400 DNA samples and arrange it in the same order as shown in the pooling tables.
2. Conduct X pool pooling by taking 2 μL of each DNA sample corresponding to each X pool row, e.g., for X2, take 2 μL of DNA from NF021 to NF040 in the example pooling table.
3. Conduct Y pool pooling by taking 2 μL of each DNA sample corresponding to each Y pool column, e.g., for Y4, take 2 μL of DNA from NF004, NF024, NF044, NF064, NF084, NF104, NF124, NF144, NF164, NF184, NF204, NF224, NF244, NF264, NF284, NF304, NF324, NF344, NF364, and NF384 in the example pooling table.
4. Use the 40 pool samples (20 X pools (X1-X20) and 20 Y pools (Y1-Y20)) that cover 400 individual DNA samples for TAIL-PCR (Subheading 3.2.3) or store samples at −20 °C.

	Y1	Y2	Y3	Y4	Y5	Y6	Y7	Y8	Y9	Y10	Y11	Y12	Y13	Y14	Y15	Y16	Y17	Y18	Y19	Y20
X1	NF001	NF002	NF003	NF004	NF005	NF006	NF007	NF008	NF009	NF010	NF011	NF012	NF013	NF014	NF015	NF016	NF017	NF018	NF019	NF020
X2	NF021	NF022	NF023	NF024	NF025	NF026	NF027	NF028	NF029	NF030	NF031	NF032	NF033	NF034	NF035	NF036	NF037	NF038	NF039	NF040
X3	NF041	NF042	NF043	NF044	NF045	NF046	NF047	NF048	NF049	NF050	NF051	NF052	NF053	NF054	NF055	NF056	NF057	NF058	NF059	NF060
X4	NF061	NF062	NF063	NF064	NF065	NF066	NF067	NF068	NF069	NF070	NF071	NF072	NF073	NF074	NF075	NF076	NF077	NF078	NF079	NF080
X5	NF081	NF082	NF083	NF084	NF085	NF086	NF087	NF088	NF089	NF090	NF091	NF092	NF093	NF094	NF095	NF096	NF097	NF098	NF099	NF100
X6	NF101	NF102	NF103	NF104	NF105	NF106	NF107	NF108	NF109	NF110	NF111	NF112	NF113	NF114	NF115	NF116	NF117	NF118	NF119	NF120
X7	NF121	NF122	NF123	NF124	NF125	NF126	NF127	NF128	NF129	NF130	NF131	NF132	NF133	NF134	NF135	NF136	NF137	NF138	NF139	NF140
X8	NF141	NF142	NF143	NF144	NF145	NF146	NF147	NF148	NF149	NF150	NF151	NF152	NF153	NF154	NF155	NF156	NF157	NF158	NF159	NF160
X9	NF161	NF162	NF163	NF164	NF165	NF166	NF167	NF168	NF169	NF170	NF171	NF172	NF173	NF174	NF175	NF176	NF177	NF178	NF179	NF180
X10	NF181	NF182	NF183	NF184	NF185	NF186	NF187	NF188	NF189	NF190	NF191	NF192	NF193	NF194	NF195	NF196	NF197	NF198	NF199	NF200
X11	NF201	NF202	NF203	NF204	NF205	NF206	NF207	NF208	NF209	NF210	NF211	NF212	NF213	NF214	NF215	NF216	NF217	NF218	NF219	NF220
X12	NF221	NF222	NF223	NF224	NF225	NF226	NF227	NF228	NF229	NF230	NF231	NF232	NF233	NF234	NF235	NF236	NF237	NF238	NF239	NF240
X13	NF241	NF242	NF243	NF244	NF245	NF246	NF247	NF248	NF249	NF250	NF251	NF252	NF253	NF254	NF255	NF256	NF257	NF258	NF259	NF260
X14	NF261	NF262	NF263	NF264	NF265	NF266	NF267	NF268	NF269	NF270	NF271	NF272	NF273	NF274	NF275	NF276	NF277	NF278	NF279	NF280
X15	NF281	NF282	NF283	NF284	NF285	NF286	NF287	NF288	NF289	NF290	NF291	NF292	NF293	NF294	NF295	NF296	NF297	NF298	NF299	NF300
X16	NF301	NF302	NF303	NF304	NF305	NF306	NF307	NF308	NF309	NF310	NF311	NF312	NF313	NF314	NF315	NF316	NF317	NF318	NF319	NF320
X17	NF321	NF322	NF323	NF324	NF325	NF326	NF327	NF328	NF329	NF330	NF331	NF332	NF333	NF334	NF335	NF336	NF337	NF338	NF339	NF340
X18	NF341	NF342	NF343	NF344	NF345	NF346	NF347	NF348	NF349	NF350	NF351	NF352	NF353	NF354	NF355	NF356	NF357	NF358	NF359	NF360
X19	NF361	NF362	NF363	NF364	NF365	NF366	NF367	NF368	NF369	NF370	NF371	NF372	NF373	NF374	NF375	NF376	NF377	NF378	NF379	NF380
X20	NF381	NF382	NF383	NF384	NF385	NF386	NF387	NF388	NF389	NF390	NF391	NF392	NF393	NF394	NF395	NF396	NF397	NF398	NF399	NF400

Fig. 1 Example of a genomic DNA pooling table for a two-dimensional pool. Twenty samples for X2 are boxed in *red line*, while 20 samples for Y4 are boxed in *purple line*

3.1.3 Primer Design and Dilution

Tnt1 is an LTR retrotransposon, which has two 610 bp long terminal repeats (LTRs) at both ends with exactly the same sequence and orientation. For this reason, the *Tnt1*-specific primer for the primary TAIL-PCR cannot be located in the LTR regions. Since Illumina sequencing reads are relatively short, to maximize the useful genomic sequence in a short Illumina read, the *Tnt1* primer for the secondary PCR should be as close to the end of *Tnt1* as possible. In *Tnt1*, the last 17 nucleotides of LTR from the forward direction are GGGGTTTATTCCCAACA, which will easily form secondary structures as an oligo. Therefore, the reverse direction to design primers for TAIL-PCR is chosen. The primary primer (*Tnt1*-Re) is 13 bp away from the LTR. Since the secondary PCR products for direct Illumina sequencing are used, each pool (X or Y) has to be bar coded. A nine-nucleotide barcode in front of the *Tnt1*-specific nested primer for each of the 40 pools is introduced (*see* Table 2).

1. Obtain primers as specified in Tables 1 and 2.
2. Dilute *Tnt1*-specific primers: primary PCR primer *Tnt1*-Re is diluted to 20 μM; secondary PCR primers (with bar codes) are diluted to 10 μM.
3. Dilute AD primers: all AD primers are diluted to 100 μM.

3.2 FST Recovery by TAIL-PCR

3.2.1 TAIL-PCR

The basis for this protocol is the original protocol of TAIL-PCR [4, 5]. However, instead of performing three rounds of PCR, only two rounds are performed (*see* **Note 1**). All TAIL-PCR reactions are in a 40 μL total volume (*see* **Note 2**). The goal is to recover as many *Tnt1* insertions as possible using TAIL-PCR. For each template, five TAIL-PCR reactions (*Tnt1*-specific primer + each of the five AD primers) are performed.

1. Prepare master mix for the primary PCR (TAIL1) for 45 reactions (enough to aliquot 20 X-pool templates and 20 Y-pool templates) for each AD primer in combination with *Tnt1*-Re primer.

TAIL1 master mix (one reaction):	
10× PCR buffer (Takara)	4 μL
dNTP (2.5 mM, Takara)	3.5 μL
Tnt1-Re primer (20 μM)	0.4 μL
AD primer (100 μM)	0.6 μL
ExTaq (Takara)	0.3 μL
Water	28.2 μL
Total volume	37 μL

2. Aliquot 37 μL of the master mix into each PCR tube that contains 3 μL of pooled DNA templates.
3. Perform TAIL1 PCR. Use heated lid. Use the following program:

1. 94 °C for 3 min
2. 94 °C for 30 s
3. 62 °C for 1 min
4. 72 °C for 2:30 min
5. Five cycles of **steps 2–4** (*see* **Note 3**)
6. 94 °C for 30 s
7. 25 °C for 3 min (50% ramp)
8. 72 °C for 2:30 min (32% ramp)
9. Two cycles of **steps 6–8** (*see* **Note 4**)
10. 94 °C for 10 s
11. 68 °C for 1 min
12. 72 °C for 2:30 min
13. 94 °C for 10 s
14. 68 °C for 1 min
15. 72 °C for 2:30 min
16. 94 °C for 10 s
17. 44 °C for 1 min
18. 72 °C for 2:30 min
19. 15 cycles of **steps 10–18** (*see* **Note 5**)
20. 72 °C for 5 min
21. 10 °C hold

4. Prepare secondary PCR (TAIL2) template dilution by mixing 2 μL of each TAIL1 PCR reaction with 98 μL of H_2O to make a 50 times dilution of the primary PCR products.
5. Transfer 2 μL of the diluted primary PCR products into fresh PCR tubes.
6. Add 1 μL of each pool-specific primer (10 μM) to the corresponding diluted template tube as each DNA pool has a pool-specific bar coded *Tnt1* primer for TAIL2.
7. Aliquot 37 μL of the following master mix to each tube:

TAIL2 master mix (one reaction)	
10× PCR buffer (Takara)	4 μL
dNTP (2.5 mM, Takara)	3.5 μL
AD primer (100 μM)	0.6 μL
ExTaq (Takara)	0.3 μL
Water	28.6 μL
Total volume	37 μL

8. Run TAIL2 PCR. Use heated lid and the following program:

1. 94 °C for 3 min
2. 94 °C for 10 s
3. 64 °C for 1 min
4. 72 °C for 2:30 min
5. Five cycles of **steps 2–4** (*see* **Note 3**)
6. 94 °C for 10 s
7. 64 °C for 1 min
8. 72 °C for 2:30 min
9. 94 °C for 10 s.
10. 64 °C for 1 min
11. 72 °C for 2:30 min
12. 94 °C for 10 s
13. 44 °C for 1 min
14. 72 °C for 2:30 min
15. 15 cycles of **steps 6–14** (*see* **Note 5**)
16. 94 °C for 10 s
17. 44 °C for 1 min
18. 72 °C for 3 min
19. Five cycles of **steps 16–18** (*see* **Note 6**)
20. 72 °C for 5 min
21. 10 °C hold

3.2.2 Electrophoresis of PCR Products

1. Transfer 5 μL of each PCR reaction from TAIL1 or TAIL2 into fresh tube.
2. Add 1 μL of 6× loading dye and mix well.
3. Load on 1% agarose gel.
4. Run gel electrophoresis for 25 min at 135 V.
5. Acquire image of DNA bands under UV light (Fig. 2a, *see* **Note 7**).

3.2.3 Pool TAIL-PCR Products

The TAIL2 PCR products are used for sequencing using the Illumina MiSeq. Products from 20 X pools and 20 Y pools (representing 400 DNA samples) are pooled into one sample for sequencing.

1. Add 2 μL of 0.5 M EDTA (pH 8.0) to each PCR tube to avoid potential extension of templates by Taq polymerase at room temperature after pooling.
2. Pool each of the five AD primer-amplified TAIL2 PCR products with the same X or Y pool number into one tube. This will result in a volume of ~200 μL; there are 20 X pools and 20 Y pools.
3. Pool 20 X pools and 20 Y pools into one Falcon tube (~8 mL).
4. Take 5 μL of the pooled PCR products for agarose gel electrophoresis to check the abundance of PCR products (Fig. 2b).

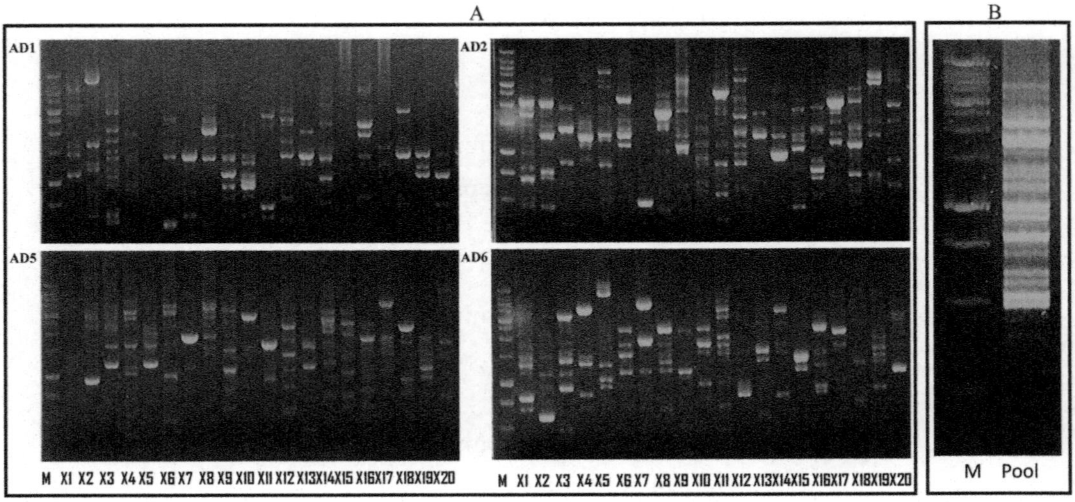

Fig. 2 A typical agarose gel electrophoresis result of TAIL2 PCR products in 20 individual X pools with different AD primers (panel **a**) and pooled PCR products from 20 X pools and 20 Y pools from all AD primers used in this protocol (panel **b**). M, DNA marker

3.2.4 PCR Product Purification

The pooled PCR products, typically about 8 mL, should be further purified before subjecting them to Illumina sequencing.

1. Transfer 4 mL of the pooled PCR products into a fresh tube.
2. Add 4 mL of phenol-chloroform-isoamyl alcohol (25:24:1, saturated with 10 mM Tris, 1 mM EDTA, pH 8.0; Sigma) to the pooled PCR products. Mix well by inverting the tubes.
3. Centrifuge at 4500 × *g* for 10 min at room temperature.
4. Carefully transfer the upper layer aqueous phase (about 4 mL) to a new 15 mL Falcon tube containing 0.45 mL of 3 M sodium acetate.
5. Add 4 mL of 2-propanol, mix well, and put at −20 °C overnight.
6. Mix and evenly divide the solution into six Eppendorf tubes and centrifuge at 16,000 × *g* at 4 °C for 15 min.
7. Discard liquid and wash and combine pellets in six tubes into one tube with 1 mL of 70% ethanol.
8. Dry pellet and dissolve in 300 μL of ddH_2O.
9. Take 150 μL of DNA solution for further purification using Agencourt AMPure XP magnetic beads (Beckman Coulter Inc.) following the manufacturer's protocol, except that the bound DNA is eluted in 150 μL of H_2O.

3.2.5 Sequencing PCR Amplicons Using Illumina MiSeq

1. Check the purified PCR products for their size on an Agilent Tape Station and quantify their concentration on a Thermo-Fisher Qubit fluorometer.
2. Construct libraries using the KAPA Biosystems Hyper Prep library preparation kit according to the manufacturer's instruc-

tion. Briefly, DNA molecules are end repaired and A tailed, ligated to Illumina-compatible adapters, and then size selected using Agencourt AMPure XP SPRI beads. Enrichment for properly ligated DNA molecules is then performed using PCR followed by an additional Agencourt AMPure XP SPRI bead cleanup step.

3. Check the individual libraries for their proper size on a Tape Station followed by Qubit quantification and library pooling.
4. Quantify the pooled set of libraries using a KAPA Biosystems Library Quantification qPCR kit on a Roche 480 LightCycler.
5. Once absolutely quantified, dilute the pooled samples to 2 nM, then denature, and then load them onto an Illumina MiSeq as per the manufacturer's protocol.
6. Sequence using paired-end, V2 chemistry to a read length of 250 bp.

3.3 Sequencing Reads Assembly and Analysis

The sequencing reads (typically 10–15 million 250 bp reads) are assembled and analyzed using the pipeline described below:

1. Demultiplex the reads using the FASTX toolkit's bar code splitter script (http://hannonlab.cshl.edu/fastx_toolkit/); this requires exact matches with the bar code sequences. Analyze the resulting files to produce a listing of all read pairs and which bar code they are labeled with.
2. Next, the bar code and adapter sequences are removed from the original, unsplit read files, and at the same time, low-quality bases are removed from the ends of each read until two consecutive bases are found with quality scores of at least 30. The entire read pair is discarded if either read was trimmed to less than 30 base pairs in length.
3. Assemble FST sequences using the complete set of trimmed reads, including those that do not contain any of the valid bar codes (*see* **Note 8**) by using Trinity (https://github.com/trinityrnaseq/trinityrnaseq/wiki) release 2013-02-25. While Trinity is designed to assemble transcriptomes rather than genomes, it produces higher quality FST sequences than any of the genome assemblers tested, as well as fewer redundant sequences or sequences not matching any locations in the *M. truncatula* genome.
4. Map the trimmed reads back to the contig sequences with Bowtie2 (http://bowtie-bio.sourceforge.net/bowtie2/index.shtml) version 2.2.3, using the "local-sensitive" alignment settings.
5. Based on the IDs of the reads mapped to each contig, the number of reads from each X and Y bar code group that contributes to that contig can be determined, and the contig can be assigned

to a specific line based on the sample pooling table. For example, if the contig has reads with X2 and Y4 bar codes, it is assigned to NF024. Each pool is represented by a specific combination of one X and one Y barcode. Two sets of pool assignments are produced: high-confidence FSTs and low-confidence FSTs. If one X bar code makes up the majority of all X-labeled reads mapped to an FST, and one Y bar code makes up the majority of all Y-labeled reads, then that FST is assigned to the pool corresponding to that X/Y combination with high confidence. Otherwise, the combination of the most common X and Y bar codes is used to assign that FST to a pool with low confidence.

6. After the FST pool assignment is complete, compile the FST sequences for each pool into separate files.
7. Scan the FST sequences for the *Tnt1* signature sequence "GACATCATCA," and if the signature is found at the 3′ end of the sequence, the FST is reverse complemented so that the signature appears at the 5′ end.
8. Number FSTs in each pool in descending order by length and write to separate files.
9. These FSTs can now be deposited into an appropriate database (like http://medicago-mutant.noble.org/mutant/).

4 Notes

1. There are multiple insertions in each mutant line. Therefore, after pooling of 20 individual DNA samples into one pool, there are possibly several hundred insertions to be amplified by TAIL-PCR. From the primary PCR, some insertions are amplified better than others due to insertion location, amplicon size, and/or AD primer composition. In the secondary PCR, there is biased amplification because of nonuniformly amplified primary PCR products – the templates for the secondary PCR. If a tertiary PCR is run, there will be even more bias in the amplification, which may cause loss of some low-abundant amplicons from the secondary PCR.
2. We tried different PCR reaction volumes and found that for complex DNA templates, low reaction volumes cause inconsistent amplification results. Reactions in 40 μL volumes produce consistent and reproducible results.
3. Five high-stringency cycles favor the production of single-stranded products, primed by the *Tnt1*-specific primer.
4. Two low-stringency cycles facilitate AD primer annealing. First, the annealing temperature is gradually decreased until it reaches 25 °C (50% ramp), and then again, gradually increased (32% ramp) until it reaches extension temperature (72 °C).

5. Fifteen super-cycles of two high-stringency cycles interlaced with one low-stringency cycle (TAIL cycling).
6. Five cycles with low annealing temperature in order to increase amount of products for sequencing. These cycles may be omitted if products are cloned and then sequenced.
7. Unlike TAIL-PCR experiments that amplify only one or two bands, our PCR products contain multiple bands for each of the five AD primers we used. Even though we run five independent TAIL-PCR reactions for each pooled template using different AD primers, there are still some insertions that are not amplified.
8. We use the Illumina paired-end sequencing approach. Theoretically, only reads from the *Tnt1* primer end have bar codes because only *Tnt1*-specific primers are bar coded. Reads from AD primers have no bar codes. Since Illumina automatically generates two files for the paired reads from the same DNA molecule, reads from AD primers are still identifiable based on their file ID.

Acknowledgments

This work was supported by The Samuel Roberts Noble Foundation and in part by NSF Plant Genome Grants (DBI 0703285 and IOS 1127155).

References

1. Young ND, Debelle F, Oldroyd GED, Geurts R, Cannon SB, Udvardi MK, Benedito VA, Mayer KFX, Gouzy J, Schoof H et al (2011) The *Medicago* genome provides insight into the evolution of rhizobial symbioses. Nature 480:520–524
2. Tadege M, Wen J, He J, Tu H, Kwak Y, Eschstruth A, Cayrel A, Endre G, Zhao PX, Chabaud M, Ratet P, Mysore KS (2008) Large-scale insertional mutagenesis using the *Tnt1* retrotransposon in the model legume *Medicago truncatula*. Plant J 54:335–347
3. Wu JL, Wu CJ, Lei CL, Baraoidan M, Bordeos A, Madamba MRS, Ramos-Pamplona M, Mauleon R, Portugal A, Ulat VJ et al (2005) Chemical- and irradiation-induced mutants of indica rice IR64 for forward and reverse genetics. Plant Mol Biol 59:85–97
4. Liu YG, Mitsukawa N, Oosumi T, Whittier RF (1995) Efficient isolation and mapping of *Arabidopsis thaliana* T-DNA insert junctions by thermal asymmetric interlaced PCR. Plant J 8:457–463
5. Liu YG, Chen Y, Zhang Q (2005) Amplification of genomic sequences flanking T-DNA insertions by thermal asymmetric interlaced polymerase chain reaction. Methods Mol Biol 286:341–348
6. Ochman H, Gerber AS, Hartl DL (1988) Genetic applications of an inverse polymerase chain reaction. Genetics 120:621–623
7. Horn C, Hansen J, Schnütgen F, Seisenberger C, Floss T, Irgang M, De-Zolt S, Wurst W, von Melchner H, Noppinger PR (2007) Splinkerette PCR for more efficient characterization of gene trap events. Nat Genet 39:933–934
8. Patel K, Cox R, Shiple J, Kiely F, Frazer K, Cox DR, Lehrach H, Sheer D (1991) A novel and rapid method for isolating sequences adjacent to rare cutting sites and their use in physical mapping. Nucleic Acids Res 19:4371–4375
9. d'Erfurth I, Cosson V, Eschstruth A, Lucas H, Kondorosi A, Ratet P (2003) Efficient transposition of the *Tnt1* tobacco retrotransposon in the model legume *Medicago truncatula*. Plant J 34:95–106

Chapter 4

The Generation of Doubled Haploid Lines for QTL Mapping

Daniele L. Filiault, Danelle K. Seymour, Ravi Maruthachalam, and Julin N. Maloof

Abstract

Recombinant inbred lines (RILs) are an essential tool for quantitative trait locus (QTL) mapping in *Arabidopsis thaliana*. Conventionally, the development of these lines is a time-consuming and tedious process requiring six to eight generations of selfing. Here, we describe an alternative approach: the rapid generation of RILs in *A. thaliana* via the creation of doubled haploids. In this method, F_1 plants are crossed to an engineered haploid inducer to produce haploid plants. The chromosomes of these haploids then spontaneously double, generating immortalized homozygous F_2 lines called doubled haploid RILs (DH RILs). Finally, DH RILs are genotyped using low-coverage whole-genome sequencing and are ready to be used for QTL mapping.

Key words Doubled haploids, Recombinant inbred lines, QTL mapping, *Arabidopsis thaliana*, Natural variation

1 Introduction

Quantitative trait locus (QTL) mapping is a powerful tool for dissecting the genetics of phenotypic variability [1]. In *Arabidopsis thaliana*, questions of genotype by environment interactions can be studied in detail by QTL mapping with recombinant inbred line (RIL) populations. The traditional construction of RIL populations involves crossing two parent accessions to shuffle the parental genomes through recombination and then fixing these recombined chromosomes in a homozygous state through many generations of selfing. The resulting immortalized homozygous lines can be genotyped once and then grown in many different conditions to assess the effects of genotype by environment interactions. Although a number of these populations are available to the community, some applications, such as verifying the results from genome-wide association mapping experiments, often require researchers to generate their own RIL populations to study particular accessions and variants of interest. Generating these

Wolfgang Busch (ed.), *Plant Genomics: Methods and Protocols*, Methods in Molecular Biology, vol. 1610,
DOI 10.1007/978-1-4939-7003-2_4, © Springer Science+Business Media LLC 2017

populations in the traditional way is a tedious and time-consuming process. In contrast, generating RIL populations via the doubled haploid method (DH RILs) described here is straightforward and much faster, allowing researchers greater flexibility to create custom RIL populations quickly and easily [2].

The key to constructing DH RILs is the generation of haploid plants from a diploid precursor. The chromosomes of haploid plants are then doubled, yielding fertile diploid plants that are homozygous at all genetic loci. These plants, referred to as doubled haploids (DH), are a powerful tool for plant breeders [3, 4]. In a few crop species, such as rice, onion, and rapeseed, DH plants at the scale required for mapping studies can be produced in vitro using tissue culture-mediated approaches [5–7]. In some cereals, such as maize, barley, and wheat, haploids can be generated in vivo using haploid inducer lines [8–10]. However, the generation of a large population of DH lines by either method is limited to selected plant species due to the recalcitrant and genotype-dependent response of plant cells to in vitro and in vivo haploid regeneration. This situation has changed with the recent engineering of a haploid inducer line in *A. thaliana* that generates haploids through seeds via a centromere-mediated process [11]. This discovery has spurred interest among researchers to develop similar haploid inducers for many crops of interest [12].

In *A. thaliana*, the haploid inducer is a Col-0 plant harboring a homozygous mutation in the centromeric variant of the histone H3 protein (CENH3) that has been complemented with a *CENH3 GFP-tailswap* transgene (*cenh3-1/cenh3-1; GFP-tailswap/GFP-tailswap*) [11]. This transgene replaces the N-terminal tail domain of CENH3 with that of histone H3.3 and adds an N-terminal GFP tag. When crossed with a wild-type plant (in this case, an F_1 between the two parents of interest), the genome of the inducer line is eliminated, resulting in a haploid F_2 plant (Fig. 1a). These haploids then spontaneously double to generate homozygous immortal F_2 lines which we call DH RILs. In this way, a DH RIL population is generated in only two generations after creation of the F_1, in comparison to the six to eight generations of selfing needed to generate traditional RILs.

This protocol will outline the basic steps involved in generating DH RILs in *A. thaliana*. First, haploid inducer plants and the F_1 of interest are obtained and prepared. Next, the F_1 is crossed with the haploid inducer, resulting in a mixture of progeny comprised of haploid, diploid, and aneuploid F_2 plants. Haploid F_2 plants are then selected based on their phenotype and grown. Each haploid plant produces a small amount of viable diploid seed which, when collected, comprises one homozygous F_2 line of the DH RIL population. Finally, each DH RIL is genotyped using high-throughput sequencing and is then ready to be used in QTL mapping.

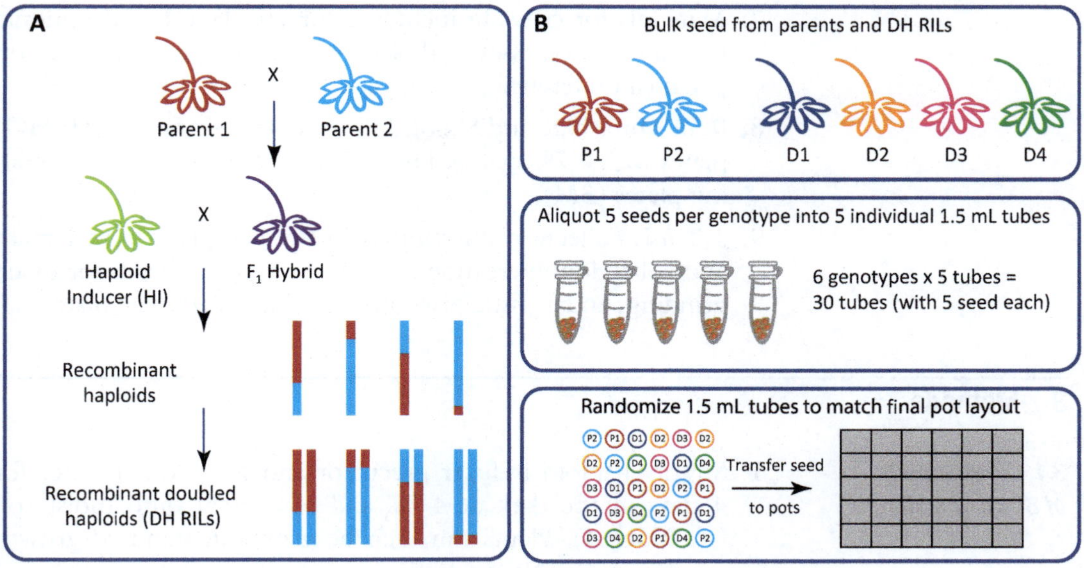

Fig. 1 Overview of DH RIL construction and experimental design. (**a**) Crossing scheme to generate DH RILs. Two parental strains are crossed to generate the desired F1 hybrid. This hybrid is then used as the male parent in a cross to the transgenic haploid inducer (HI). Haploid progeny resulting from the elimination of the HI chromosomes will be recombinant. Spontaneous chromosome doubling in these haploids will lead to the desired doubled haploid recombinant inbred lines (DH RIL). (**b**) Diagram illustrating a standard experimental design for phenotyping the DH RILs. Seed from parents and DH RILs is bulked. In this example, five seeds of each line are aliquoted into a 1.5 mL Eppendorf tube for each genotype–replicate combination. Before sowing each genotype–replicate combination onto soil, the 1.5 mL tubes are randomized into the appropriate design. *See* Subheading 3.4.1 for more details

2 Materials

1. Seed to generate the haploid inducer line, which can be ordered from either the Arabidopsis Biological Resource Center (CS66982) or the European Arabidopsis Stock Center NASC (N66982). This seed stock is the segregating progeny of *cenh3-1/CENH3-1; GFP-tailswap/GFP-tailswap* plants. The haploid inducer plants (*cenh3-1/cenh3-1; GFP-tailswap/GFP-tailswap*) used for generating DH RILs will be selected from these progeny (*see* **Note 1**).
2. Sufficient seed of the F_1 of interest to grow at least three F_1 plants (*see* **Note 2**).
3. Pots or growing tray inserts measuring ≥7 cm in diameter, soil, and humidity domes for growing plants.
4. Plastic sleeves for plant isolation (*see* **Note 3**) and stakes and ties for plant support.
5. 1.5 mL Eppendorf tubes for seed stratification and storage.
6. Desiccator for drying seeds.

7. Materials for crossing including forceps, labels, and optional magnification glasses or dissecting microscope, depending on personal preference.
8. 0.5× Murashige and Skoog Minimal Organic media (MSMO) plates with 0.7% agar and no sucrose. Micropore tape for sealing plates (3M).
9. 1.2 mL collection microtubes in 96-well plate format racks and collection microtube caps, 2.4 mm metal beads for tissue grinding, and a tissue disruptor suitable for 96-well plates.

3 Methods

3.1 Preparation of Starting Materials

1. Stratify haploid inducer precursor and F_1 seeds in water for 4 days in the dark at 4 °C and sow them onto moist soil (*see* **Note 4**). Plants can then be grown in standard growth conditions (*see* **Note 5**).
2. Sowing should be timed so that the flowering periods of both the haploid inducer and the F_1 overlap substantially. The haploid inducer has an extended flowering period and will produce flowers for many weeks, but has no vernalization requirement. *A. thaliana* natural accessions, however, have a shorter flowering period and many require extensive vernalization. Depending on the parents chosen, F_1 seed may therefore need to be sown well before haploid inducer seed in order to synchronize flowering. The length of the vernalization requirement (and the required offset in sowing times between the haploid inducer and the F_1) will depend on the parental accessions used and the vernalization temperature. Assess the vernalization requirements of the parental accessions when growing parental lines for the original F_1 cross and time sowing and vernalization of the F_1 seed appropriately.
3. Select haploid inducer plants. Since haploid inducer plants (*cenh3-1/cenh3-1 GFP-tailswap/GFP-tailswap)* produce few fertile offspring, seeds obtained from the stock center are the segregating progeny of a selfed *cenh3-1/CENH3-1 GFP-tailswap/GFP-tailswap* plant. Once these seeds are grown, haploid inducer plants can be distinguished from other genotypes by their stunted size, compact form, and bushy growth habit. This distinct morphology is demonstrated in Fig. 2e.
4. Confirm the genotypes of both the haploid inducer and the F_1 of interest (*see* **Note 6**).
5. Ten haploid inducer plants and three F_1 plants should be sufficient for the crosses needed to generate about 250 DH RILs (*see* **Note 7**).

Fig. 2 Phenotypic differentiation of haploids from aneuploid and diploid siblings. (**a**–**d**) Rosette phenotypes of month-old progeny derived from a haploid inducer (Col-0) x *Ler* cross. Aneuploids (Anp) display diverse phenotypes affecting the leaf size, shape, color, and overall plant morphology. Haploid (Hap) plants are similar to the wild-type diploids (Dip) except for a reduction in size. (**e**) Progeny from *cenh3-1/CENH3; GFP-tailswap/GFP-tailswap* showing both the WT phenotype and the compact, bushy, stunted phenotype unique to haploid inducer (*cenh3-1/cenh3-1; GFP-tailswap/GFP-tailswap*) plants. (**f**, **g**) Flowering phenotypes of plants derived from a haploid inducer x Col-0 cross. Haploid (Hap) plants phenotypically resemble diploids (Dip) except for a proportional reduction in plant and flower size. Most of the aneuploids are stunted and small (*arrows*). WT wild type, HI haploid inducer

6. The haploid inducer line, as well as the haploids produced from crosses with the inducer line, has low sterility and is therefore prone to outcrossing. These plants also have a distinctly wide and bushy growth habit when flowering, further increasing the chances of outcrossing through accidental contact with other plants. Take careful measures to avoid contamination by outcrossing throughout the entire DH RIL construction process, such as using plastic plant sleeves, choosing large pots (≥ 7 cm in diameter), avoiding overcrowding, and growing plants in an undisturbed location. A dedicated plant chamber or growth room is ideal.

3.2 Making Crosses between the Haploid Inducer Line and the F_1

1. When both the haploid inducer and the F_1 plants are flowering, crossing can begin. As the haploid inducer line is largely male sterile, it is used as the female parent and the F_1 is used as the male parent (*see* **Note 8**). Due to this low male fertility, no emasculation of the haploid inducer flowers is required, so many researchers will find crosses easy to perform even without the aid of additional magnification.
2. Remove mature flowers that are too old to have viable stigmas from one stem of a haploid inducer plant. Mark the location of the oldest flower crossed with a small tape flag or string tag describing the cross and date (*see* **Note 9**).
3. Remove a flower with mature pollen from an F_1 plant using forceps (mature pollen will be bright yellow and appear powdery). Use the pollen-laden anthers of this flower to liberally "paint" pollen onto the stigma of a haploid inducer flower. One F_1 flower will usually be sufficient to pollinate at least three or four inducer flowers. Flowers at many different developmental stages along the stem can be female parents. Even flowers whose sepals have barely started to open and whose stigma is not yet protruding can be successfully crossed. If the stigma of a flower is not yet visible, simply place pollen in and around the opening formed by the separating sepals. Pollinating all the receptive flowers on a haploid inducer stem may require several F_1 flowers.
4. Mark the location of the last cross attempted by removing the next several unopened flower buds on the stem.
5. Repeat **steps 2** through **4** for many more haploid inducer stems. Because only some of the resulting seed will be true haploids, one should conservatively aim for about 300 siliques to generate about 250 DH RILs (*see* **Note 10**).
6. Monitor crosses carefully and remove the resulting siliques when they are dry and mature, but before they shatter.
7. After harvesting, dry seeds in a desiccator before storing them in dry conditions (*see* **Note 11**).

3.3 Identification of DH RILs

3.3.1 Grow Plants for Haploid Selection

1. Sterilize seeds from crosses (*see* **Note 12**) and sprinkle on 0.5× MSMO (no sucrose) + 0.7% agar plates. Seal plates with porous-sealing tape and stratify at 4 °C in the dark for 4 days before transferring to standard plant growth conditions (*see* **Note 13**).
2. The percentage of viable seeds will be low, and germination may occur over a period of up to 4 weeks (*see* **Note 14**).
3. As soon as a seedling has four true leaves, carefully transfer it to a large (≥7 cm diameter) pot filled with moist soil. Plates will need to be repeatedly opened and resealed to remove seedlings for transplantation; use good sterile technique when preparing and handling plates to avoid contamination during this process. Haploid plants may appear small and weak at this stage and may have delayed germination, so transfer all seedlings regardless of their appearance or germination time. Expect an average of one to two transferred seedlings per silique.

3.3.2 Select Haploid Plants

1. Plants transferred in the previous step will be a mixture of haploids, diploids, and aneuploids. Examples of each of these three genotypes can be seen in Fig. 2a–d, f, g.
2. The following phenotypes can be used to distinguish these three genotypes:
   (a) *Plant size*: Haploid plants are noticeably smaller than diploids, usually about half the size. The size of aneuploid plants can vary greatly, depending on which chromosomes have abnormal copy numbers (*see* **Note 15**).
   (b) *Flower size*: The flowers of haploid plants are typically about half the size of those from diploid plants. Aneuploid plants can have a range of flower sizes spanning from very small to larger than those of diploids.
   (c) *Developmental defects*: Haploid plants look like smaller versions of diploid plants, with no large differences in leaf shape or color. Aneuploids, on the other hand, may have a wide spectrum of developmental abnormalities, such as color differences and curled or misshapen leaves, among others. The exact developmental defects observed will depend on which chromosomes have abnormal copy numbers, so although many aneuploid plants may display similar phenotypes, not every aneuploid plant will look exactly the same.
   (d) *Fertility*: Diploid plants have normal fertility levels and produce many full siliques. Despite their low fertility, haploid plants may sometimes produce a few siliques and seeds. Occasionally, an entire stalk of a haploid plant will be fertile due to spontaneous somatic genome duplication (diploidization). Aneuploid plants display a range of

fertility levels, with some plants producing siliques normally and others producing few or no siliques.

3. Using these phenotypes for selection, discard diploids and aneuploids and continue growing haploid plants only. Support and tie plants as needed and use plastic sleeves around each plant to prevent cross-contamination between lines. Roughly 2/3 of the plants transferred from plates will be DH RILs after genotyping.

3.3.3 Harvest DH RIL Seed

1. Although mostly sterile, haploid plants occasionally produce fertile doubled haploid seed. This occurs because occasional gametes will have a complete haploid chromosome complement; when two such gametes meet, fertilization results in a doubled haploid seed. In other cases, a plant can experience a spontaneous somatic doubling event, leading to entire siliques or branches of doubled haploid seed. In general, a haploid plant will produce between 20 and 200 viable doubled haploid seed, depending on the plant size (*see* **Note 16**).
2. These seeds are the putative DH RILs; each plant will produce one independent recombinant DH line. Collect seeds when siliques are mature and dry but before siliques shatter. Collect and store seeds from each plant separately and work cleanly to prevent cross-contamination of seeds between plants.
3. After harvesting, dry seeds in a desiccator before storing them in dry conditions (*see* **Note 11**).

3.4 Confirmation of DH RILs via High-Throughput Genotyping

3.4.1 Grow DH RILs for Genotyping and Phenotyping

1. Ensure that enough seed from each putative DH RIL has been generated in order to perform a replicated, randomized phenotyping experiment. Replicate number will depend on the phenotype of interest. Sow approximately five seeds per genotype per pot in case of low germination success. If additional seed is needed, genotypes can be self-fertilized in order to bulk seed for each DH RIL.
2. Calculate the number of pots needed for the phenotyping experiment (number of genotypes × number of replicates). Make sure to include parental genotypes in the experiment (Fig. 1b).
3. Aliquot five seeds per genotype into a 1.5 mL Eppendorf tube for each genotype–replicate combination. For each pot in the experiment, there will be one 1.5 mL Eppendorf tube (Fig. 1b).
4. Arrange the 1.5 mL tubes in a randomized design using a layout that facilitates quick and accurate transfer of the seeds to their final pot location. Randomization is used to ensure homogeneity of phenotypic variance across genotypes (Fig. 1b).

5. Stratify seeds in the dark at 4 °C for 3–10 days after adding 500 μL of sterilized water to each 1.5 mL Eppendorf tube (*see* **Note 4**).
6. Sow all five seeds of each genotype–replicate combination by pipetting the contents of each 1.5 mL tube onto a pot filled with moist soil.
7. After sowing, cover trays containing pots with a clear plastic humidity dome and place trays into the desired experimental growth conditions (*see* **Note 17**).
8. Remove clear plastic humidity domes after approximately 1–2 weeks or when plants have initiated a few true leaves.
9. Score germination success of each genotype–replicate combination and thin plants until only one plant resides in each pot. This step can be ignored in cases where the experimenter desires multiple plants per pot.
10. Once plants are at the sixth leaf stage or later, leaves can be sacrificed for genotyping. For each plant, collect a single leaf (~100–200 mg) for genotyping into a 1.2 mL collection microtube (in 96-well plate format) containing a metal tissue disruptor bead. In cases where destructive sampling may affect the phenotype of interest, plants should be grown separately for genotyping and phenotyping. If possible, collect replicates of the tissue to use as a backup in case of downstream errors (*see* **Note 18**).
11. Quickly freeze the tissue after collection by placing completed (and sealed) 96-well plate into a −80 °C freezer.

3.4.2 Extract DNA for High-Throughput Genotyping

1. Grind frozen tissue in 96-well plate using a Qiagen TissueLyser II (or equivalent). Quickly proceed to the first DNA extraction step and do not let the tissue thaw after grinding. Alternatively, tissue can be lyophilized before grinding and stored at room temperature.
2. Perform DNA extraction using a DNA extraction strategy that generates DNA of sufficient quality for high-throughput genotyping (*see* **Note 19**). In general, DNA should be clean, undegraded, and of high quality. The amount and concentration of DNA required for subsequent sequencing library construction will be determined by the protocol chosen in Subheading 3.4.3.

3.4.3 Construct and Sequence High-Throughput Genotyping Libraries

1. Low-coverage whole-genome sequencing is the most straightforward method for genotyping *A. thaliana* DH RILs (*see* **Note 20**). Coverage of at least 2–3× per sample, which can be obtained through multiplexing 96 libraries in an Illumina HiSeq sequencing lane, should be sufficient for most experiments.

2. Prepare sequencing libraries for each DH RIL individual using a protocol that is approved by your sequencing facility.
3. The parents of the F_1 of interest also need to be genotyped. If no high-throughput sequencing data is available from other sources (*see* **Note 21**), then sequencing libraries should also be prepared for the parents. Although parents can be sequenced at the same depth as the DH RILs, sequencing the parents at higher depth will provide more high-quality SNPs, making the genotyping of DH RILs easier and more accurate.
4. Generate high-throughput sequencing data for each library. For all subsequent steps, it is assumed the sequencing was performed on an Illumina instrument.

3.5 Generating Genotype Calls from High-Throughput Sequencing Data

3.5.1 Download and Quality Filter Sequencing Data

1. Download raw sequencing data from your genome facility. From this point in the protocol, we will demonstrate all steps using an example data set (SRR407382 and SRR407833) that can be downloaded as fastq files from NCBI's Short Read Archive (http://www.ncbi.nlm.nih.gov/sra) using the SRA toolkit and the following commands (*see* **Note 22**):

```
fastq-dump SRR407382
fastq-dump SRR407383
```

2. If samples have been multiplexed, individuals are identified in the raw sequencing data by a unique nucleotide barcode. Typically, the Illumina software will automatically sort the reads by barcode, and the user will obtain demultiplexed reads from the sequencing center. If this is not the case, custom scripts can be used to demultiplex raw data to extract the reads corresponding to individual samples. Our example uses the script `00.extract_barcode_of_interest.pl` (available at https://github.com/MaloofLab/Filiault-MMB-2016) to extract the sequences corresponding to the original parents (Parent 1 and Parent 2) as well as to a single DH RIL library.

```
#Parent 1 - 6944
perl 00.extract_barcode_of_interest.pl SRR407383.fastq AGGAC
Parent1.fastq
#Parent 2 - 6967
perl 00.extract_barcode_of_interest.pl SRR407382.fastq GATTT
Parent2.fastq
#DH example - DHRIL-16
perl 00.extract_barcode_of_interest.pl SRR407383.fastq GATTT
DH_example.fastq
```

3. Remove low-quality reads from each library. Low-quality bases at the beginning and end of each read are removed as well as reads that are shorter than a minimum length threshold. The following command provides the default parameters used to trim single-end reads using trimmomatic v0.36 [13] (*see* **Note 23**).

```
java -jar trimmomatic-0.36.jar SE Parent1.fastq
Parent1_trimmed.fastq LEADING:3 TRAILING:3 SLIDINGWINDOW:4:15:
MINLEN:20

java -jar trimmomatic-0.36.jar SE Parent2.fastq
Parent2_trimmed.fastq LEADING:3 TRAILING:3 SLIDINGWINDOW:4:15:
MINLEN:20

java -jar trimmomatic-0.36.jar SE DH_example.fastq
DH_example_trimmed.fastq LEADING:3 TRAILING:3 SLIDINGWINDOW:4:15:
MINLEN:20
```

3.5.2 Align Quality-Filtered Reads to the Reference Genome

1. Download the fasta file of the *Arabidopsis thaliana* reference genome sequence from Phytozome (https://phytozome.jgi.doe.gov) (*see* **Note 24**).
2. Generate reference genome index for the chosen alignment tool, in this case, bwa v0.7.8 [14].

```
bwa index -p Athaliana_167_TAIR9 -a is Athaliana_167_TAIR9.fa.gz
```

3. Align quality-trimmed reads to the reference genome. In this case, one mismatch (−n 1) is allowed between the read and the reference (*see* **Note 25**).

```
bwa aln -n 1 Athaliana_167_TAIR9 Parent1_trimmed.fastq >
Parent1.sai
bwa aln -n 1 Athaliana_167_TAIR9 Parent2_trimmed.fastq >
Parent2.sai
bwa aln -n 1 Athaliana_167_TAIR9 DH_example_trimmed.fastq >
DH_example.sai
```

4. Use sequence alignment index (.sai) files to create sequence alignment/map (.sam) files.

```
bwa samse Athaliana_167_TAIR9 Parent1.sai Parent1_trimmed.fastq >
Parent1.sam

bwa samse Athaliana_167_TAIR9 Parent2.sai Parent2_trimmed.fastq >
Parent2.sam

bwa samse Athaliana_167_TAIR9 DH_example.sai
DH_example_trimmed.fastq > DH_example.sam
```

3.5.3 Identify Polymorphisms in Parents and DH RILs Relative to the Reference Genome

1. Convert sam file to bam file using samtools v1.3 [15].

```
samtools view -b Parent1.sam > Parent1.bam
samtools view -b Parent2.sam > Parent2.bam
samtools view -b DH_example.sam > DH_example.bam
```

2. Sort bam file by chromosome and position using samtools v1.3.

```
samtools sort Parent1.bam > Parent1_sorted.bam
samtools sort Parent2.bam > Parent2_sorted.bam
samtools sort DH_example.bam > DH_example_sorted.bam
```

3. Identify polymorphisms using the samtools v1.3 mpileup command. mpileup requires a samtools-specific index of the reference genome. This can be generated using the faidx command.

```
samtools faidx Athaliana_167_TAIR9.fa
samtools mpileup -f Athaliana_167_TAIR9.fa Parent1_sorted.bam -o Parent1.pileup
samtools mpileup -f Athaliana_167_TAIR9.fa Parent2_sorted.bam -o Parent2.pileup
samtools mpileup -f Athaliana_167_TAIR9.fa DH_example_sorted.bam -o DH_example.pileup
```

4. Filter mpileup files to identify single nucleotide polymorphisms (SNPs) that are segregating between the two parents. A custom script was used to identify positions segregating between the two parental genotypes. Additionally, this script outputs the putative genotype call of each segregating position (Parent 1 = 0, Parent 2 = 1). In this case, the SNPs are called relative to the reference genome. The mpileup file contains both non-polymorphic (reference) and polymorphic positions for each genotype.

```
perl 01.extract_SNPs_from_mpileup.pl Parent1.pileup Parent2.pileup DH_example.pileup DH_example.geno
```

3.5.4 Generate Final Genotype Calls for each DH RIL

1. Remove spurious genotype calls using an HMM algorithm. Since the sequencing coverage is low for this data set, a number of incorrect genotype calls are generated. These spurious calls are likely the result of sequencing or alignment errors and they are often enriched in centromeric regions. Since the DH RILs are expected to be recombinant inbred lines with few recombination events occurring per line, an HMM can quickly predict the expected genotypic state at each chromosomal position, removing incorrect genotype calls. Although there are a number of available HMM packages, the one demonstrated below can be easily implemented in the R statistical package [16, 17] (*see* **Note 26**). In this case, the probability of transitioning between parental genotypes on a chromosome is 5%, while the probability of remaining in the current genotypic state is 95% (i.e., transition probabilities). Additionally, given an observed genotype call, the probability that call is correct is 85%, while 15% of the time the genotype will be incorrect (i.e., emission probability). We have found that these are the best probability settings for this data, but these values are likely to vary by population type (i.e., F_2 versus RIL) and sequencing depth. Figure 3 shows the raw genotype

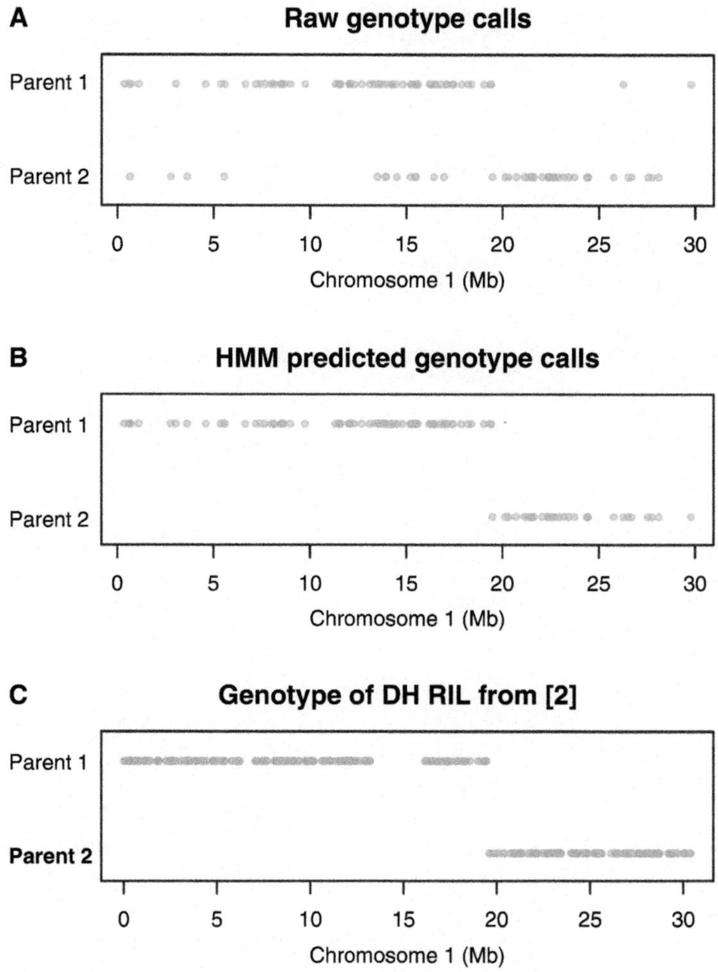

Fig. 3 Prediction of DH RIL genotypes from low-coverage sequencing. (**a**) The raw genotype output for a representative DH RIL (from Subheading 3.5.3). Here, each point represents a SNP segregating between the two parents of the population. (**b**) The HMM predicted genotypes (from Subheading 3.5.4) for each position in (**a**). Low-coverage sequencing can lead to unlikely genotype calls, and these are corrected by the HMM. (**c**) The genotype of this DH RIL from [2]. There are a larger number of segregating SNPs in (**c**) because high-coverage parental data was used to identify polymorphisms (*see* **Note 27**)

calls for the example DH RIL (Fig. 3a), as well as the HMM genotypes (Fig. 3b), and the genotypes observed in the original analyses [2] for a representative chromosome (Fig. 3c) (*see* **Note 27**).

```
install.packages("HMM")
library(HMM)

snps <- read.table("DH_example.geno",sep="\t",fill=TRUE)
snps <- snps[!is.na(snps[,6]),]
head(snps)
dim(snps)

states <- c("A","B") #Parent 1 = A; Parent 2 = B
symbols <- c(0,1)    #Parent 1 = 0; Parent 2 = 1

transProbs <- matrix(c(0.95,0.05,0.05,0.95),
              ncol=length(states),
              nrow=length(states),
              byrow=TRUE)
emissionProbs <- matrix(c(0.85,0.15,0.15,0.85),
                 nrow=length(states),
                 ncol=length(symbols))

hmm <- initHMM(states,symbols,
               transProbs=transProbs,
               emissionProbs=emissionProbs)

vit <- viterbi(hmm,as.character(snps[,6]))
out <- cbind(snps,vit)
colnames(out) <-
c("Chromosome","Position","Reference_base","Coverage",
            "DH_base","DH_genotype","HMM_prediction")
head(out)
```

2. Remove genotypes from the data set that are not true DH RILs. These lines could be aneuploid, contain fragments of the haploid inducer genome, or be contaminants from cross-pollination. To cull genotypes containing contamination from the Col-0 haploid inducer, identify SNPs that are shared between the two parents, but that differ from Col-0. If there is an excess of Col-0 SNPs in a putative DH, particularly in centromeric regions, then these lines may have entire or fragmented chromosomes derived from the haploid inducer and should be excluded from QTL mapping. Expect approximately 8% of lines to contain Col-0 fragments [2]. As DH RILs are expected to be completely homozygous, aneuploids and individuals generated via cross-contamination will exhibit an excess of heterozygosity, and these lines should also be removed from subsequent analyses.
3. Combine genotype calls for each DH RIL and proceed with QTL mapping using the guidelines outlined in the R/qtl package.

```
@Article{Broman2003,

title = {R/qtl: {QTL} mapping in experimental
crosses},
```

```
author = {Karl W. Broman and Hao Wu and Saunak
Sen and Gary A. Churchill},
journal = {Bioinformatics},
year = {2003},
volume = {19},
pages = {889-890},
}.
```

4 Notes

1. In this method, we use *cenh3-1/cenh3-1 GFP-tailswap/GFP-tailswap* plants (sometimes known as GFP-tailswap) as the haploid inducer. One disadvantage of using this inducer is that haploid plants must be selected by phenotype once grown (*see* Subheading 3.3.2). Another option is to use the haploid inducer known as SeedGFP-HI. These plants have an additional GFP construct that is expressed exclusively in the embryo and endosperm of seeds, making it possible to select haploid seeds prior to sowing [18]. If interested in using SeedGFP-HI, more details can be found in [19].
2. In this protocol, we will use F_1 plants as the starting material. Plants from the F_2 (or higher generation crosses) can also be used if additional recombination events are desired.
3. Custom plant sleeves can be made from thin clear polypropylene sheets commonly used to package flowers. Trim the sheet into a rectangle about as tall as the predicted flower stalks of the plant you would like to cover and about 1.5 times as wide as the desired sleeve circumference. Roll this rectangle into a tube (the overlapping ends will provide some additional structural support) and secure with adhesive tape. Slide a pre-rolled plastic sleeve over each plant when it begins to flower and leave in place until siliques are mature.
4. For this and all future stratification steps, the choice of parents may affect the stratification requirement. Assess whether this stratification regime is appropriate for your parents and F_1, and if not, modify this and all subsequent stratification steps as required.
5. Unless otherwise stated, the standard long-day *A. thaliana* growth conditions of your facility will be appropriate for all plant growth steps in this protocol. For those growing *A. thaliana* for the first time, The Arabidopsis Biological Resource Center provides a good overview of standard plant growth procedures (https://abrc.osu.edu/seed-handling).

6. It is highly recommended that the genotypes of all starting materials be verified before proceeding. A detailed protocol for genotyping the haploid inducer line is available in [19]. Parents and F_1 plants can be confirmed either by PCR genotyping or through next-generation sequencing. Markers suitable for single-marker genotyping of many *A. thaliana* natural accessions can be developed from the 1001 Genomes Project data set (http://1001genomes.org/) [20]. The Strain ID Tool of the 1001 Genomes Project (http://tools.1001genomes.org/strain_id/#strain-id) can be used to identify a genotype from next-generation sequencing data. Any parent not included in the 1001 Genomes Project data set should be resequenced using next-generation sequencing, since this data will be necessary for genotyping the resulting DH RILs.
7. In this protocol, we give estimates of the number of plants and crosses needed to produce about 250 DH RILs. Since growing conditions may vary from lab to lab, these numbers are intended as a guideline rather than a guarantee.
8. Since the inducer line is not completely male sterile, the direction of the cross can be reversed if desired. This reversal has two disadvantages. First, the flowers of the female parent would need to be emasculated, greatly increasing the crossing time and effort. Secondly, many more crosses would be needed to compensate for the low male sterility of the inducer line. For these reasons, we have not attempted to construct DH RILs using the inducer line as the male parent, although it is theoretically feasible.
9. Since haploid inducer flowers are largely male sterile, it is acceptable to include flowers that are older than one would normally consider as female parents. In short, any flower with petals and an intact stigma can be used as a female parent in these crosses.
10. The number of plants suggested in **step 5** of Subheading 3.1 assumes that the necessary crosses will be made over the course of several weeks as additional flowers mature.
11. The optimal length of dry storage can vary from accession to accession. Assess the length of time needed to give optimal germination results for the parents of the F_1 used.
12. We use chlorine gas sterilization for our seeds (outlined as "vapor-phase sterilization" in [21]). Other surface sterilization methods should also work, but we have not tested them.
13. To make the subsequent work of transplanting and identifying haploids more manageable, seeds can be plated in several rounds spaced 3–4 weeks apart.

14. Haploid germination can be accelerated by treatment with gibberellic acid, either by soaking the seeds in 100 μM GA_3 before plating or by adding GA_3 to the MSMO plates to a final concentration of 25 μM [19].
15. It may take a little bit of experience to become confident at spotting differences in size and developmental phenotypes. In addition to using the photographs in Fig. 1 as a reference, it can also help to have example plants of both parents and/or an F_1 growing at the same time for comparison purposes.
16. Although not generally needed, haploid plants can be treated with colchicine to induce chromosome doubling. See the protocol outlined in [11].
17. Plants can also be grown in conditions other than those outlined here, for example, on MS agar plates. If alternative conditions are desired, ensure that the tissue can be harvested from each plant without jeopardizing the phenotype of interest. Alternatively, DH RILs can be grown separately for genotyping and phenotyping.
18. If experiment is blocked by replicate (i.e., genotypes are independently randomized within each replicate), then the tissue does not need to be harvested from all replicates.
19. A number of protocols will be appropriate for this step, including CTAB, Qiagen DNeasy, and many others.
20. There are other options for genotyping, including high-coverage reduced-representation sequencing, as used in [2]. Your choice of method will depend on the capabilities of your sequencing facility, the costs per individual (and marker), the coverage desired, and the expected rate of polymorphism between the original parents. For whole-genome sequencing, marker coverage is expected to be low, but many polymorphic positions are interrogated per individual. In contrast, the per site coverage of reduced-representation libraries is higher, depending on the chosen restriction enzyme, but far fewer polymorphic sites are interrogated. More discussion of the methods and example protocols for each method are available in [22].
21. High-coverage sequencing and genotyping data for a large number of *A. thaliana* accessions is available through the 1001 Genomes Project (http://1001genomes.org/) [20].
22. If working in a single working directory in a Unix or Linux environment, the example code in Subheading 3.5 can be pasted into your terminal. Note that the software packages must be properly installed and added to your PATH for the code to run.

23. If using paired-end data, this command must be altered. See the trimmomatic manual for more details (http://www.usadellab.org/cms/uploads/supplementary/Trimmomatic/TrimmomaticManual_V0.32.pdf).
24. Although the most current *A. thaliana* annotation is TAIR10, the sequences of the chromosomes in the genome fasta file have not changed since TAIR9. Therefore, Athaliana_167_TAIR9.fa.gz corresponds to TAIR10 and is the current genome version at this time. The TAIR10 fasta files for each chromosome (TAIR10_chr1.fas, etc) can also be downloaded directly from ftp://ftp.arabidopsis.org/home/tair/Sequences/whole_chromosomes/. If this option is used, the chromosome fasta files will then need to be concatenated manually to produce a whole-genome fasta file suitable for this analysis.
25. The data provided for the example is 40 bp single-end data. It would not be uncommon for data produced in your sequencing facility or downloaded from the 1001 Genomes Project to be 100-150 bp paired-end data. If this is the case, these mapping commands should be modified, either by using bwa sampe in place of bwa samse in Subheading 3.5.2, **step 4** or by using bwa mem in Subheading 3.5.2, **step 3** and removing Subheading 3.5.2, **step 4**.
26. The commands in Subheading 3.5.4 should be run directly in R after setting the proper current working directory. The working directory should contain the DH_example.geno file.
27. The results in Fig. 3c differ from those obtained with the example data set since the original analysis used whole-genome parental data to identify segregating polymorphisms and therefore had more power to detect recombination events.

Acknowledgments

We thank Anne Britt for helpful discussions. The DH RIL project in the Maloof Lab was funded by NSF grants IOS-0923752 and IOS-0820854. RM acknowledges Ramalingaswami Fellowship awarded by the Department of Biotechnology (DBT), Government of India.

References

1. Lynch M, Walsh B (1997) Mapping and characterizing QTLs: inbred line crosses. In: Genetics and analysis of quantitative traits. Sinauer Associates, Sunderland, MA, pp 431–489
2. Seymour DK, Filiault DL, Henry IM, Monson-Miller J, Ravi M, Pang A, Comai L, SWL C, Maloof JN (2012) Rapid creation of Arabidopsis doubled haploid lines for quantitative trait locus mapping. Proc Natl Acad Sci

U S A 109:4227–4232. doi:10.1073/pnas.1117277109

3. Dunwell JM (2010) Haploids in flowering plants: origins and exploitation. Plant Biotechnol J 8:377–424. doi:10.1111/j.1467-7652.2009.00498.x
4. Forster BP, Thomas WTB (2010) Doubled haploids in genetics and plant breeding. In: Janick J (ed) Plant breeding reviews, vol 25. John Wiley & Sons, Inc., Hoboken, NJ, pp 57–88. http://onlinelibrary.wiley.com/book/10.1002/9780470650301
5. Guiderdoni E, Galinato E, Luistro J, Vergara G (1992) Anther culture of tropical japonica indica hybrids of rice (*Oryza sativa* L.) Euphytica 62:219–224. doi:10.1007/bf00041756
6. Alan AR, Mutschler MA, Aigars B, Ed C, Earle ED (2003) Production of gynogenic plants from hybrids of *Allium cepa* L. and *A. roylei* Stearn. Plant Sci 165:1201–1211. doi:10.1016/s0168-9452(03)00327-3
7. Ferreira ME, Williams PH, Osborn TC (1994) RFLP mapping of *Brassica napus* using doubled haploid lines. Theor Appl Genet 89:615–621. doi:10.1007/BF00222456
8. Coe EH (1959) A line of maize with high haploid frequency. Am Nat 93:381–382. doi:10.1086/282098
9. Kasha KJ, Kao KN (1970) High frequency haploid production in barley (*Hordeum vulgare* L.) Nature 225:874–876. doi:10.1038/225874a0
10. Barclay IR (1975) High frequencies of haploid production in wheat (*Triticum aestivum*) by chromosome elimination. Nature 256:410–411. doi:10.1038/256410a0
11. Ravi M, Chan SWL (2010) Haploid plants produced by centromere-mediated genome elimination. Nature 464:615–618. doi:10.1038/nature08842
12. Tek AL, Stupar RM, Kiyotaka N (2015) Modification of centromere structure: a promising approach for haploid line production in plant breeding. Turk J Agric For 39:557–562. doi:10.3906/tar-1405-137
13. Bolger AM, Lohse M, Usadel B (2014) Trimmomatic: a flexible trimmer for Illumina sequence data. Bioinformatics 30:2114–2120. doi:10.1093/bioinformatics/btu170
14. Li H, Durbin R (2009) Fast and accurate short read alignment with Burrows-Wheeler transform. Bioinformatics 25:1754–1760. doi:10.1093/bioinformatics/btp324
15. Li H, Handsaker B, Wysoker A, Fennell T, Ruan J, Homer N, Marth G, Abecasis G, Durbin R, 1000 Genome Project Data Processing Subgroup (2009) The sequence alignment/map format and SAMtools. Bioinformatics 25:2078–2079. doi:10.1093/bioinformatics/btp352
16. R Core Team (2016) R: A language and environment for statistical computing. R Foundation for Statistical Computing, Vienna, Austria. http://www.R-project.org/
17. Scientific Software Development - Dr. Lin Himmelmann and www.linhi.com (2010) HMM: HMM - Hidden Markov Models. R package version 1.0 http://CRAN.R-project.org/package=HMM
18. Ravi M, Marimuthu MPA, Tan EH, Maheshwari S, Henry IM, Marin-Rodriguez B, Urtecho G, Tan J, Thornhill K, Zhu F, Panoli A, Sundaresan V, Britt AB, Comai L, Chan SWL (2014) A haploid genetics toolbox for *Arabidopsis thaliana*. Nat Commun 5:5334. doi:10.1038/ncomms6334
19. Ravi M, Bondada R (2016) Genome elimination by tailswap CenH3: in vivo haploid production in *Arabidopsis thaliana*. In: Murata M (ed) Chromosome and genomic engineering in plants. Springer, New York, pp 77–99
20. The 1001 Genomes Consortium (2016) 1,135 genomes reveal the global pattern of polymorphism in *Arabidopsis thaliana*. Cell 166(2):481–491
21. Clough SJ, Bent AF (1998) Floral dip: a simplified method for *Agrobacterium*-mediated transformation of *Arabidopsis thaliana*. Plant J 16:735–743. doi:10.1046/j.1365-313x.1998.00343.x
22. Rowan BA, Seymour DK, Chae E, Lundberg DS, Weigel D (2017) Methods for genotyping-by-sequencing. In: White SJ, Cantsilieris S (eds) Genotyping: methods and protocols. Springer, New York. pp 221–242

Part II

Phenotypes

Chapter 5

Assessing Distribution and Variation of Genome-Wide DNA Methylation Using Short-Read Sequencing

Jörg Hagmann and Claude Becker

Abstract

DNA methylation is one of the most prominent epigenetic marks and is particularly complex in plant genomes. Whole-genome bisulfite sequencing (WGBS) using short reads has become the standard tool to study genome-wide patterns of DNA methylation. The goal of the present protocol is to enable readers to perform WGBS on both the wet lab and the computational side. We briefly outline important steps in bisulfite library preparation, then focus on the different aspects of DNA methylation analysis, from read mapping to identifying biologically relevant differential methylation between different samples.

Key words DNA methylation, Short-read sequencing, Bisulfite, Differential methylation, Differentially methylated region

1 Introduction

DNA methylation generally refers to the covalent attachment of a methyl group to the 5 position of the cytosine pyrimidine ring (5-methylcytosine). It is the most widely studied of all epigenetic marks, which shape the plant genome and play important roles in plant development, stress response, and transposon silencing (out of the many review articles on the topic, *see*, e.g., [1]). In plants, the investigation of whole-genome DNA methylation patterns is particularly challenging, because DNA methylation can occur in three different sequence contexts: CG, CHG, and CHH (where H represents any base but G). Each sequence context has a different underlying molecular machinery responsible for the establishment and maintenance of the methylation; in consequence, depending on the biological question that is addressed, it is important to be able to discern DNA methylation by sequence context. While genome-wide methylation patterns can be studied by using either chip-based approaches (methylation-sensitive tiling arrays) or immunoprecipitation of methylated DNA, followed by sequencing

Wolfgang Busch (ed.), *Plant Genomics: Methods and Protocols*, Methods in Molecular Biology, vol. 1610,
DOI 10.1007/978-1-4939-7003-2_5, © Springer Science+Business Media LLC 2017

(MeDIP-seq) [2, 3], these methods lack both single-base resolution and differentiation by sequence context.

With the advent of short-read sequencing technology, the method of choice to study site-specific DNA methylation has been whole-genome sequencing of bisulfite-treated DNA. Treatment of genomic DNA with sodium bisulfite leads to deamination of unmethylated cytosines to uracil, whereas methylated cytosines remain unchanged [4]. After PCR amplification, uracil will appear as thymine in the sequencing read. It is this C-to-T change with regard to the reference genome that informs the user about the original methylation status of the cytosine in question. *Arabidopsis thaliana* was the first organism of which a genome-wide DNA methylation map was generated using this technology [5, 6], and WGBS has seen a rapid increase in usage over the last years.

While determining and quantifying the methylation status of each cytosine in the genome have become standard practice, differential analysis of DNA methylation between samples (e.g., control vs. treated; wild type vs. mutant) is still challenging and has led to the development of a multitude of statistical tools [7]. As the large majority of these was developed with mammalian DNA methylation patterns in mind, and as mammalian DNA methylation is restricted almost exclusively to cytosines in the CG sequence context [8], these tools are not always amenable to plant genomes.

In the current protocol, we provide a step-by-step guideline on how to analyze DNA methylation in plants, from library preparation to sequencing data analysis and testing for differential methylation between multiple samples.

2 Material

2.1 Library Preparation

1. Column-based kit for the extraction of genomic DNA (gDNA).
2. Bisulfite treatment kit (*see* **Note 1**).
3. Library preparation kit or reagents (*see* **Note 1**).
4. DNA polymerase suitable for uracil-containing templates (*see* **Note 2**).

2.2 Computational Analysis

1. Linux workstation (*see* **Note 3**).
2. Software packages from https://sourceforge.net/projects/methpipeline (*see* **Note 4**).

3 Methods

3.1 Library Preparation

1. Isolate gDNA using a column-based purification from frozen plant tissue. We discourage from using cetyltrimethylammonium bromide (CTAB)-based protocols, as residuals of this chemical might interfere with the efficiency of BS treatment.

Libraries can be prepared from as little as 10 ng of gDNA, although we recommend using 100 ng starting material whenever material is not limited.

2. Prepare sequencing libraries according to kit manufacturers' instructions as if preparing standard genomic DNA libraries, and stop after adapter ligation and subsequent cleanup.
3. Prepare adapter-ligated DNA in a total volume of 40 μL for treatment with sodium bisulfite (*see* **Note 5**). Several kits are available to perform this protocol; we follow the protocol for the EpiTect Plus Bisulfite kit (Qiagen, Hilden, Germany) here.
4. Add 85 μL of freshly prepared sodium bisulfite solution and 15 μL protection buffer, mix by pipetting (do not vortex in this or any of the following steps).
5. Incubate in a thermal cycler with a 100 μL heating block. Smaller heating blocks, as they are frequent in older PCR machines, lead to drastic decrease in conversion efficiency and should be avoided. Do not expose samples to temperatures below 20 °C while they are in BS solution, either before or after the thermal cycle. However, place samples on ice immediately after the ensuing cleanup. Run the thermal cycle as follows (this takes approximately 10 h and can be run overnight):

Step	Time (min)	Temperature (°C)
Denaturation	5	95
Incubation	25	60
Denaturation	5	95
Incubation	85	60
Denaturation	5	95
Incubation	175	60
Incubation	25	60
Denaturation	5	95
Incubation	85	60
Denaturation	5	95
Incubation	175	60
Hold	Infinite	20

6. Perform cleanup according to the manual and elute in 15 μL of elution buffer.
7. PCR-amplify the library using 7.5 μL of the elution as template. Add 25 μL Kapa HiFi Uracil+ HotStart polymerase ReadyMix (Kapa Biosystems), 5 μL of Illumina-compatible

primers, and 2.5 μL water. Amplify using the following PCR protocol for 14 cycles:

Initial denaturation	98 °C	45 s
Denaturation	98 °C	15 s
Annealing	65 °C	30 s
Extension	72 °C	30 s
Final extension	72 °C	60 s

8. Continue with post-PCR library cleanup as indicated in the library preparation instructions.

3.2 Mapping of BS Short Reads

1. Analyze the general properties of the bisulfite-treated reads using quality control tools such as FastQC (http://www.bioinformatics.babraham.ac.uk/projects/fastqc/) and, if necessary, remove overall low quality bases at both read ends (e.g., using fastq-mcf; https://github.com/ExpressionAnalysis/ea-utils/blob/wiki/FastqMcf.md).
2. Compute an index of the reference genome using bismark (*see* **Notes 6** and 7; http://www.bioinformatics.babraham.ac.uk/projects/bismark/):
```
bismark_genome_preparation <path_of_reference genome>
```
3. Align reads against this index by the following commands with minimum parameters. For the full list of bismark's parameters, we refer readers to the bismark manual: http://www.bioinformatics.babraham.ac.uk/projects/bismark/Bismark_User_Guide.pdf. Depending on the nature of the sequencing reads, use:
```
bismark <genome_folder> -1 <reads1_file> -2 <reads2_file>
```
if the library was sequenced paired end (make sure that the read files are sorted in a way such that pairs of reads occur at identical line numbers in both files) or
```
bismark <genome_folder> <single_read_file>
```
if the library was sequenced single end.

3.3 Quantifying Per-Position Methylation Levels and Building the Genome Matrix

1. Retrieve methylation information for each individual sample from the mapping files using the script bismark_methylation_extractor, provided with the bismark package:
```
bismark_methylation_extractor <-s or -p> --comprehensive --CX_context --bedGraph <bismark_sam_file>
```
Specify -s for single-end libraries and -p for paired-end libraries.

2. Sort the bedGraph output file <*bismark.cov> of bismark_methylation_extractor by genomic position:
```
sort -k1,1d -k2,2g <bismark_outfile> > <bis-
mark_ordered_outfile>
```
3. Create a tab-separated sample file with the following columns: (1) unique sample name and (2) full path of the sample-specific bismark sam file (*_bismark.sam).
4. Generate a genome matrix containing per-position information of all samples (*see* **Note 8**):
```
perl generate_genome_matrix.pl -s <sample_
file> -r <reference_genome_fasta> -i bismark
-o <genome_matrix_file>
```
Depending on the number of samples and the genome size, the matrix can be quite large. To ease computation and reduce disk space, the parameter -c restricts matrix entries to those that show a minimum read depth of <c> in at least one sample (recommended: three or higher).

3.4 Determining Statistically Significant Methylation by Position

1. Test each position in every sample for statistically significant methylation (*see* **Note 9**). Start by extracting positions from the genome matrix that correspond to genomic sequences known to be unmethylated. Due to sequencing errors and incomplete bisulfite conversion, genomic positions should be tested for a statistically significant methylation signal. This is commonly done by simply testing the methylation rate against the nonconversion rate of the bisulfite treatment mentioned (*see* **Note 5**), followed by multiple testing correction.
2. Extract the positions of the unmethylated genome sequences from the genome matrix, e.g., for the chloroplast sequence "ChrC" by using the grep command in a terminal window:
```
grep "^ChrC" <genome_matrix> > <genome_ma-
trix.unmeth>
```
3. Calculate the false methylation rates on the unmethylated genome sequence:
```
perl false_methylation_rates.pl <genome_ma-
trix_unmeth> 5 > FMRs.txt
```
4. Perform the statistical test and retrieve methylated positions per sample:
```
perl binomial_test_for_methylation.pl -g
<genome_matrix> FMRs.txt > <genome_matrix_
with_MPs>
```
The output is the input genome matrix with an additional column attached. It contains one character per sample, ordered the same way as the samples in the matrix:

"o": not methylated.

"m": statistically significantly methylated.

"Z": zero coverage.

"z": below minimum coverage (this can be adjusted by -c in the script binomial_test_for_methylation.pl).

Thus, methylated positions of a sample can be retrieved by parsing through the genome matrix and gathering sites where the quality string contains "m" in the respective sample column, e.g., for sample 10:

```
awk '{ split($NF, a, ""); if (a[10]=="m")
print }' <genome_matrix_with_MPs>
```

3.5 Determining Differentially Methylated Positions (DMPs)

Although DMPs are good indicators of overall variation in a dataset, the biological relevance of differential methylation at single cytosines is questionable. Furthermore, DMP detection is heavily biased by coverage and sequence context. We will therefore not indulge in DMP analysis and rather focus on the detection of differentially methylated regions (DMRs). For full descriptions on how to detect and statistically validate DMPs in a multi-sample dataset, we refer readers to refs. [9, 10].

3.6 Segmenting the Genome into Methylated Regions (MRs)

1. For each sample, run the HMM segmentation algorithm for MR detection using the following command (*see* **Notes 10** and **11**):

   ```
   hmm_mrs -o <output_MR_file> -p <output_param-
   eter_file> -x 1 <input_genome_matrix>
   ```

   hmm_mrs reports MRs in a simple BED file format. The columns are (1) chromosome, (2) start position (included), (3) end position (included), (4) "MR" followed by a consecutive number as unique region identifier, (5) the number of cytosines within the MR, and (6) a strand identifier "+" for compatibility reasons.

2. If there are two or more replicates per group, merge MRs from these replicates into one BED file, such that overlapping MRs start with the most upstream (5′) and end with the most downstream (3′) position (e.g., by using bedtools: http://bedtools.readthedocs.io/en/latest/index.html):

   ```
   sort k1,1 -k2,2n -k3,3n <MR_file_1> <MR_file_2>
   | bedtools merge -i stdin > <merged_MR_file>)
   ```

3.7 Determining Differentially Methylated Regions (DMRs)

3.7.1 Selecting Segments to Test for Differential Methylation

1. Prepare a sample list file with the following content for each sample (sample here refers to grouped replicates, or individual samples, depending on what pairwise tests the user wants to investigate):

Column 1: sample ID (has to be identical to the one in the MR BED file).

Column 2: comma-separated list of columns of the sample's replicates in the genome matrix file (0-based; specify "4" for sample 1 in the matrix, "5" for sample 2, etc.).

Column 3: path to the MR file (in case of merged replicates, this path should refer to the merged MR file of that replicate group).

2. Define the testable segments of all pairwise comparisons between samples listed in the sample file (*see* **Notes 12** and **13**):

```
perl select_regions_to_test.pl -s <sample_list_file> -m <genome_matrix> -o <output_folder>
```

3.7.2 Statistical Test of Differential Methylation

1. Perform the statistical test for differential methylation by running perform_differential_tests.pl for each batch of segments and for each pairwise comparison of samples (i.e., for each output file from Subheading 3.6, **step 1**; *see* **Note 14**). A command including all mandatory parameters looks like this:

```
perl perform_differential_tests.pl -s <sample_list_file> -m <genome_matrix> -i <regular_expression_for_input_files
```

(It has to be double quoted! e.g., "./*.[0-9]*" would submit a job for all files in the current folder that end with a dot followed by a digit).

The statistical test produces one output file (with the appendix ".DIFF") for each input file.

3.7.3 Grouping Strains into Epialleles

1. Concatenate and position sort all output files from the statistical files of the previous step into one combined file:

```
sort -g -m -k1 -k2 <all_test_files>.DIFF > all_pairwise_tests.txt.DIFF
```

2. Run the `group_samples_and_test_groups.pl` script as follows (*see* **Note 15**):

```
perl group_samples_and_test_groups.pl pl -s <sample_list_file> -m <genome_matrix> -d all_pairwise_tests.txt.DIFF -i 2
```

The output of this command is a series of files that each represents an iteration of the grouping and processing. Grouped but still overlapping DMRs between samples can be found in the file `all_pairwise_tests.txt.DIFF.qv.GROUPS`, DMRs between sample groups in the file all_pairwise_tests.txt.DIFF.qv.GROUPS.grouptests.qv.GROUPS.

3.7.4 Resolving Overlapping DMRs

1. To resolve overlapping regions (*see* **Note 16**) and obtain the final list of DMRs, run the following command:

```
perl resolve_overlapping_dmrs.pl -f <overlapping_DMR_file> -m < 'max_signif_samples'/'min_qvalue' > <final_DMRs>
```

DMRs and hDMRs are treated equally in the ordering process. If you are interested only in hDMRs, extract hDMRs from all

DMRs (e.g., by `awk '$NF!="-"' <DMR_file> > <hDMR_file>`) and run the script resolve_overlapping_dmrs.pl only on the hDMR file. *See* **Note 17** for explanations on the output file.

4 Notes

1. Several manufacturers offer kits for preparing bisulfite (BS)-treated libraries. Alternatively, kits for the preparation of genomic libraries can be used, in combination with a bisulfite treatment kit. We typically use the TruSeq Nano DNA sample prep kit (Illumina, CA, USA) in combination with the EpiTect Plus Bisulfite kit (Qiagen, Hilden, Germany). When using other library prep kits, users should verify that the adapters for Illumina sequencing contain methylated cytosines only, as otherwise cytosines in the adapters will be deaminated in the BS treatment, preventing amplification of the library and consequently preventing binding to the flow cell.
2. Most commercially available polymerases are not able to read uracil efficiently. We recommend the use of KAPA HiFi HotStart Uracil+ polymerase (Kapa Biosystems, MA, USA) for library amplification, but other commercial polymerases are available.
3. Although all computational processes described here can be run on a standard personal computer (Linux), we highly recommend the use of a multicore computational server or, better still, a computational cluster, particularly when sample numbers are large and many comparisons need to be integrated in the differential analysis.
4. Due to space limitations, we cannot list and explain all parameters of every tool mentioned in this protocol. In this chapter we indicate commands with standard or default parameter values; we refer readers to the manuals of the respective published packages as well as to refs. 9, 10 for further details.
5. BS treatment of DNA is a chemical process, and as such it is never 100% efficient. During the data analysis, it is necessary to quantify the efficiency of the BS conversion and to assess the amount of nonconverted unmethylated cytosines. If you are working with *Arabidopsis thaliana*, it is safe to use cytosines in the chloroplast genome as a reference, as *A. thaliana* chloroplast DNA is unmethylated and every unconverted C can be accounted for by incomplete conversion. If you are working with other species of which you do not know the chloroplast sequence or of which it is known that the plastome is (completely or partially) duplicated in the nuclear genome (as is the case in maize, e.g., but also in many other plant species), it is necessary to include a standard DNA in your library. In this

case, add 0.1 % or lambda-phage DNA to your sample before you start the library preparation. Reads mapping to lambda can ultimately be used to assess the degree of nonconversion in your sample.

6. At this point we suppose that there is a reference genome sequence for the plant species that you are working with. For BS sequencing data analysis without a reference genome, we recommend ref. 11.

7. Other tools for aligning BS reads to a reference genome are available. The software to identify differentially methylated regions explained here is compatible with bismark output files. When using other mappers, users need to adjust the pipeline accordingly.

8. All subsequent steps for retrieving methylation data described here rely on a specific file format, which we refer to as a genome matrix, which is composed like this:

1	3	CG	C	40/0/0/3	40/20/1/4	40/100/4/0
1	8	CHG	G	40/10/1/9	40/0/0/7	40/60/3/2

The columns are:

1. Chromosome ID.
2. Position (1 based).
3. Sequence context.
4. Strand (C, forward; G, reverse).
5. Quadruple for each sample: quality value/methylation rate/number of methylated reads/number of unmethylated reads. When using bismark for read mapping, the quality value is always set to the maximum value 40.

The genome matrix contains the raw read counts for all genomic cytosine sites in all samples.

9. Due to sequencing errors and incomplete bisulfite conversion, genomic positions should be tested for a statistically significant methylation signal. This is commonly done by simply testing the methylation rate against the nonconversion rate of the bisulfite treatment mentioned above, followed by multiple testing correction.

10. This step uses a two-state hidden Markov model implementation to segment the genome into highly and lowly methylated regions. The theoretical basis of this approach has been explained in detail in Molaro et al. [12] and has been adapted to the specificities of plant DNA methylation [10]. In brief, the model learns methylation rate distributions for both an unmethylated and a methylated state for each sequence context separately

(CG, CHG, and CHH) while simultaneously estimating transition probabilities between the two states from genome-wide data. On the trained model, the most probable path of the HMM along the genome is then used to define regions of high and low methylation.

11. For each of the following command lines, we refer readers to the manual provided with the software for full information on all available parameters.
12. For details on how segments to be tested for differential methylation are selected, see the full manual (https://sourceforge.net/projects/methpipeline/files/v0.1/DMR_manual.pdf). In brief, identical segments (same start and end positions) between two samples with (1) differing methylation levels or (2) with high methylation state in all samples are tested.
13. Explaining the exact content of the output files generated by this command exceeds the scope of this protocol. It is sufficient to know that these files constitute the input for the subsequent statistical test (Subheading 3.7.2).
14. Since elaborate numerical optimization is applied (there is no exact training formula), this step is computationally intense, and it is recommended to execute it highly parallelized, e.g., on a cluster. The preset mode of job submission to a cluster in our script is using Sun Grid Engine's (SGE) *qsub* command. If other cluster software systems are used, the script qsub.pl has to be manually adapted accordingly. We refer readers to the complete manual on https://sourceforge.net/projects/methpipeline/files/v0.1/DMR_manual.pdf for adjusting qsub.pl to their needs and computational environment. An alternative to using a cluster is to specify -l, which generates (hundreds/thousands of) bash scripts of the individual processes that can be locally executed on any server(s).
15. Since the statistical test can be very sensitive for subtle methylation changes (especially for longer segments), we stringently correct the *p*-values of all pairwise comparisons for multiple testing at once. We use Storey's method [13] to calculate a false discovery rate (FDR). To retrieve epiallele frequencies, we group the strains based on their statistical differences to each other. To optionally assess the grouping, and to further stringently filter "obvious" signals, groups are statistically tested against each other with the same test as for the previous step (if option -i is larger than 1). Finally, another filter for "strong" and potentially biologically meaningful methylation differences is applied to obtain "highly differentially methylated" regions (hDMRs). These include only segments in which the mean methylation rate differs at least by a factor of 3 between any two groups.

16. Since the region selection step (Subheading 3.6, **step 1**) conservatively considered many overlapping segments to perform tests for differential methylation, DMRs are likely to overlap each other. We provide two strategies to generate a set of non-overlapping DMRs: All DMRs from a set of overlapping DMRs are ordered either by the descending number of samples that show significant differential methylation for the segments or by the ascending value of the corrected *p*-value of the segments. The first DMR in that list is reported, as well as subsequent DMRs down the sorted list, if they do not overlap with the previously selected segments. The script resolve_overlapping_dmrs.pl provides the two modes termed max_signif_samples and min_qvalue.
17. The output file contains following relevant columns: (1) chromosome; (2) start position; (3) DMR length; (4) number of groups; (5) sample groups, separated by "///," samples in parentheses are not significantly different to any other sample; (6) average methylation levels. Samples are separated by ',', contexts by "/", sample order as in sample sheet; (7) statistically significant sequence contexts of DMR; and (8) statistically significant sequence contexts of hDMR or "-" for no hDMR).

Acknowledgments

This work was supported by the Max Planck Society.

References

1. Pikaard CS, Mittelsten Scheid O (2014) Epigenetic regulation in plants. Cold Spring Harb Perspect Biol 6(12):a019315. doi:10.1101/cshperspect.a019315
2. Zhang X, Yazaki J, Sundaresan A, Cokus S, Chan SW, Chen H, Henderson IR, Shinn P, Pellegrini M, Jacobsen SE, Ecker JR (2006) Genome-wide high-resolution mapping and functional analysis of DNA methylation in arabidopsis. Cell 126(6):1189–1201. doi:10.1016/j.cell.2006.08.003
3. Zilberman D, Gehring M, Tran RK, Ballinger T, Henikoff S (2007) Genome-wide analysis of Arabidopsis thaliana DNA methylation uncovers an interdependence between methylation and transcription. Nat Genet 39(1):61–69. doi:10.1038/ng1929
4. Frommer M, McDonald LE, Millar DS, Collis CM, Watt F, Grigg GW, Molloy PL, Paul CL (1992) A genomic sequencing protocol that yields a positive display of 5-methylcytosine residues in individual DNA strands. Proc Natl Acad Sci U S A 89(5):1827–1831
5. Cokus SJ, Feng SH, Zhang XY, Chen ZG, Merriman B, Haudenschild CD, Pradhan S, Nelson SF, Pellegrini M, Jacobsen SE (2008) Shotgun bisulfite sequencing of the Arabidopsis genome reveals DNA methylation patterning. Nature 452(7184):215–219. doi:10.1038/nature06745
6. Lister R, O'Malley RC, Tonti-Filippini J, Gregory BD, Berry CC, Millar AH, Ecker JR (2008) Highly integrated single-base resolution maps of the epigenome in Arabidopsis. Cell 133(3):523–536. doi:10.1016/j.cell.2008.03.029
7. Robinson MD, Kahraman A, Law CW, Lindsay H, Nowicka M, Weber LM, Zhou X (2014) Statistical methods for detecting differentially methylated loci and regions. Front Genet 5:324. doi:10.3389/fgene.2014.00324
8. Lister R, Pelizzola M, Dowen RH, Hawkins RD, Hon G, Tonti-Filippini J, Nery JR, Lee

L, Ye Z, Ngo QM, Edsall L, Antosiewicz-Bourget J, Stewart R, Ruotti V, Millar AH, Thomson JA, Ren B, Ecker JR (2009) Human DNA methylomes at base resolution show widespread epigenomic differences. Nature 462(7271):315–322. doi:10.1038/nature08514

9. Becker C, Hagmann J, Muller J, Koenig D, Stegle O, Borgwardt K, Weigel D (2011) Spontaneous epigenetic variation in the Arabidopsis thaliana methylome. Nature 480(7376):245–249. doi:10.1038/nature10555

10. Hagmann J, Becker C, Muller J, Stegle O, Meyer RC, Wang G, Schneeberger K, Fitz J, Altmann T, Bergelson J, Borgwardt K, Weigel D (2015) Century-scale methylome stability in a recently diverged Arabidopsis thaliana lineage. PLoS Genet 11(1):e1004920. doi:10.1371/journal.pgen.1004920

11. van Gurp TP, Wagemaker NC, Wouters B, Vergeer P, Ouborg JN, Verhoeven KJ (2016) epiGBS: reference-free reduced representation bisulfite sequencing. Nat Methods 13(4):322–324. doi:10.1038/nmeth.3763

12. Molaro A, Hodges E, Fang F, Song Q, McCombie WR, Hannon GJ, Smith AD (2011) Sperm methylation profiles reveal features of epigenetic inheritance and evolution in primates. Cell 146(6):1029–1041. doi:10.1016/j.cell.2011.08.016

13. Storey JD, Tibshirani R (2003) Statistical significance for genomewide studies. Proc Natl Acad Sci U S A 100(16):9440–9445. doi:10.1073/pnas.1530509100

Chapter 6

Circular Chromosome Conformation Capture in Plants

Stefan Grob

Abstract

The study of nuclear architecture promises novel insights into genome function and regulation. Hereby, quantitative methods based on chromosome conformation capture (3C) revolutionized the field, as they allow accurate and unbiased characterization of 3D genome organization of genomic regions of interest. The choice of the appropriate 3C derivate is crucial to acquire results suited for a specific research question. Circular 3C (4C) is the method of choice to study the genome-wide 3D architecture of a specific genomic region of interest. Here, I present a robust 4C protocol, established in *Arabidopsis thaliana*, which can be employed by any experienced molecular biologist and is applicable in various other plant species.

Key words 3C, 4C, 5C, Hi-C, *Arabidopsis*, Chromatin folding, Chromatin organization, Chromosomal interactions, Chromatin loops, Higher-order organization, Nuclear architecture

1 Introduction

In 2002, Job Dekker and colleagues published a novel method to study chromosomal architecture, which in the following years revolutionized the field [1]. Chromosome conformation capture (3C) allows for analyzing interaction frequencies and, thus, the relative spatial proximity between two specific genomic regions. Soon after, several groups further developed the basic 3C concept to establish various 3C derivate technologies, creating a family of 3C-based methods. Whereas 3C represents the *one-to-one* approach to study specific chromatin interactions, chromosome conformation capture carbon copy (5C) [2], the *many-to-many* 3C version, is used to simultaneously assess interaction frequencies of numerous sets of genomic regions. Circular chromosome conformation capture (4C) [3, 4] is the *one-to-all* approach and allows to study the genome-wide interaction frequencies relative to a specific genomic region of interest (the viewpoint). Finally, Hi-C, the *all-to-all* approach and newest member of the 3C family, is used to map all chromosome interactions at once and is the method of choice to characterize overall nuclear architecture [5]. All 3C-based

Wolfgang Busch (ed.), *Plant Genomics: Methods and Protocols*, Methods in Molecular Biology, vol. 1610, DOI 10.1007/978-1-4939-7003-2_6, © Springer Science+Business Media LLC 2017

technologies are based on a similar protocol. Spatial proximity between genomic regions is captured by formaldehyde cross-linking of chromatin in intact nuclei, leading to the covalent binding of regions in close spatial proximity. The cross-linked chromatin is subsequently digested and religated. Hence, the previously cross-linked interaction partners end up on a single circular molecule, termed the 3C template. Finally the measurement of the abundance of the 3C templates will determine the interaction frequency and, thus, the spatial proximity of a given set of interacting partners. The various 3C derivate methods vary in the readout of the interaction frequencies:

- *3C* (one-to-one) measures interaction frequencies between two genomic regions semiquantitatively (by PCR) or quantitatively (by qPCR) using primers specific to the two interacting regions of interest.
- *5C* (many-to-many) employs the same quantification method as 3C; however, multiplex PCR followed by high-throughput sequencing allows simultaneous analysis of interaction frequencies of many genomic regions.
- *4C* (one-to-all) involves a secondary digestion and religation step. 3C templates representing interaction frequencies of the genomic region of interest to all other genomic regions are amplified by inverse PCR, using primers specific to the viewpoint. The resulting PCR products are subsequently analyzed by high-throughput sequencing or microarray hybridization.
- *Hi-C* (all-to-all) simultaneously measures all genome-wide interaction frequencies by high-throughput sequencing of the entire 3C template pool.

Hi-C may appear as the most appealing method to perform as it reveals all chromosomal interaction frequencies at once. However, for various biological questions, other 3C derivate methods are preferred. By avoiding the data complexity of Hi-C, especially 4C represents a powerful method to assess the 3D interactome of specific loci, as it produces high-resolution interaction data at low cost and time requirements. 4C interaction data of several viewpoints (up to 12 for *Arabidopsis thaliana*) can be generated by using only one Illumina sequencing lane, still providing significantly higher resolution than can be achieved using Hi-C. Thus, if the research question concerns a specific or several genomic loci, 4C is preferred over Hi-C.

Initially, 4C follows the standard 3C workflow to generate 3C templates (*see* Fig. 1). Subsequently, the 3C templates are digested once more by a secondary restriction enzyme. Whereas the primary restriction enzyme is usually a six cutter, the secondary restriction enzyme then recognizes four base pair restriction sites. Thus, the secondary restriction enzyme exhibits a high likelihood to cut within the primary 3C template. Subsequently, another intramolecular ligation step is performed to yield 4C tem-

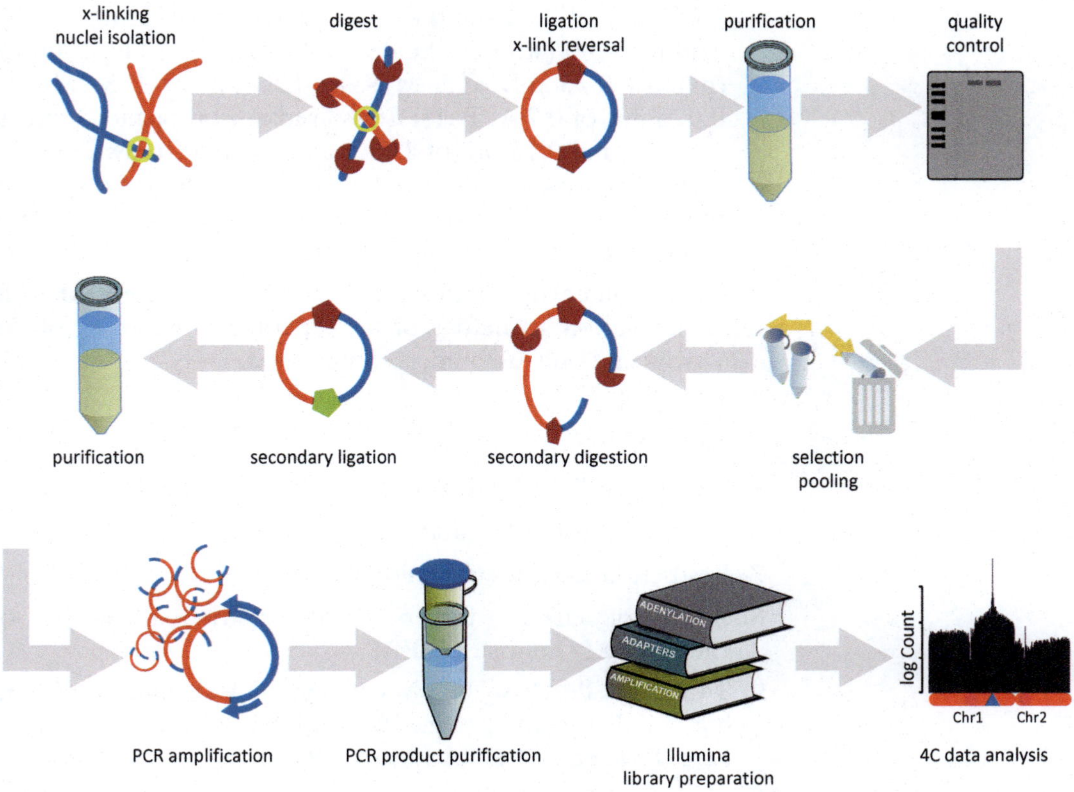

Fig. 1 4C workflow. *Top row*: 3C template generation (1° restriction digest). *Middle row*: quality assessment and 4C template generation (2° restriction digest). *Bottom row*: PCR amplification and Illumina sequencing library preparation

plates (*see* Fig. 1). An inverse PCR, using "outward"-facing primers, which flank the primary and secondary restriction site, will subsequently amplify a viewpoint-specific interactome. The here presented 4C protocol is robust and can be performed by any experienced molecular biologist. Completion of the entire protocol takes approximately 2 weeks.

2 Materials

2.1 Reagents

Prepare reagents using ultrapure water (ddH_2O; e.g., Milli-Q) and store at room temperature (RT) if not otherwise stated.

1. Nuclei isolation buffer (NIB): 20 mM Hepes (pH 8), 250 mM sucrose, 1 mM $MgCl_2$, 0.5 mM KCl, 40% glycerol, 0.25% Triton X-100, 0.1 mM phenylmethanesulfonyl fluoride (PMSF), and 0.1% 2-mercaptoethanol.

   Preparation: prepare freshly on the day of the experiment. For 150 mL NIB, pour 60 mL of glycerol into a measuring cylinder (do not use a pipette), and add 63.375 mL of ddH_2O. Transfer the entire mix to a 250 mL glass bottle. Add

18.75 mL of 2 M sucrose (sterile filtered), 3.75 mL of 10% Triton X-100 (store in the dark), 3 mL of 1 M Hepes buffer (pH 8), 0.75 mL of 1 M KCl, 0.15 mL of 1 M $MgCl_2$, 0.075 mL of 0.2 M PMSF (dissolved in isopropanol, store at –20 °C), and 0.15 mL of 2-mercaptoethanol (be aware of the toxicity of 2-mercaptoethanol, work under a fume hood). Mix the NIB by gentle swirling avoiding the formation of bubbles and keep at RT under a fume hood.

2. 4% formaldehyde solution. Dilute from 36% formaldehyde stock solution at the day of the experiment. Be aware of the toxicity and only use under a fume hood.
3. 2 M glycine.
4. 20% SDS.
5. 20% Triton X-100. Store at RT, and keep in the dark.
6. Complete protease inhibitor tablets (Roche).
7. Restriction enzymes of choice.
8. Restriction enzyme buffers for the restriction enzymes of choice (e.g., CutSmart, NEB).
9. T4 ligase. Purchase highly concentrated enzyme displaying high catalytic activity (e.g., 10–20 μL Weiss units (WU)/μL—here: Promega T4 DNA Ligase (HC)).
10. High-fidelity DNA polymerase (here: Phusion High-Fidelity DNA polymerase, NEB).
11. 50 mM $MgCl_2$.
12. PCR buffer (here: 5× Phusion HF/GC PCR buffer).
13. 10 mM dNTP mix.
14. Ligation buffer 10×: 0.5 M Tris-Cl, 0.1 M $MgCl_2$, 0.1 M DTT, and pH 7.5. Although commercial ligation buffer can be used, it is advisable to prepare your own ligation buffer due to large quantities needed.
15. 100 mM ATP.
16. Molecular biology grade bovine serum albumin (BSA): 10 mg/mL stock. Store at –20 °C.
17. Proteinase K: 10 mg/mL stock, dissolved in ddH_2O, store aliquots at –20 °C.
18. DNase-free RNase A. To prepare a 10 mg/mL stock, weigh 30 mg of lyophilized RNase A powder and add 30 μL of 1 M Tris–HCl (pH 8), 45 μL of 5 M NaCl, and 2925 μL of ddH_2O. Boil for 10 min, cool down to RT, and aliquot. Store at –20 °C.
19. Phenol (liquefied), be aware of the toxicity and only use under a fume hood.
20. Chloroform, be aware of the toxicity and only use under a fume hood.

21. Isoamyl alcohol.
22. 3 M sodium acetate.
23. Glycogen: 20 mg/mL.
24. 5 M sodium chloride.
25. 0.5 M EDTA.
26. 1 M Tris–HCl.
27. Ethanol: 100 and 70%.
28. Agarose.
29. Ethidium bromide, be aware of the toxicity and handle correspondingly.
30. Illumina library preparation kit.
31. AMPure XP PCR purification beads (Agencourt).

2.2 Equipment

1. Vacuum pump.
2. Mortar and pestle.
3. Whatman filter paper.
4. Miracloth.
5. 15 and 50 mL conical tubes (Falcon tubes).
6. 1.5 mL reaction tubes.
7. 0.2 mL reaction tubes.
8. Netting material (e.g., nylon mesh, mesh size approx. 1–2 mm).
9. Thermomixer, ideally equipped for 15 mL reaction tubes.
10. Cooled water bath.
11. Refrigerated microcentrifuge (for 1.5 mL tubes).
12. Refrigerated centrifuge equipped for 15 and 50 mL conical tubes.
13. Thermal cycler.
14. Fume hood.
15. Q-bit spectrometer.
16. Magnetic rack.

3 Methods

3.1 Cross-Linking and Isolation of Plant Nuclei

3.1.1 Collection of Plant Material and Vacuum Infiltration (1.5–2 h)

- *Work under a fume hood at room temperature (RT).*
- *50 mL conical tube, 250 mL glass bottle, measuring cylinder, vacuum pump, desiccator, nylon mesh, and ring.*
- *36% formaldehyde, NIB, protease inhibitor tablets, and 2 M glycine.*

1. Harvest 5–10 g of plant material of your choice (*see* **Note 1**), and equally distribute it to two 50 mL conical tubes (i.e., Falcon tubes).

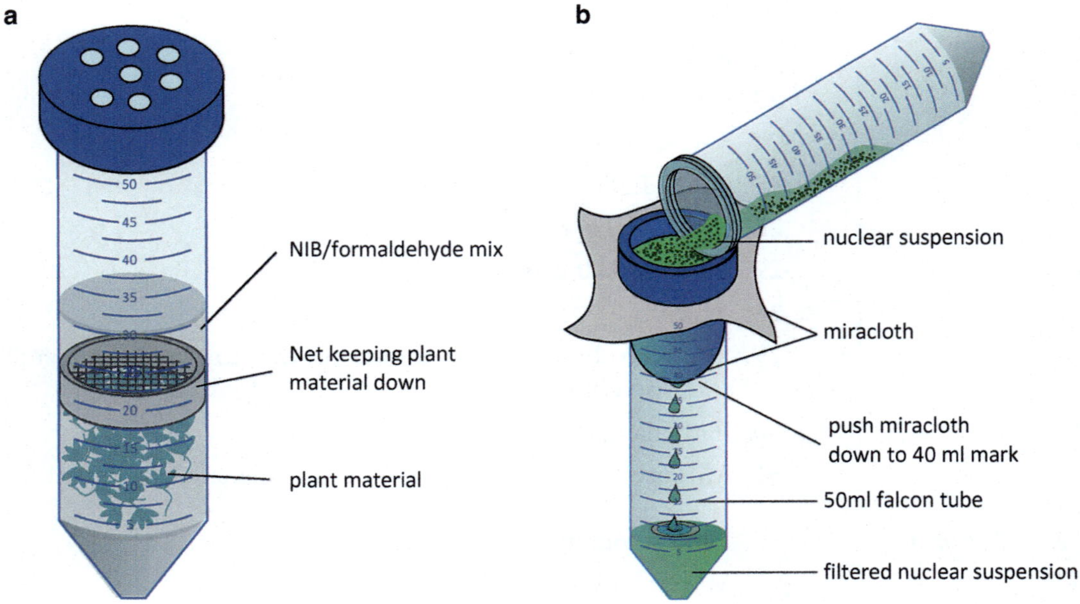

Fig. 2 Illustrations on how to set up (**a**) vacuum infiltration and (**b**) filtration of nuclei

2. To each tube, add 15 mL of NIB and 15 mL of 4% formaldehyde solution.
3. Place the bottle containing the residual NIB on ice, and add two tablets of protease inhibitor (complete protease inhibitor, Roche).
4. Use a ring holding a nylon mesh to fully submerge the plant material (*see* Fig. 2a).
5. Pierce holes into the lid of the conical tube and place them under vacuum for 1 h. Turning on and off the vacuum pump a few times during the first 5 min significantly enhances vacuum infiltration.
6. Add 1.9 mL of 2 M glycine to quench the formaldehyde, and apply the vacuum for another 5 min.
7. Discard the NIB/formaldehyde mixture according to the local chemical waste regulations.
8. Wash the plant material three times with ddH_2O, to remove residual NIB/formaldehyde mix.

3.1.2 Grinding and Filtration of Plant Material (1–1.5 h)

- *Work on bench/fume hood at RT/on ice.*
- *Paper towel, mortar and pestle, 50 mL conical tubes, liquid nitrogen, and Miracloth.*
- *NIB.*

1. Place the washed plant material between paper towels, and apply slight pressure to dry the plant tissue (*see* **Note 2**).

2. Place the plant material from both conical tubes into a pre-cooled mortar, add liquid nitrogen, and thoroughly grind the material to a fine powder (*see* **Note 3**).
3. Transfer to powder evenly into two 50 mL conical tubes, and immediately place them on ice.
4. Working under a fume hood on ice, add 20 mL of NIB to each conical tube and resuspend by gentle swirling/rotating the tubes.
5. Place an approx. 7 × 7 cm-sized piece of Miracloth on two fresh 50 mL conical tubes (*see* Fig. 2b), and filter the suspension through the Miracloth (*see* **Note 4**).
6. Repeat the filtration once.

3.1.3 Purification of Nuclei (1 h)

- *Work at bench on ice.*
- *Refrigerated centrifuge(s) suitable for 50 mL conical tubes and 1.5 mL reaction tubes.*
- *NIB, 1.2× restriction enzyme buffer.*

1. Pellet the nuclei by spinning at 3000 × *g* at 4 °C for 15 min.
2. Discard the supernatant according to the local chemical waste disposal regulations.
3. Add to each tube 2 mL NIB and resuspend the pellets by gentle swirling. If necessary, slowly pipette up and down using a cutoff pipette tip to force the pellet into suspension. Minimize pipetting to ensure the integrity of the nuclei.
4. Distribute the suspension of both conical tubes into a total of four 1.5 mL reaction tubes (*see* **Note 5**).
5. Centrifuge at 1900 × *g* at 4 °C for 5 min. Remove the supernatant and resuspend the nuclei in 1 mL of NIB.
6. Repeat the centrifugation and resuspension step twice or more times (total of two to three times, *see* **Note 6**).
7. When the nuclei reached satisfactory purification, remove the supernatant, and add 400 μL of 1.2× restriction enzyme buffer (e.g., CutSmart® (NEB)).
8. Resuspend the nuclei and spin at 1900 × *g* at 4 °C for 5 min.
9. Repeat this step once (total two times).
10. Add 500 μL of 1.2× restriction enzyme buffer, and slowly pipette up and down to bring nuclei into suspension.

3.2 Digestion, Fill-In Reaction, and Ligation

3.2.1 Permeabilization of the Nuclear Membrane (2 h)

- *Work at bench at RT.*
- *Thermal mixer.*
- *20% SDS, 20% Triton X-100.*

1. Permeabilize the nuclear membranes by addition of 5 μL of 20% SDS, and incubate for 40 min at 65 °C under constant shaking (900 rpm).

2. Incubation at 37 °C for 20 min under constant shaking.
3. Add 50 μL of 20% Triton X-100 to quench SDS. This is needed for efficient chromatin digestion.
4. Incubate for 1 h at 37 °C under constant shaking (*see* **Note 7**).

3.2.2 Digestion (2 h + Overnight + 2 h)

- *Work at bench at RT.*
- *Thermal mixer, 1.5 mL reaction tubes.*
- *Restriction enzyme of choice, 10× restriction enzyme buffer.*

1. To control the initial chromatin integrity, transfer 60 μL of each of the four nuclear suspensions in a fresh 1.5 mL reaction tube, and store at 4 °C until further use (*see* Subheading 3.4.2).
2. Add 15 μL of 10× restriction enzyme buffer and 115 μL of ddH_2O to each tube.
3. For digestion, add 100 U of restriction enzyme of choice (*see* **Note 8**), and incubate at 37 °C for 2 h under constant shaking (900 rpm).
4. Add another 200 U of restriction enzyme, and incubate overnight at 37 °C under constant shaking.
5. A final 100 U of restriction enzyme is added, and the samples are incubated 1–2 h at 37 °C under constant shaking.
6. To inactivate the restriction enzyme, add 20 μL of 20% SDS, and incubate for 25 min at 65 °C.
7. To later control for the digestion efficiency, 60 μL from each subsample is transferred to a fresh 1.5 mL reaction tube and stored at 4 °C until further use (*see* Subheading 3.4.2) (*see* **Notes 9** and **10**).

3.2.3 Ligation (6 h)

- *Work at bench on ice.*
- *15 mL conical tubes, thermal mixer equipped for 15 mL conical tubes.*
- *10% Triton X-100, 10× ligation buffer, 10 mg/mL molecular biology grade BSA, 100 mM ATP, and highly concentrated T4 ligase.*

1. Transfer the four subsamples to four 15 mL conical tubes, and place them on ice.
2. To prepare the ligation reaction and to sequester the SDS, add to each tube 750 μL 10% Triton X-100, 750 mL 10× ligation buffer, 75 μL of 10 mg/mL molecular biology grade BSA, and 5.3 mL of ddH_2O, and incubate for 1 h at 37 °C.
3. To start the ligation reaction, add 75 μL of 100 mM ATP and 10 WU of T4 ligase (*see* **Note 11**). Incubate 2.5 h at 16 °C.
4. Add another 10 WU of T4 ligase and incubate for another 2.5 h at 16 °C.
5. Incubate 45 min at RT.

3.3 Reversal of the Cross-Linking and RNase Treatment (Overnight + 3 h)

- *Work at bench at RT.*
- *Thermal mixer equipped for 15 mL conical tubes and 1.5 mL reaction tubes.*
- *10 mg/mL proteinase K, 10 mg/mL RNase A.*

1. The formaldehyde cross-linking is reversed by the addition of 50 μL of 10 mg/mL proteinase K.
2. To the previously stored pre- and post-digests control samples, add 5 μL of proteinase K and 130 μL of ddH_2O.
3. Incubate all samples overnight at 65 °C under constant shaking (900 rpm).
4. To complete the reversal of the cross-linking, another 50 μL (5 μL for the controls, respectively) of 10 mg/mL proteinase K is added, and the reactions are incubated for another 2 h at 65 °C under constant shaking (*see* **Note 12**).
5. Perform RNase treatment by adding 30 μL (5 μL for the controls, respectively) of 10 mg/mL RNase A, and incubate for 45 min at 37 °C.

3.4 DNA Purification (1–2 h)

- *Work under fume hood at RT.*
- *15 mL conical tubes, 50 mL conical tubes, 1.5 mL reaction tubes, −80 °C freezer, centrifuges equipped for 15 mL conical tubes, and 1.5 mL reaction tubes.*
- *Phenol, chloroform, isoamyl alcohol, 3 M NaOAc, and 20 mg/mL glycogen.*

3.4.1 3C Samples

1. Add 7.5 mL of phenol/chloroform/isoamyl alcohol (25:24:1 v/v/v) mixture to each tube, and shake vigorously for 40 s.
2. Centrifuge 15 min at 4500 × *g* at RT.
3. Remove the aqueous (upper!) phase without disturbing the debris-containing interphase, and transfer it to a fresh 15 mL conical tube.
4. To remove residual phenol, add 7.5 mL of chloroform/isoamyl alcohol (24:1 v/v), and mix by vigorously shaking the tubes for 40 s.
5. Centrifuge 15 min at 4500 × *g* at RT.
6. Transfer the aqueous phase without disturbing the interphase to a fresh 50 mL conical tube.
7. To precipitate the 3C template DNA, add 7 mL of ddH_2O, 1.4 mL of 3 M NaOAc, 30 μL of 20 mg/mL glycogen (*see* **Note 13**), and fill the 50 mL conical tube with ice-cold ethanol up to the 50 mL mark.
8. Mix by inverting the tubes, and incubate overnight at −80 °C.

9. Pellet the DNA by spinning the tubes at >3000 × *g* for 1 h at 4 °C, discard the supernatant, and wash the pellet once with 70% ethanol.
10. Air-dry the pellets and resuspend in 150 μL of ddH_2O.

3.4.2 Control Samples (Pre- and Post-digestion Control)

1. Adjust the working volume to a final 200 μL by adding 35 μL of ddH_2O to each control sample.
2. Add 200 μL of phenol/chloroform/isoamyl alcohol (25:24:1 v/v/v), and mix by vigorously shaking for 40 s.
3. Separate the phases by spinning at >11,000 × *g* for 5 min.
4. Transfer the aqueous phase into a fresh 1.5 mL reaction tube.
5. Repeat the purification adding 200 μL of chloroform/isoamyl alcohol (24:1 v/v).
6. Transfer the upper phase to a fresh 1.5 mL reaction tube.
7. Add 2 μL of 20 mg/mL glycogen, 20 μL of 3 M NaOAc, and 550 μL of ice-cold ethanol.
8. Store the controls alongside with the 3C samples at –80 °C.
9. Pellet the DNA of the control samples by spinning at 4 °C at >11,000 × *g* for 45 min.
10. Wash once with 70% ethanol, air-dry, and finally resuspend the control samples in 30 μL of ddH_2O.

3.5 Quality Controls

3.5.1 Assessing the Digestion and Ligation Efficiency (2 h)

1. Measure DNA concentrations of pre-digest controls, post-digest controls, and 3C samples. This needs to be done accurately and can be best achieved with a Q-bit measuring device (*see* **Note 14**).
2. Load 180 ng of DNA from each 3C and control samples onto a 1% agarose gel, containing 4 μL of ethidium bromide (medium sized gel).
3. Run for 60–90 min at 100 V.
4. Assess the quality of chromatin (pre-digest control), the efficiency of the restriction digest (post-digest control), and the ligation efficiency (3C sample) (*see* **Note 15** and Fig. 3).

3.6 4C Template Generation

3.6.1 Secondary Restriction Digest (4 h)

- *Work at bench at RT.*
- *Thermal mixer.*
- *Secondary restriction enzyme of choice), 10× restriction enzyme buffer.*

1. Pool all 3C templates that passed the quality control assay. Note the number of satisfactory 3C templates will determine the volumes of all reagents hereafter. In the following steps, we assume that the pooled volume is 150 μL; however, if there is a larger volume available after pooling (e.g. 300 μL), volumes of the reagents below have to be adjusted (e.g., doubled).

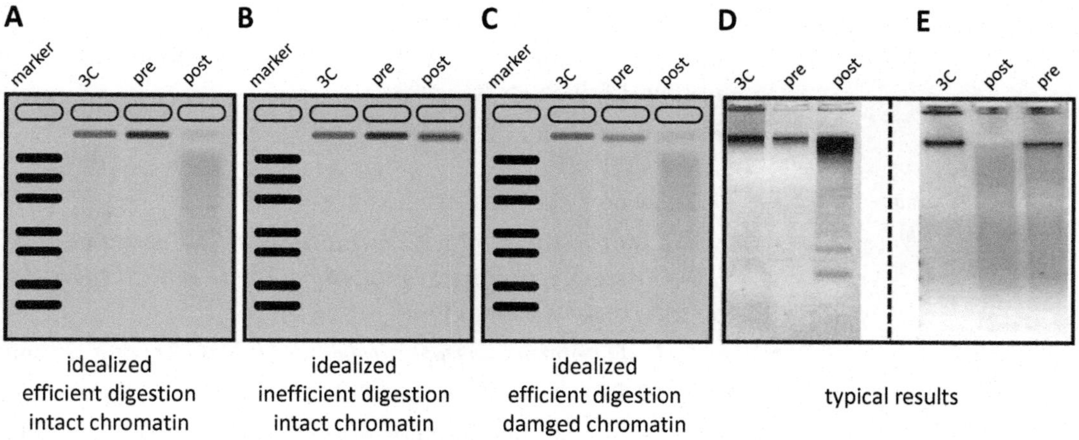

Fig. 3 3C quality controls. (**a**) Successful 3C template generation. The cross-linked chromatin is intact prior digestion (*see* lane "pre") and is subsequently efficiently digested (*see* lane "post"). The majority of digested chromatin is religated (compare lanes "post" to "3C"). (**b**) Unsatisfactory 3C template generation. Although the cross-linked chromatin is intact, the sample should be discarded, as the digestion efficiency is not satisfactory (*see* lane "post"). (**c**) Unsatisfactory 3C template generation. Chromatin integrity is compromised prior digestion (*see* lane "pre"); hence, the sample should be discarded. Note that (**a**–**c**) show idealized gel pictures and where produced using a drawing software. (**d** and **e**) Typical 3C results with satisfactory (**d**) and borderline (**e**) chromatin integrity (*see* lanes "pre"), efficient digestion (*see* lanes "post"), and efficient ligation (*see* lane "3C"), as the "smear" in the lane "post" nearly disappeared in lane "3C" (**d**). Note that it is not always possible to acquire a gel image, which clearly shows the smaller digested fragments ("smear") and at the same time is not overexposed for longer fragments. To estimate both it is recommended to acquire several pictures with different exposure times

2. Mix 150 μL of 3C templates, 20 μL of 10× restriction enzyme buffer, and 25 U of secondary restriction enzyme (*see* **Note 8**).
3. Fill with ddH_2O to reach a final volume of 200 μL.
4. Incubate at 37 °C for 1.5 h (adjust temperature according to manufacturer's protocol for the respective restriction enzyme).
5. Add another 25 U of restriction enzyme and incubate 1.5 h at 37 °C.
6. Heat inactivates the restriction enzyme by incubating at 65 °C for 30 min (or according to manufacturer's protocol).
7. Check the digestion efficiency of an aliquot (approx. 180 ng) on a 1% agarose gel.

3.6.2 Secondary Ligation (5 h)

- *Work at bench on ice.*
- *Cooled water bath.*
- *10× ligation buffer, 100 mM ATP, highly concentrated T4 ligase.*

1. Transfer the digested 3C templates to a 15 mL conical tube, and add 700 μL of 10× ligation buffer, 70 μL of 100 mM ATP, and 10 WU of T4 ligase.

2. Incubate 2.5 h at 16 °C.
3. Add an additional 10 WU of T4 ligase, and incubate for another 2.5 h at 16 °C.
4. Incubate 45 min at RT.

3.6.3 4C Template Purification (1–2 h)

- *Work under fume hood at RT.*
- *15 mL conical tubes, 50 mL conical tubes, 1.5 mL reaction tubes, −80 °C freezer, centrifuges equipped for 15 mL conical tubes, and 1.5 mL reaction tubes.*
- *Phenol, chloroform, isoamyl alcohol, 3 M NaOAc, and 20 mg/mL glycogen.*

1. Add 7.5 mL of phenol/chloroform/isoamyl alcohol (25:24:1 v/v) mixture to each tube, and shake vigorously for 40 s.
2. Centrifuge 15 min at 4500 × *g* at RT.
3. Remove the aqueous (upper!) phase without disturbing the debris-containing interphase, and transfer it to a fresh 15 mL conical tube.
4. To remove residual phenol, add 7.5 mL of chloroform/isoamyl alcohol (24:1 v/v), and mix by vigorously shaking the tubes for 40 s.
5. Centrifuge 15 min at 4500 × *g* at RT.
6. Transfer the aqueous phase without disturbing the interphase to a fresh 50 mL conical tube.
7. To precipitate the 3C template DNA, add 7 mL of ddH_2O, 1.4 mL of 3 M NaOAc, and 30 μL of 20 mg/mL glycogen, and fill the 50 mL conical tube with ice-cold ethanol up to the 50 mL mark.
8. Mix by inverting the tubes, and incubate overnight at −80 °C.
9. Pellet the DNA by spinning the tubes at >3000 × *g* for 1 h at 4 °C.
10. Discard the supernatant, and wash the pellet once with 70% ethanol.
11. Air-dry the pellets and resuspend in 100 μL of ddH_2O.

3.7 4C PCR

3.7.1 PCR Setup (1 h)

- *Work at bench on ice.*
- *Thermal cycler.*
- *Phusion DNA polymerase, 5× HF PCR buffer, 50 mM $MgCl_2$, 10 mM dNTP, and viewpoint primers.*

1. To amplify a viewpoint-specific interactome (*see* **Notes 16** and **17**), set up twelve 50 μL PCR reactions in parallel (*see* **Note 18**).
2. For each reaction mix 31 μL ddH_2O, 10 μL 5× PCR buffer, 0.5 μL 50 mM $MgCl_2$, 1 μL 10 mM dNTP mix, 2.5 μL of each

10 μM viewpoint primers, 1.5 μL (approx. 50 ng) 4C template, and 0.5 μL Phusion polymerase.

3. Run trial PCR programs and assess different cycle numbers and annealing temperatures to ensure efficient and linear amplification of the PCR (*see* **Note 19**).
4. Run the 4C PCR program with the previously optimized parameters.

3.7.2 4C PCR Quality Control (3 h)

- *Work at bench at RT.*
- *Gel running equipment.*
- *Agarose, ethidium bromide.*

1. Run a 1% agarose gel to check the efficiency and specificity of the 4C PCR (*see* **Note 19**).

3.7.3 4C Sample Purification (2 h)

- *Work at bench at RT.*
- *Table top centrifuge.*
- *PCR purification kit.*

1. Pool all satisfactory 4C PCR products, and purify them using a commercial PCR purification kit according to the manufacturer's protocol. Your 4C sample is now ready for sequencing library preparation.

3.8 Sequencing Library Preparation

For most projects involving next-generation sequencing, sequencing library preparation is carried out by specialized facilities. Standard genomic DNA library protocols (here Illumina *TruSeq Nano* DNA Library Preparation Kit) are usually suitable to prepare 4C libraries. Other library protocols (commercial and customized) may work as efficiently. The necessary steps are adapter ligation and library amplification. Note that the library preparation must not include fragmentation and size selection.

3.8.1 End Repair (1 h: Not Mandatory)

- *Work at bench at RT/on ice.*
- *Magnetic stand, thermal mixer.*
- *Illumina End Repair Mix 2 (ERP), Resuspension Buffer (RSB), AMPure XP beads, and freshly prepared 80% EtOH.*

1. Transfer 60 μL of 4C PCR product into a 200 μL reaction tube.
2. Add 40 μL of ERP and mix by pipetting up and down ten times.
3. Incubate for 30 min at 30 °C and subsequently store them on ice.
4. Transfer the samples to a fresh 1.5 mL reaction tube.

5. Add 160 μL of well-dispersed AMPure beads, and mix by pipetting up and down ten times.
6. Incubate 15 min at RT.
7. Place the tubes on a magnetic stand for 15 min, and subsequently remove the supernatant without disturbing the beads.
8. Add 400 μL of 80% EtOH, incubate 30 s, and subsequently remove the supernatant.
9. Repeat this washing step once.
10. Let the beads dry at RT or 37 °C (do not over dry!).
11. Resuspend beads in 20 μL of RSB.

3.8.2 Adenylation and Adapter Ligation (1 h)

- *Work at bench at RT.*
- *Thermal cycler, magnetic stand.*
- *A-tailing mix (ATL), RSB, DNA ligation mix (LIG), DNA adapter indexes, Stop ligation mix (STL), AMPure XP beads, and 80% EtOH.*

1. Add 12.5 μL of ATL, and incubate in thermal cycler (program: 37 °C for 30 min/70 °C for 5 min/4 °C for 5 min).
2. To ligate Illumina paired-end sequencing adapters, add 2.5 μL of RSB, 2.5 μL of DNA adapters, and 2.5 μL of LIG, and mix gently by slowly pipetting up and down ten times.
3. Incubate the mixture for 10 min at 30 °C.
4. To stop the ligation reaction, add 5 μL STL, and mix thoroughly by pipetting up and down.
5. Transfer the sample (42.5 μL) to a fresh 1.5 mL reaction tube.

3.8.3 Cleanup (1 h)

- *Work at bench at RT.*
- *1.5 mL reaction tubes, 0.2 mL reaction tubes.*
- *AMPure beads, EtOH, and RSB.*

1. Add 42.5 μL of well-dispersed AMPure beads, and mix by pipetting up and down ten times.
2. Incubate 15 min at RT.
3. Place the tubes on a magnetic stand for 15 min, and subsequently remove the supernatant without disturbing the beads.
4. Add 200 μL of 80% EtOH, incubate 30 s, and subsequently remove the supernatant.
5. Repeat this washing step once.
6. Let the beads dry at RT or 37 °C (do not overdry!).
7. Resuspend beads in 52.5 μL of RSB.
8. Incubate at RT for 2 min, and then place the tubes on a magnetic stand for 2 min.

9. Transfer 50 μL of the supernatant to a fresh 1.5 mL reaction tube.
10. Perform a second cleanup by adding 50 μL of AMPure beads, and incubate the mixture for 15 min at RT.
11. Place the reaction tubes on a magnetic rack for 2 min, and remove supernatant without disturbing the beads.
12. Wash twice with 80% EtOH and let the beads dry. Finally resuspend in 27.5 μL RSB and incubate for 2 min.
13. Place the tubes on a magnetic stand for 2 min and transfer the DNA containing supernatant to a fresh 0.2 mL reaction tube.

3.8.4 Amplification (1 h)

- *Work at bench at RT.*
- *Thermal cycler.*
- *Enhanced PCR Master Mix (ePMM), PCR Primer Cocktail (PPC), RSB, AMPure beads, and 80% EtOH.*

1. Add 5 μL of PPC and 20 μL of ePMM, and mix thoroughly by pipetting up and down.
2. Run a PCR program to amplify your libraries. The following program gave robust amplification in my hands:

1.	95 °C	3 min
2.	98 °C	20 s
3.	60 °C	15 s
4.	72 °C	30 s
5.	Go to **step 2**, eight cycles total	
6.	72 °C	5 min
7.	4 °C	On hold

3. Following the PCR, transfer the sample to a 1.5 mL reaction tube. Note, to avoid contamination, the following steps are usually performed in a "post-PCR" lab.
4. Add 50 μL of AMPure beads (37 μL for paired-end sequencing), and mix thoroughly by pipetting up and down.
5. Incubate 15 min at RT.
6. Place the tubes on a magnetic stand and remove the supernatant.
7. Wash twice with 200 μL of 80% EtOH without disturbing the beads.
8. Air-dry the bead pellet and resuspend in 20 μL RSB.
9. Mix thoroughly and incubate for 2 min at RT.
10. Place sample again on a magnetic rack for 2 min.
11. Transfer the supernatant containing your library to a fresh 1.5 mL reaction tube. Your library is now ready for sequencing!

4 Notes

1. If the experimental setup does not require a specific tissue, I recommend using 14-day-old whole seedlings as a starting material. Seedling tissue exhibits a large nuclei/tissue ratio, as the cells are rather small. Furthermore, seedlings are less likely to accumulate cell wall components and secondary metabolites, which might interfere with the 3C template preparation. Grow the seedlings in plastic petri dishes containing standard Murashige and Skoog (MS). For *Arabidopsis*, ca. 50 μL of seeds distributed over three standard petri dishes should yield enough material for your experiment. When harvesting prevent contamination of the plant material with pieces of culture media.
2. Wet seedlings will make the grinding step considerably harder. Thus, place the seedlings onto Whatman filter paper, cover the plant material with another filter paper, and subsequently place the two filter papers holding the cross-linked plant material in between several layers of absorbent paper towels. By applying slight pressure (e.g., putting a plastic rack on top of the stack), the water will quickly be absorbed by the paper towels. Eventually, exchange the paper towels, leaving the filter paper stack intact.
3. Do not rush the grinding step. Properly ground plant material will make the later filtration step considerably easier and will significantly increase the final 3C template yield. I advise to repeat the grinding several times, adding new liquid nitrogen in between.
4. Due to the large amount of plant material used as an input, the Miracloth is likely to clog. Complete resuspension of the ground plant material in NIB is crucial: leave the tubes with the (incompletely) resuspended plant material on ice for a while (approx. 15 min). In case the Miracloth clogs, add additional NIB (usually 5 mL will easily fit in the filtering funnel). If this measure is not sufficient to unclog the Miracloth, remove the suspension by pipetting slowly using a 1 mL pipette tip, from which the tip was cut off. Then filter the suspension through a fresh piece of Miracloth, attached to the same 50 mL tube as was used before, and add additional NIB (e.g., 5 mL). In total, up to 35 mL of NIB can be used per individual conical tube. As parts of the suspension will stay on the piece of Miracloth, expect about 25–30 mL of filtrate in the conical tube after using a total of 35 mL of NIB for filtration.
5. Prior to this point, the plant material was separated in different tubes to "facilitate the work flow (balancing of centrifuge, efficient handling of tubes, preventing Miracloth from clogging, etc.). From this point on, the four 1.5 mL reaction tubes have

to be kept separate for experimental reasons, as we aim at obtaining four independent subsamples. Do not expect all four subsamples to reach satisfactory quality, as digestion and relegation steps, are not always efficient. All high-quality subsamples will later be pooled.

6. Cellular debris that is left due to insufficient nuclei purification can inhibit the restriction enzyme. With some experience, the sufficient number of nuclei purification steps can be estimated by the eye: finally the suspension should appear light green or (ideally) whitish. As a trade-off, an excess of purification steps will decrease the final yield.

7. SDS will not only permeabilize the nuclear membrane but also efficiently denature enzymes, including restriction enzymes and ligases. Using Triton X-100, the SDS is quenched to minimize negative effects on the enzymes later used during the protocol. I advise to minimize the use of SDS. The importance of permeabilizing the nuclear membrane is debated in the field; thus, SDS can possibly be also omitted throughout the protocol (and hence also the Triton X-100).

8. The choice of the primary and secondary restriction enzyme is crucial for a successful 4C experiment. Try to use enzymes that are not blocked by methylation, can be purchased in high concentrations, and are known to digest efficiently. I recommend using *HindIII* or *BamHI* as primary restriction enzymes and *DpnII* or *NlaIII* as secondary restriction enzymes. For your viewpoint of interest, the primary and secondary restriction enzyme should be sufficiently spaced (minimum of 200 bp) to allow bending of the DNA, which is crucial for efficient religation.

   Furthermore, if various viewpoints should be analyzed in parallel, it is advisable to use the same set of primary and secondary restriction for all viewpoints. This allows using the same 4C template for all viewpoints and thus minimizes potential artifacts arising from the generation of several different 4C templates. Using this protocol, up to 6 μg of 4C template can be generated, which is sufficient to amplify the 4C interactomes of up to ten viewpoints. Please also refer to **Note 16** for 4C primer design.

9. Satisfactory digestion efficiency can be difficult to achieve. For trouble shooting, refer to these points:
   - Type and amount of restriction enzyme.
   - Number of washing steps with both NIB and restriction enzyme buffer.
   - Concentration of SDS used to permeabilize the nuclear membrane.

10. The pre- and the post-digestion controls will later be important to assess the initial chromatin quality (pre-digestion control), the digestion efficiency (post-digestion control), and the ligation efficiency (3C template).
11. Be aware of the different types of units that are used by various manufacturers to describe the concentration of ligases: for example, Promega ligases are sold in "Weiss units" (WU), whereas New England Biolabs (NEB) uses "cohesive end units" (CEU). The final concentration of active ligase varies significantly for the two unit definitions. In this protocol WU are relevant.
12. The final 3C template yield is affected by the efficiency of the de-cross-linking and the proteinase K treatment. Protein-bound DNA may not be efficiently isolated as it may stay in the phenol:chloroform phase.
13. DNA precipitation is inhibited by the DTT, which is added to the ligation buffer [6]. To lower the DTT concentration, ddH_2O is added.
14. A Nanodrop spectrometer will most likely produce erroneous results on the DNA concentration, as traces of phenol and other impurities will interfere with the spectrometry. Using a measuring device that relies on double-stranded DNA specific probes, such as a Q-bit, will yield more robust results.
15. The pre-digest control is used to assess the initial chromatin integrity. A clear single high molecular band represents good chromatin quality. Fragmented initial chromatin will be visible as a smear and the corresponding subsamples should be discarded (*see* Fig. 3). Ideally the post-digestion sample will produce a smear on an agarose gel, and no high molecular band should be detectable. Due to highly repeated sequences, a ladderlike pattern within the smear can be observed. Discard to corresponding 3C subsample if the post-digest control does not give appropriate results. The 3C template can be used to assess the ligation efficiency: completely religated 3C samples will resemble the pre-digest controls, with a somewhat lower high molecular band.
16. Appropriate primer design is crucial for a successful 4C experiment. Primers should flank either restriction site, facing "outward" (i.e., one primer flanks the primary restriction site, the other flanks the secondary restriction site). Thus, primers are inversely designed compared to canonical PCR primers (*see* Fig. 4b). To avoid losing sequencing information, the primers should be located within 50 bp of the respective restriction site (usually 50 bp of a 100 bp sequencing read is sufficient to unambiguously map the read). The primers have to uniquely bind to the viewpoint of choice; thus, I highly recommend to

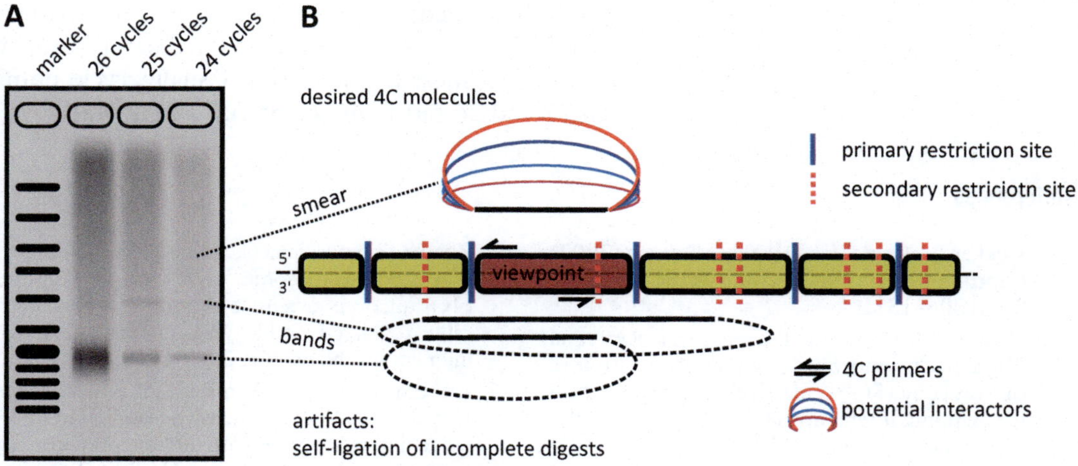

Fig. 4 4C PCR. (**a**) A typical 4C PCR analyzed on an agarose gel. The well-defined bands usually represent artifacts arising from incomplete digestion of chromatin. (**b**) *Top*: desired 4C templates using the viewpoint (from 1° restriction site to 2° restriction site as backbone). Middle: representation of the 4C primer localization and orientation. *Bottom*: artifacts, arising from incomplete digestion of DNA. By skipping on 1° restriction site, the following 2° restriction site can engage in self-ligation of the backbone

thoroughly check the primer's specificity, using BLAST or other DNA alignment programs. Unspecific primer binding can significantly distort the 4C results as they can produce putative interaction peaks, which, however, stem from the amplification of another unspecific viewpoint.

17. Analyzing the band pattern produced by incomplete digestion can assess the specificity of the 4C PCR. The observed band sizes can be compared to theoretical values, which are determined by analyzing size of restriction digest fragments of your genomic region of interest (*see* Fig. 4a, b). When the primary restriction site of the viewpoint's restriction fragment was partially skipped, the following secondary restriction site and the original secondary restriction site of the viewpoint's restriction fragment can ligate to each other, forming self-circulation of the viewpoint's genomic region.
18. To balance the 4C interactome amplification, twelve PCR reactions are set up in parallel. If the pattern on the agarose gel exhibits prominent differences, redesign the primers to obtain a more robust and reproducible PCR amplification.
19. Linear amplification is difficult to define as the templates in your pre-PCR 4C sample occur in different concentrations. Typically, templates representing self-ligation of incompletely digested DNA are most abundant and, hence, give the most prominent bands. They are not necessarily representative for the overall linearity of the PCR. Try various cycle numbers between 20 and 30 cycles, and choose the lowest cycle number,

which, subjectively judged, yields a satisfactory ratio between "smear" (your 4C interactome) and well-defined bands (usually artifacts) and confers linear amplification. Usually, cycle numbers between 23 and 26 satisfy these criteria (*see* Fig. 4a).

References

1. Dekker J, Rippe K, Dekker M et al (2002) Capturing chromosome conformation. Science 295:1306–1311
2. Dostie J, Richmond TA, Arnaout RA et al (2006) Chromosome conformation capture carbon copy (5C): a massively parallel solution for mapping interactions between genomic elements. Genome Res 16:1299–1309
3. Simonis M, Klous P, Splinter E et al (2006) Nuclear organization of active and inactive chromatin domains uncovered by chromosome conformation capture-on-chip (4C). Nat Genet 38:1348–1354
4. Zhao Z, Tavoosidana G, Sjölinder M et al (2006) Circular chromosome conformation capture (4C) uncovers extensive networks of epigenetically regulated intra- and interchromosomal interactions. Nat Genet 38:1341–1347
5. Lieberman-Aiden E, Van Berkum NL, Williams L et al (2009) Comprehensive mapping of long-range interactions reveals folding principles of the human genome. Science 326:289–293
6. Louwers M, Splinter E, Van Driel R et al (2009) Studying physical chromatin interactions in plants using chromosome conformation capture (3C). Nat Protoc 4:1216–1229

Chapter 7

Genome-Wide Profiling of Histone Modifications and Histone Variants in *Arabidopsis thaliana* and *Marchantia polymorpha*

Ramesh Yelagandula, Akihisa Osakabe, Elin Axelsson, Frederic Berger, and Tomokazu Kawashima

Abstract

Histone modifications and histone variants barcode the genome and play major roles in epigenetic regulations. Chromatin immunoprecipitation (ChIP) coupled with next-generation sequencing (NGS) is a well-established method to investigate the landscape of epigenetic marks at a genomic level. Here, we describe procedures for conducting ChIP, subsequent NGS library construction, and data analysis on histone modifications and histone variants in *Arabidopsis thaliana*. We also describe an optimized nuclear isolation procedure to prepare chromatin for ChIP in the liverwort, *Marchantia polymorpha*, which is the emerging model plant ideal for evolutionary studies.

Key words Chromatin immunoprecipitation, Histone modifications, Histone variants, *Arabidopsis thaliana*, *Marchantia polymorpha*

1 Introduction

In eukaryotes, the core histones H2A, H2B, H3, and H4 compose a series of nucleosomes and pack the DNA into chromatin [1]. Histones also provide the foundation of epigenetic control, regulating gene expression through histone modification and histone variants [2–4]. For example, in *Arabidopsis thaliana*, lysine 9 di-methylation of histone H3 (H3K9me2) is associated with gene silence; on the other hand, lysine 4 tri-methylation of histone H3 (H3K4me3) marks active genes [5, 6]. Histone variants refer to paralogous genes of a histone family (e.g., H2A) that encode related but distinct protein isoforms (e.g., H2A.X and H2A.Z) [7]. Knowledge of histone variants, particularly in H2A and H3, has been recently developed and shows that histone variants play crucial roles in regulating various processes such as transcription, chromosome segregation, DNA repair and recombination, chromatin remodeling and chromatin

Wolfgang Busch (ed.), *Plant Genomics: Methods and Protocols*, Methods in Molecular Biology, vol. 1610,
DOI 10.1007/978-1-4939-7003-2_7,

compaction, and germ line-specific DNA packaging and zygotic activation [8–12].

Chromatin immunoprecipitation (ChIP) is widely used as a method to investigate the abundance of each histone mark or histone variant at a locus or genome-wide level. ChIP protocols for the model plant *Arabidopsis* have been established [13–16]. However, a ChIP protocol should be modified and optimized when used for a different species, such as the emerging model species *Marchantia polymorpha*, to ensure successful data acquisition. *Marchantia* is a liverwort, one of the earliest diverging land plants, which has also evolved histone variants similar to what have been identified in other land plants [17]. Its genome sequence is already available to the public (www.phytozome.jgi.org), and many molecular tools such as homologous recombination and the CRISPR/Cas9 system are available in *Marchantia* [18]. Due to its unique evolutionary position between green algae and the rest of the land plants [19], *Marchantia* is an excellent model plant not only for molecular but also for evolutionary studies. It is thus critical and beneficial for the research community to establish an optimized ChIP protocol for *Marchantia*, allowing the investigation of the epigenetic landscape of *Marchantia* as well as the evolution of gene regulation through epigenetics in land plants.

In this chapter, we first describe a ChIP protocol for histone modification and histone variants in *Arabidopsis*. We also describe the *Marchantia* ChIP protocol in which the nuclei isolation step has been optimized [20]. This optimization is necessary due to the lower density of nuclei in *Marchantia* vegetative tissue in comparison to *Arabidopsis*. By starting with four times *Marchantia* vegetative tissue, we have successfully obtained the similar amount of chromatin to that of *Arabidopsis*. Next-generation sequencing (NGS) library construction as well as NGS data analysis is also described with helpful tips.

2 Materials

Use molecular biology grade chemicals for making all reagents. All the reagents are prepared using sterile double-distilled Mono Q water. Toxic and volatile chemicals should always be handled in a chemical fume hood and be discarded according to safety regulations.

2.1 Cross-linking and Nuclei Isolation

1. ½ MS medium agar plates: Dissolve 2.205 g of MS medium (catalog no. M0231 from DUCHEFA BIOCHEMIE B.V) and 10 g of sucrose into 900 mL water. Add 8 g of plant agar and adjust pH to 5.7 with 1 N KOH while stirring. Make it up to 1 L with water. Sterilize by autoclave and pour into petri plates in the sterile hood.

2. ½ Gamborg B5 medium agar plates: Dissolve 1.5 g of Gamborg B5 medium (catalog no. G0209 from DUCHEFA BIOCHEMIE B.V) and 0.5 g of MES (2-(N-morpholino)ethanesulfonic acid) into 900 mL water. Add 10 g of plant agar and adjust pH to 5.5 with 1 N KOH while stirring. Make it up to 1 L with water. Sterilize by autoclave and pour into petri plates in the sterile hood.
3. Protease inhibitor cocktail cOmplete™ Protease Inhibitor Cocktail (Roche). Dissolve one tablet in 2 mL of sterile water to obtain 25× stock, which can be stored at −20 °C. Only add to buffers immediately before use.
4. 1 M phosphate buffer, pH 7.0: 61 mL of 1 M Na_2HPO_4, and 39 mL of 1 M NaH_2PO_4 stock solution. Sterilize by filtering and store at room temperature.
5. MC buffer: 10 mM phosphate buffer pH 7.0, 100 mM NaCl, 0.1 M sucrose. Sterilize by filtering and store at 4 °C.
6. *Arabidopsis* cross-linking buffer: 1% of formaldehyde in MC buffer. Always freshly prepare before use. For making 25 mL cross-linking buffer, add 676 μL of 37% formaldehyde stock solution to 25 mL of MC buffer.
7. 1× PBS: 137 mM NaCl, 2.7 mM KCl, 10 mM Na_2HPO_4, and 1.8 mM KH_2PO_4.
8. *Marchantia* cross-linking buffer: 1% of formaldehyde in 1× PBS buffer.
9. 2 M glycine solution: 0.15 g glycine in 1 mL water.
10. Vacuum pump and exsiccator unit.
11. M1 buffer: 10 mM phosphate buffer pH 7.0, 100 mM NaCl, 10 mM beta mercaptoethanol, 1 M hexylene glycol, 1× cOmplete protease inhibitor cocktail. Sterilize by filtering and store at 4 °C.
12. M2 buffer: 10 mM phosphate buffer pH 7.0, 100 mM NaCl, 10 mM beta mercaptoethanol, 10 mM $MgCl_2$, 0.5% Triton X-100, 1 M hexylene glycol, 1× cOmplete protease inhibitor cocktail. Sterilize by filtering and store at 4 °C.
13. M3 buffer: 10 mM phosphate buffer pH 7.0, 100 mM NaCl, 10 mM beta mercaptoethanol, 1× cOmplete protease inhibitor cocktail. Sterilize by filtering and store at 4 °C.
14. 200 mM MES-KOH buffer, pH 5.3: Dissolve 19.5 g of MES in 400 mL of water, adjust pH with KOH to 5.3, then make it up to 500 mL with water. Filter sterilize and store at room temperature.
15. MP1 buffer: 10 mM MES-KOH buffer pH 5.3, 10 mM NaCl, 10 mM KCl, 0.4 M sucrose, 2% (w/v) PVP-10, 10 mM $MgCl_2$, 10 mM 2-mercaptoethanol, 6 mM EGTA, 1× cOmplete protease inhibitor cocktail. Prepare this buffer just before use.

16. MP2 buffer: 10 mM MES-KOH buffer pH 5.3, 10 mM NaCl, 10 mM KCl, 0.25 M sucrose, 10 mM $MgCl_2$, 10 mM 2-mercaptoethanol, 0.2% Triton-X 100, 1× cOmplete protease inhibitor cocktail.
17. MP3 buffer: 10 mM MES-KOH buffer pH 5.3, 10 mM NaCl, 10 mM KCl, 1.7 M sucrose, 2 mM $MgCl_2$, 10 mM 2-mercaptoethanol, 1× cOmplete protease inhibitor cocktail.
18. 50 mL falcon tubes.
19. Miracloth (Calbiochem).
20. 40 μm cell strainers (VWR).
21. 10 μm filter mesh (Buddeberg laboratory technology).
22. Glass funnels.
23. Cooling centrifuge.
24. Rotating stand: Test tube rotator.

2.2 Chromatin Preparation

1. Nuclei lysis buffer: 1% SDS, 10 mM EDTA pH 8.0, 50 mM Tris–HCl pH 8.0, 1× cOmplete protease inhibitor cocktail. Sterilize by filtering and store at room temperature.
2. ChIP dilution buffer: 0.01% SDS, 1.1% Triton X-100, 1.2 mM EDTA, 16.7 mM Tris–HCl pH 8.0, 167 mM NaCl. Sterilize by filtering and store at 4 °C.
3. Bioruptor (Diagenode).
4. Bioruptor compatible tubes 1.5 mL (Diagenode).

2.3 Immunoprecipitation

1. Protein A/G magnetic beads (Life Technologies): Dynabeads® Protein A or Dynabeads® Protein G.
2. Magnetic separation rack: MagnaRack™ (Invitrogen).
3. Antibodies: Native antibodies raised against histone variant or histone modification of interest. Epitope tag-specific antibodies can be used if the histone of interest is tagged with the epitope tag (*see* **Note 1**).
4. DNA LoBind 1.5 mL microcentrifuge tubes (Eppendorf®).
5. Low-salt wash buffer: 0.1% SDS, 1% Triton X-100, 2 mM EDTA pH 8.0, 20 mM Tris–HCl pH 8.0, 150 mM NaCl. Sterilize by filtering and store at 4 °C.
6. High-salt wash buffer: 0.1% SDS, 1% Triton X-100, 2 mM EDTA pH 8.0, 20 mM Tris–HCl pH 8.0, 500 mM NaCl. Sterilize by filtering and store at 4 °C.
7. LiCl wash Buffer: 0.25 M LiCl, 1% IGEPAL CA630, 1% deoxycholic acid, 1 mM EDTA, 10 mM Tris–HCl pH 8.0. Sterilize by filtering and store at 4 °C.
8. 1× TE buffer: 10 mM Tris–HCl pH 8.0, 1 mM EDTA pH 8.0. Sterilize by filtering and store at 4 °C.

9. Elution Buffer: 1% SDS (2 mL 10% SDS), 0.1 M $NaHCO_3$ (2 mL 1 M $NaHCO_3$).
10. 10× reverse cross-link buffer: 2 M NaCl, 0.1 M EDTA, 0.4 M Tris–HCl pH 8.0, 0.4 mg/mL proteinase K.
11. Proteinase K.
12. RNase A (DNase and protease-free) 10 mg/mL.
13. 3 M sodium acetate pH 5.2.
14. Thermomixer.
15. MinElute PCR Purification Kit (Qiagen).
16. qPCR reagents, e.g., FastStart essential DNA green master (Roche).

2.4 Library Preparation

1. Library preparation kit: TrueSeq ChIP library preparation kit (Illumina) or Ovation Ultralow System V2 1-16 (Nugen).
2. DNF-474 High-Sensitivity NGS Fragment Analysis Kit (Advanced Analytical Technologies).

3 Methods

Carry out all the steps on ice or in cold room unless otherwise noted.

3.1 Tissue Collection and Cross-Linking

3.1.1 Tissue Collection and Cross-Linking for *Arabidopsis thaliana*

1. Harvest tissue either from 10-day seedlings grown on a ½ MS plate or young rosette leaves from plants grown on soil into a 50 mL falcon tube. Approximately 300 mg of tissue is sufficient for each antibody ChIP. In the current protocol, we start with 2 g of tissues, which is sufficient for six ChIP reactions.
2. Cross-link harvested tissue by adding 25 mL of freshly prepared cross-linking buffer (1% formaldehyde in MC buffer) to the 50 mL falcon tube on ice. Apply vacuum for 15 min to assist penetration of cross-linking buffer into the tissues (*see* **Note 2**).
3. Stop cross-link reaction by adding 1.565 mL of 2 M glycine solution. Mix by inverting tubes ten times and apply vacuum for 5 min.
4. Wash the tissues with 25 mL MC buffer three times. Remove the buffer residuals on the tissues by blotting with paper towels.

   Stop point: Cross-linked tissues can be snap frozen in liquid nitrogen and stored at −80 °C for up to 6 months.

3.1.2 Tissue Collection and Cross-linking for *Marchantia polymorpha*

1. Harvest tissue from 14-day thalli grown on a ½ Gamborg B5 plate into a glass beaker. For each antibody ChIP, approximately 1.3 g of tissue is sufficient for good results. In the current protocol, we start with 8 g of tissue, which is sufficient for six ChIP reactions.

2. Cross-link harvested tissue by adding 67 mL of freshly prepared cross-linking buffer (1% formaldehyde in 1× PBS buffer) to the glass beaker on ice. Apply vacuum for 5 min and release the air pressure (total twice) to assist penetration of cross-linking buffer into the tissues (*see* **Note 2**).
3. Stop cross-link reaction by adding 4.5 mL of 2 M glycine solution (to 0.125 M) to 67 mL of cross-linking buffer, mix sample well, and apply vacuum for 10 min.
4. Remove the leftover buffer on the tissues by blotting with paper towels.

   Stop point: Cross-linked tissues can be snap frozen in liquid nitrogen and stored at −80 °C.

3.2 Nuclei Isolation

3.2.1 Nuclei Isolation from *Arabidopsis thaliana*

1. Snap-freeze cross-linked tissues in liquid nitrogen and grind to fine powder using pre-cooled mortar and pestle in liquid nitrogen.
2. Carefully transfer powder to a precooled 50 mL falcon tube and suspend in 25 mL of M1 buffer.
3. Filter the suspension through two layers of miracloth using a glass funnel into a new 50 mL falcon tube. Add an additional 10 mL of M1 buffer to collect leftover nuclei in the residue.
4. Spin the filtrate at 2000 × *g* at 4 °C for 5 min and discard the supernatant.
5. Resuspend the pellet in 5 mL of M2 buffer and spin at 2000 × *g* at 4 °C for 5 min. The supernatant color should be green due to chloroplast burst.
6. Repeat M2 buffer wash four more times. After the final wash, the supernatant color should be clear (chloroplast removal).
7. Resuspend the nuclei pellet in 5 mL of M3 buffer and spin at 2000 × *g* at 4 °C for 5 min. Discard the supernatant. Go to Subheading 3.3.

3.2.2 Nuclei Isolation from *Marchantia polymorpha*

1. Snap-freeze cross-linked tissues in liquid nitrogen and grind to fine powder using precooled mortar and pestle in liquid nitrogen.
2. Carefully transfer powder to a precooled 50 mL falcon tube and suspend in 40 mL of MP1 buffer.
3. Filter the suspension through one layer of miracloth into a new 50 mL falcon tube.
4. Filter again through one layer of miracloth into a new 50 mL of falcon tube.
5. Filter the sample through a 40 μm filter into a new 50 mL falcon tube.

6. Filter the sample through a 10 μm filter into a new 50 mL falcon tube using a glass funnel.
7. Filter again through a 10 μm filter into a new 50 mL falcon tube using a glass funnel.
8. Spin the filtrate at 3000 × *g* at 4 °C for 10 min and discard the supernatant.
9. Resuspend the pellet in 15 mL of MP2 buffer and incubate on ice for 15 min. The supernatant color should be green due to chloroplast burst.
10. Spin the resuspension at 3000 × *g* at 4 °C for 10 min. Repeat MP2 buffer wash two more times. After the final wash, the supernatant color should be clear (chloroplast removal).
11. Resuspend the pellet in 5 mL of MP3 buffer and transfer the samples to 5 mL Eppendorf tubes.
12. Spin the resuspension at 16,000 × *g* at 4 °C for 1 h and discard the supernatant.

3.3 Chromatin Preparation

From here, *Arabidopsis* and *Marchantia* samples are treated in the same manner.

1. Resuspend the purified nuclear pellet in 350 μL nuclei lysis buffer (with 1× cOmplete proteinase inhibitor), transfer the nuclei suspension to 1.5 mL Eppendorf tube, and incubate on ice for 20 min with occasional mixing.
2. Add 1050 μL of ChIP dilution buffer (with 1× cOmplete proteinase inhibitor) to lower the SDS concentration to 0.25%. After mixing, split the suspension into five 1.5 mL tubes (Diagenode) equally (~280 μL each). Avoid making air bubbles. Also set aside a 10 μL aliquot as a pre-sonication sample.
3. Shear the chromatin using the Bioruptor in a cold room. Settings: Power high, 15 cycles – 30 s on and 30 s off. After eight cycles, stop and change water in the Bioruptor with precooled water to the suggested level. We highly recommend optimizing the shearing conditions depending on the instrument and tubes used.
4. Spin tubes at (>20,000 × *g*) at 4 °C for 15 min to pellet down the debris. Pool the soluble chromatin in one tube and repeat the spin for ten more minutes to get rid of any remaining debris and transfer the supernatant to a fresh tube. From the total chromatin, set aside 140 μL (1/10 of total) for input sample. *Stop point*: Chromatin can be snap frozen in liquid nitrogen and stored at −80 °C for up to 6 months. This can be thawed and used for the next steps in the experiments.
5. Optional: check chromatin shearing (*see* **Note 3**).

3.4 Chromatin Immunoprecipitation

3.4.1 Washing Magnetic Beads

1. Resuspend protein A/G magnetic beads by vortexing.
2. Transfer the required volume of magnetic beads (70 μL × six samples + extra = 450 μL for this protocol) into a 1.5 mL Eppendorf tube, and place it on the MagnaRack for 5 min to pellet the beads.
3. Discard the supernatant without disturbing the pellet.
4. Resuspend the beads in 1 mL ChIP dilution buffer by vortexing or inverting the tube several times.
5. Place the tube on the MagnaRack for 5 min to pellet the beads.
6. Discard the supernatant without disturbing the pellet.
7. Repeat the ChIP dilution buffer wash (**steps 4–6**) once more.
8. Resuspend beads in ChIP dilution buffer in the initial volume (450 μL).

3.4.2 Preclearing the Chromatin Solution

1. If using stored chromatin samples, spin at 15,000 rpm at 4 °C for 10 min to clear the chromatin, and transfer the supernatant to a fresh tube. Otherwise, go to **step 2**.
2. Add ChIP dilution buffer to the sheared chromatin sample (~1400 μL) to make the volume up to 3.6 mL to reduce the concentration of SDS to <0.1%.
3. Split the chromatin into six tubes (600 μL each), and add 30 μL of prewashed magnetic beads protein A or G depending on the antibody used in the experiment to each tube. Store remaining beads at 4 °C for later step.
4. Incubate at 4 °C for 1 h, rotating.
5. Quick spin the tubes for a few seconds and put them on a MagnaRack for 10 min to pellet the magnetic beads.
6. Transfer the supernatant to fresh tubes and repeat the step to pellet any remaining beads left in the supernatant.
7. Transfer the chromatin-containing supernatant to new 1.5 mL DNA LoBind tubes (*see* **Note 4**).

3.4.3 Immunoprecipitation

1. Add the respective antibody to each tube. We generally use 5–10 μg of antibody (*see* **Note 5**). Close the tubes properly and incubate at 4 °C overnight, rotating.
2. Add 40 μL of prewashed protein A or protein G magnetic beads, depending on the antibody used (*see* **Note 6**). Incubate at 4 °C for 3 h, rotating.
3. Quick spin the tubes for a few seconds and place them on a MagnaRack for 5 min to pellet the beads.
4. Carefully transfer the supernatant without disturbing the pellet to new tubes, and store them at −80 °C. These stored supernatants can help when troubleshooting. Do not let the beads dry during this time.

5. Add 1 mL of low-salt wash buffer to each tube, mix them well, and incubate at 4 °C for 5 min, rotating.
6. Quick spin the tubes for a few seconds and place them on a MagnaRack for 5 min to pellet the beads.
7. Discard the supernatant without disturbing the pellet.
9. Repeat **steps 5–7** once more.
10. Add 1 mL of high-salt wash buffer to each tube, mix them well, and incubate at 4 °C for 5 min, rotating.
11. Quick spin the tubes for a few seconds and place them on a MagnaRack for 5 min to pellet the beads.
12. Discard the supernatant without disturbing the pellet.
13. Add 1 mL of LiCl wash buffer to each tube, mix them well, and incubate at 4 °C for 5 min, rotating.
14. Quick spin the tubes for a few seconds and place them on a MagnaRack for 5 min to pellet the beads.
15. Discard the supernatant without disturbing the pellet.
16. Add 1 mL of 1× TE buffer to each tube, mix them well, and incubate at 4 °C for 5 min, rotating.
17. Quick spin the tubes for a few seconds and place them on a MagnaRack for 5 min to pellet the beads.
18. Resuspend the beads in 1 mL of 1× TE buffer and transfer total suspension to a new 1.5 mL DNA LoBind Eppendorf tube. Transferring beads to a new tube at this step reduces the background.
19. Incubate the tubes at 4 °C for 5 min, rotating.
20. Quick spin the tubes for a few seconds and place them on a MagnaRack for 5 min to pellet the beads.
21. Discard the supernatant without disturbing the pellet.

3.5 Elution, Reverse Cross-Linking, and DNA Purification

1. For elution, resuspend the washed beads in 250 μL of freshly prepared elution buffer. Vortex briefly to mix and incubate at 65 °C for 15 min on a thermomixer with shaking.
2. Spin down at 15,000 rpm speed at 4 °C for 5 min and carefully transfer the supernatant to a fresh Eppendorf tube.
3. Repeat elution with another 250 μL of elution buffer and combine elutes to have 500 μL.
4. For reverse cross-linking, add 51 μL of 10× reverse cross-link mix to 500 μL samples and incubate for 3 h at 50 °C and overnight at 65 °C using a thermomixer with shaking. Make sure to use safe lock tubes to avoid evaporation.
5. The next day, add 10 μg RNase A and incubate at room temperature for 30 min.

6. For DNA purification, adjust the pH of the elute solution by adding 55 μL (1/10 volume) of sodium acetate pH 5.2 to each sample and purify the DNA with a Qiagen MinElute PCR Purification Kit. At the final step, elute the DNA in 15 μL into new 1.5 mL Eppendorf LoBind tubes.

3.6 NGS Library Generation

1. For ChIP enrichment analysis, dilute the ChIP DNA by five times and use 1 μL for each qPCR reaction. ChIP enrichment is calculated by normalizing either by % input method or fold enrichment method (*see* **Note 7**).
2. Successfully enriched samples are converted to next-generation sequencing libraries using an Illumina TrueSeq ChIP library preparation kit according to supplier's instructions (*see* **Note 8**).
3. Analyze the quality of the libraries by using the fragment analyzer. A good library shows fragments between 300 and 400 bp.
4. To validate the library quality, qPCR is carried out using diluted library DNA. Libraries, which show satisfactory enrichment, can be sequenced at a local NGS facility service.

3.7 ChIP Libraries Sequencing and Analysis

1. Sequence *Arabidopsis thaliana* or *Marchantia polymorpha* using 50 bp single-end reads and a sequencing depth of five million uniquely mapped reads per sample. This generally provides enough coverage for global identification of histone marks/variants; however, the required sequencing depth depends on the scope of the experiment.
2. Investigate the sequencing quality using tools such as fastQC [21] that provide a comprehensive report including statistics for per sequence quality, duplication rates, and GC sequence content.
3. Map the sequence reads to the genome of interest. For ChIP-seq data, Bowtie2 [22] software is a popular alignment tool (*see* **Notes 9–11**).
4. Control the alignment and duplication rates. BioBloom [23] provides a fast and easy way to identify potential contaminations.
5. Analyze the data. The exact steps of the analysis greatly depend on questions of interest, but there are a few steps that are often included: plotting of *chromosome profiles* and *averaged profiles* and heatmaps over genes, TEs, and other regions of interest using deepTools [24]. *Peak calling* by MACS2 [25] can be used for histone mark datasets. For histone variants nucleosome positioning, software such as PING [26] should be used.

4 Notes

1. Generating specific antibodies to highly conserved histone variants, such as H3 variants, is quite challenging, and sometimes it may not be possible. In that case, a genome-wide distribution of histone variants can be carried out using epitope tagging. Because most of the important modifications of histone H3 are present on N-terminal tails of histones, N-terminal epitope tag might result in artificial phenotypes and C-terminal epitope tag may reduce such problems.
2. It is essential to make sure that tissues are submerged in the cross-linking buffer for proper cross-linking. Formaldehyde is volatile and toxic, so carry out this step in a chemical fume hood, avoid contact with skin, and discard waste carefully according to safety regulations.
3. We suggest checking chromatin shearing before proceeding to the next step of the experiment. During this time, chromatin can be stored at 4 °C up to 1 week without affecting its integrity. Isolate DNA from pre-sonication and input samples by reverse cross-linking as mentioned in Subheading 3.5, **steps 4–6**. Analyze DNA fragmentation by running ~15 ng DNA on 1.5% agarose gel with SYBR gold. Majority of DNA fragments should be distributed between 200 and 300 bp (Fig. 1). Efficient fragmentation of chromatin is important for the resolution and success of histones and histone modifications ChIP experiment.

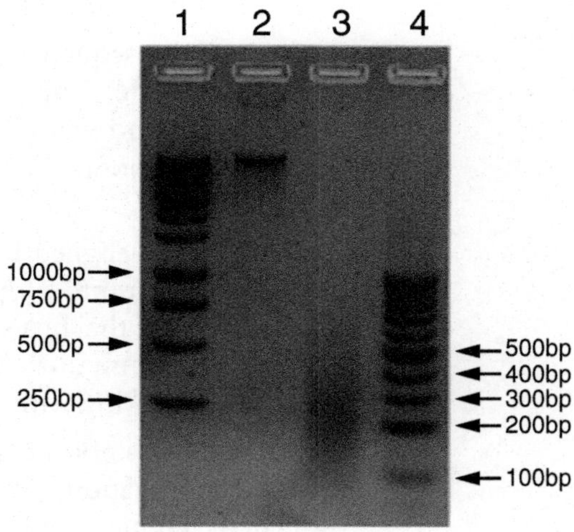

Fig. 1 Distribution pattern of DNA fragments isolated from sonicated chromatin. In each lane, DNA is electrophoresed on 1.5% agarose gel and stained with SYBR gold. *Lanes 2* and *3* are loaded with DNA isolated from pre-sonicated and post-sonicated *Marchantia polymorpha* chromatin, respectively. *Lanes 1* and *4* are loaded with 1 kb and 100 bp DNA ladders respectively

4. Preclearing chromatin solution step allows to remove components that bind to protein A/G magnetic beads prior to antibody-chromatin pulldown. This step usually improves the signal-to-noise ratio of the ChIP data.
5. It is advisable to standardize the concentration of each antibody to be used. We check commercial antibody activity of each lot either by Western blot or immuno-staining using H3 as a positive control and IgG as a negative control.
6. Some antibodies may give significant nonspecific background; blocking magnetic beads with 1% BSA prior to use may reduce the background in some cases. Incubate prewashed protein A/G magnetic beads in Subheading 3.4.1 overnight in 1% BSA at 4 °C, rotating, before antibody-chromatin pulldown in Subheading 3.4.3.
7. If a native antibody is used for ChIP, use IgG with the same plant material as the negative control. If an epitope tag antibody is used to profile a histone variant, the wild-type (untagged line) plant material instead of the transgenic tagged line plant material is used as the negative control. We use at least three independent loci to confirm the ChIP enrichment when its genome distribution is known. When characterizing a novel histone variant distribution, we try multiple genomic loci including gene coding regions, intergenic regions, and TE regions to investigate any presence of the ChIP enrichment.
8. We successfully generated good-quality libraries from the NuGEN Ovation Ultralow System kit with ChIP DNA as low as 1 ng. To avoid kit-specific amplification biases, use a single kit to generate all the libraries in one experiment.
9. To inspect the sequencing result, use of a so-called genome browser is useful. The browser will allow you to load your aligned reads and will then visualize the reads aligned to the genome. Commonly used genome browsers include IGV [27] and IBG [28].
10. Converting the aligned bam file into the bigWig format [29] before loading it into a genome browser will reduce the memory needed by the browser and will speed up the genome browser's response time. This can save you a lot of time, especially when working with many samples at the same time.
11. To achieve a smoother and more representative genome browser visualization, one can extend the aligned reads to the estimated fragment size before loading the data into the browser.

Acknowledgments

Thanks to S. Akimcheva and J. S. Russell for technical assistance and critical reading, respectively. AO was supported by a Grant-in-Aid for the Japan Society for the Promotion of Science (JSPS) Fellows. RY, TK, EA, and FB were funded by GMI and FWF (grants 2163-B16 linked to ERA-CAPS project EVOREPRO).

References

1. Luger K, Dechassa ML, Tremethick DJ (2012) New insights into nucleosome and chromatin structure: an ordered state or a disordered affair? Nat Rev Mol Cell Biol 13:436–447
2. Strahl BD, Allis CD (2000) The language of covalent histone modifications. Nature 403: 41–45
3. Hake SB, Allis CD (2006) Histone H3 variants and their potential role in indexing mammalian genomes: the "H3 barcode hypothesis". Proc Natl Acad Sci U S A 103:6428–6435
4. Deal RB, Henikoff S (2011) Histone variants and modifications in plant gene regulation. Curr Opin Plant Biol 14:116–122
5. Roudier F et al (2011) Integrative epigenomic mapping defines four main chromatin states in Arabidopsis. EMBO J 30:1928–1938
6. Feng S, Jacobsen SE (2011) Epigenetic modifications in plants: an evolutionary perspective. Curr Opin Plant Biol 14:179–186
7. Talbert PB et al (2012) A unified phylogeny-based nomenclature for histone variants. Epigenetics Chromatin 5:1–19
8. Talbert PB, Henikoff S (2014) Environmental responses mediated by histone variants. Trends Cell Biol 24:642–650
9. Filipescu D, Szenker E, Almouzni G (2013) Developmental roles of histone H3 variants and their chaperones. Trends Genet 29:630–640
10. Maze I, Noh K-M, Soshnev AA, Allis CD (2014) Every amino acid matters: essential contributions of histone variants to mammalian development and disease. Nat Rev Genet 15:259–271
11. Otero S, Desvoyes B, Gutierrez C (2014) Histone H3 dynamics in plant cell cycle and development. Cytogenet Genome Res 143:114–124
12. Yelagandula R et al (2014) The histone variant H2A.W defines heterochromatin and promotes chromatin condensation in Arabidopsis. Cell 158:98–109
13. Kaufmann K et al (2010) Chromatin immunoprecipitation (ChIP) of plant transcription factors followed by sequencing (ChIP-SEQ) or hybridization to whole genome arrays (ChIP-CHIP). Nat Protoc 5:457–472
14. Saleh A, Alvarez-Venegas R, Avramova Z (2008) An efficient chromatin immunoprecipitation (ChIP) protocol for studying histone modifications in Arabidopsis plants. Nat Protoc 3:1018–1025
15. Ito T, Takahashi N, Shimura Y, Okada K (1997) A serine/threonine protein kinase gene isolated by an in vivo binding procedure using the Arabidopsis floral homeotic gene product, AGAMOUS. Plant Cell Physiol 38:248–258
16. Lippman Z, Gendrel A-V, Colot V, Martienssen R (2005) Profiling DNA methylation patterns using genomic tiling microarrays. Nat Methods 2:219–224
17. Kawashima T et al (2015) Diversification of histone H2A variants during plant evolution. Trends Plant Sci 20:419–425
18. Berger F, Bowman JL, Kohchi T (2016) Marchantia. Curr Biol 26:R186–R187
19. Shimamura M (2016) Marchantia polymorpha: taxonomy. Phylogeny and morphology of a model system. Plant Cell Physiol 57:230–256
20. Saxena PK, Fowke LC, King J (1985) An efficient procedure for isolation of nuclei from plant protoplasts. Protoplasma 128:184–189
21. Andrews S FastQC A quality control tool for high throughput sequence data. http://www.bioinformatics.babraham.ac.uk/projects/fastqc/
22. Langmead B, Salzberg SL (2012) Fast gapped-read alignment with Bowtie 2. Nat Methods 9:357–359
23. Chu J et al (2014) BioBloom tools: fast, accurate and memory-efficient host species sequence screening using bloom filters. Bioinformatics 30:3402–3404
24. Ramírez F et al (2016) deepTools2: a next generation web server for deep-sequencing data analysis. Nucleic Acids Res 44(W1):W160–W165
25. Zhang Y et al (2008) Model-based analysis of ChIP-Seq (MACS). Genome Biol 9:1–9
26. Zhang X, Robertson G, Woo S, Hoffman BG, Gottardo R (2012) Probabilistic inference for

nucleosome positioning with MNase-based or sonicated short-read data. PLoS One 7:e32095

27. Robinson JT et al (2011) Integrative genomics viewer. Nat Biotechnol 29:24–26

28. Nicol JW, Helt GA, Blanchard SG, Raja A, Loraine AE (2009) The integrated genome browser: free software for distribution and exploration of genome-scale datasets. Bioinformatics 25:2730–2731

29. Kent WJ, Zweig AS, Barber G, Hinrichs AS, Karolchik D (2010) BigWig and BigBed: enabling browsing of large distributed datasets. Bioinformatics 26:2204–2207

Chapter 8

Tissue-Specific Transcriptome Profiling in *Arabidopsis* Roots

Erin E. Sparks and Philip N. Benfey

Abstract

Spatiotemporal transcriptome profiles from specific tissues are critical for understanding plant development and responses to the environment. One approach to isolate specific tissues is fluorescence-activated cell sorting (FACS). In this chapter, we outline methods for the FACS isolation of root protoplasts followed by transcriptome profiling using RNA sequencing.

Key words Fluorescence-activated cell sorting, Roots, Protoplast, Transcriptomics, RNAseq

1 Introduction

The rapid advancement in genome-wide analyses has dramatically expanded the amount of information available on gene expression (transcription and translation), DNA methylation, histone modification, and more. The data obtained from these approaches is important for our understanding of how plants interpret signals to enact developmental programs or respond to stimuli. However, genome-wide information is often obtained at the whole organ or whole plant scale, which can obscure tissue-level changes that are important contributors to physiology [1–4]. Individual tissue profiles are key to understanding how spatiotemporal changes in gene expression influence plant outcomes.

In *Arabidopsis thaliana*, tissue-specific analyses have been used to identify key regulators of root development [1, 5], abiotic stress responses [2, 6–8], and shoot apical meristem identity [9, 10]. Isolation of specific tissues is achieved through two main approaches, laser capture microdissection (LCM) and fluorescence-activated cell sorting (FACS) (for a comprehensive list of analyses, *see* ref. [11]). For plant species in which transformation is inefficient and/or fluorescent reporter lines do not exist, LCM can be used to isolate specific cell populations (e.g., [12, 13]). The advantage of LCM is that, theoretically, any population that can be

Wolfgang Busch (ed.), *Plant Genomics: Methods and Protocols*, Methods in Molecular Biology, vol. 1610,
DOI 10.1007/978-1-4939-7003-2_8, © Springer Science+Business Media LLC 2017

visualized under a microscope can be isolated. However, LCM isolation is usually performed on fixed tissues, which may compromise the quality of the material captured for subsequent analysis (*see* ref. [14] for additional information). This chapter focuses on the isolation of fluorescently labeled cell populations using FACS. Many tissue-specific marker lines already exist in *Arabidopsis*, and new lines are readily generated due to the ease of plant transformation. The advantages of FACS include rapid sample acquisition and high-quality material obtained from living tissues.

Here we describe an update and expansion of previous methods [15, 16] for tissue-specific transcriptome profiling in *Arabidopsis* roots. This protocol relies on the enzymatic digestion of plant cell walls to release the protoplasts (termed "protoplasting"), for subsequent FACS enrichment of a fluorescently labeled cell population, isolation of RNA, and transcriptome profiling by RNA sequencing (RNAseq). This isolation protocol has been successfully used to analyze tissue-specific root transcriptomes (mRNAs, miRNAs, and natural antisense transcripts) [1, 5, 17, 18], proteomes [19], DNA methylation [20], and metabolomes [4].

2 Materials

2.1 Plant Materials

1. Seeds from a fluorescently labeled marker line (*see* **Note 1**).
2. 50 mL conical tubes.
3. 50% bleach +0.1% Tween20 for seed sterilization.
4. Square plates (100 × 100 × 15 mm) containing nutritive gel: 1% agar, 1× Murashige Skoog (MS), and 1% sucrose.
5. Autoclaved (20 min gravity cycle) nylon mesh (Nitex Cat 03-100/44, Sefar), cut into the size of a Post-It note (*see* **Note 2**).
6. Forceps.
7. Alcohol lamp filled with 190 or 200 proof ethanol.
8. 3.5 mL sterile graduated transfer pipette.
9. Parafilm.
10. Standard growth chamber (e.g., Percival or Conviron).

2.2 Enzymatic Digestion (Protoplasting)

1. Solution A (1 mL per sort, *see* **Note 3**): For 200 mL

600 mM Mannitol	21.86 g
2 mM $MgCl_2$	2 mL of 0.2 M $MgCl_2 \cdot 6H_2O$ stock solution
0.1% BSA	0.2 g (Sigma A-3912)
2 mM $CaCl_2$	2 mL of 0.2 M $CaCl_2 \cdot 2H_2O$ stock solution
2 mM MES	2 mL of 0.2 M MES·hydrate stock solution
10 mM KCL	2 mL of 1 M KCl stock solution
pH to 5.5 with 1 M Tris	

2. Solution B (7 mL per Sort, *see* **Note 3**):
   50 mL Solution A.
   0.75 g cellulase (Calbiochem or Sigma).
   0.05 g pectolyase (Sigma).
3. #21 carbon steel surgical blades (one per sort).
4. 70 μM cell strainers (e.g., Thermo Fisher Scientific) (two per sort).
5. 40 μM cell strainers (e.g., Thermo Fisher Scientific) (one per sort).
6. 35 × 10 mm petri dishes (one per sort).
7. 50 mL conical tubes (two per sort).
   3.5 mL graduated transfer pipette (one per sort).
8. 15 mL conical tubes (one per sort).
9. 5 mL polystyrene round-bottom tubes (or other as accommodated by the sorting machine) (one per sort).
10. Orbital shaker.
11. Centrifuge for 15 mL conical tubes.
12. P1000 micropipette and tips.

2.3 Fluorescence-Activated Cell Sorting

1. Cell sorter equipped with a laser specific for the fluorescent protein of interest (e.g., Beckman Coulter MoFlo Astrios).
2. Eppendorf tube for collection containing RNA isolation buffer (e.g., 100 μL Qiagen RLT Buffer with 10 μL β-mercaptoethanol added per manufacturer's instructions).
3. Dry ice for storing collected cells after sorting.

2.4 RNA Isolation and Analysis

1. RNA isolation kit (e.g., Qiagen RNeasy Micro or Mini kit).
2. DNase cleanup kit (e.g., Qiagen RNase-free DNase, which is included in RNeasy kit).
3. Qubit fluorometer (Thermo Fisher Scientific).
4. Qubit RNA HS assay kit (Thermo Fisher Scientific).
5. 2100 Bioanalyzer (Agilent Technologies).
6. Bioanalyzer RNA 6000 Nano or Pico kit (Agilent Technologies).

2.5 RNAseq Library Preparation

1. Poly-A selection kit (e.g., Dynabeads mRNA Purification Kit, Thermo Fisher Scientific) or ribo-depletion kit (e.g., Plant Ribo-Zero rRNA Removal Kit, Illumina).
2. Magnetic stand that can accommodate one or more Eppendorf tubes.
3. RNAseq library preparation kit (e.g., TruSeq mRNA Library Prep Kit, Illumina).
4. Bioanalyzer RNA 6000 Pico kit (Agilent Technologies).
5. Thermocycler.
6. Agencourt AMPure XP beads.

7. Bioanalyzer High Sensitivity DNA Kit (Agilent Technologies).
8. Core facility or company with a next-generation sequencer (e.g., Illumina HiSeq 2500).

2.6 RNAseq Analysis

1. Macintosh- or Linux-enabled computer.
2. FASTX-toolkit (http://hannonlab.cshl.edu/fastx_toolkit/index.html).
3. Tophat (https://ccb.jhu.edu/software/tophat/index.shtml).

3 Methods

3.1 Seed Sterilization and Plating

1. Aliquot seeds from a fluorescently labeled marker line into a 50 mL conical tube for sterilization (*see* **Note 4**). Table 1 contains a list of the commonly used root marker lines and the

Table 1
Frequently used root marker lines

Marker line	Population marked	Number of plates per sort	Approximate number of cells [a]	RNA amounts (#Cells)[b]
APL	Protophloem, phloem and phloem companion cells	4–6	50,000+	254.8 ng (200,000)
CO2	Meristematic cortex	4–6	50,000+	390 ng (50,000)
COBL9	Lateral root cap and trichoblast	10–12	100,000+	767 ng (50,000)
CORTEX	Differentiated cortex	6–8	100,000+	520 ng (46,000)
E30	Differentiated endodermis	4–6	30,000+	468 ng (150,000)
GL2	Atrichoblast	4–6	50,000+	650 ng (50,000)
PET111	Columella	4–6	50,000+	241.8 ng (100,000)
S17	Phloem pole pericycle	4–6	50,000+	192.4 ng (50,000)
S18	Differentiated xylem	4–6	20,000+	83.2 ng (60,000)
S32	Protophloem and metaphloem	4–6	20,000+	338 ng (80,000)
S4	Meristematic xylem	4–6	20,000+	390 ng (60,000)
SCR	Quiescent center and endodermis	4–6	50,000+	364 ng (50,000)
WER	Lateral root cap and atrichoblast	3–6	250,000+	468 ng (50,000)
WOL	Stele	4–6	150,000+	520 ng (50,000)
WOX5	Quiescent center and initial cells	4–6	10,000+	78 ng (30,000)

[a]Approximate number of cells refers to the minimum number sorted and does not reflect the total number of cells that can be collected

[b]These numbers are from one experiment. Numbers should be used for reference only and do not reflect the amount of RNA that will be isolated from every experiment

approximate number of cells recovered from one set of pooled plates (i.e., one sort). The total weight of seeds should be calculated based on the target number of cells and the number of replicates (*see* **Note 1**). Here "one sort" refers to 4–12 plates combined, depending on the marker line.

2. Add water and stratify seeds at 4 °C overnight for 2 days, protected from light (*see* **Note 5**).
3. Decant water from the seeds. Add 50% bleach and 0.1% Tween20 to the tube for surface sterilization. Incubate for 5 min on a rotating platform to ensure thorough sterilization. Rinse the seeds 5–6 times by filling the 50 mL conical tube with sterile water under sterile conditions. Allow the seeds to settle or briefly spin down between washes.
4. While seeds are being sterilized, add nylon mesh to each of the plates (Fig. 1a, b). In a sterile hood, flame sterilize a pair of forceps with an alcohol lamp. Using the sterilized forceps, carefully remove a piece of mesh from the autoclaved package and place onto the surface of a 1× MS, 1% sucrose, and 1% agar plate. Add mesh to all of the plates, sterilizing the forceps between every plate to minimize bacterial or fungal contamination.
5. After the final wash, remove the majority of water (to ~5 mm above the seeds) with a 3.5 mL sterile graduated transfer pipette. Use a different pipette for each of the genotypes to avoid cross-contamination of the seeds.
6. Using the same transfer pipette, carefully paint two lines of seeds across the mesh on the prepared plates. These lines should be 3–5 seeds thick across the plate (Fig. 1c).
7. Seal plates with Parafilm (*see* **Note 6**), and transfer to a growth chamber set at 22 °C and 16 h light and 8 h dark conditions for 5–6 days. Position plates vertically in the growth chamber so that roots grow along the surface of the mesh-covered agar plate (Fig. 1d).

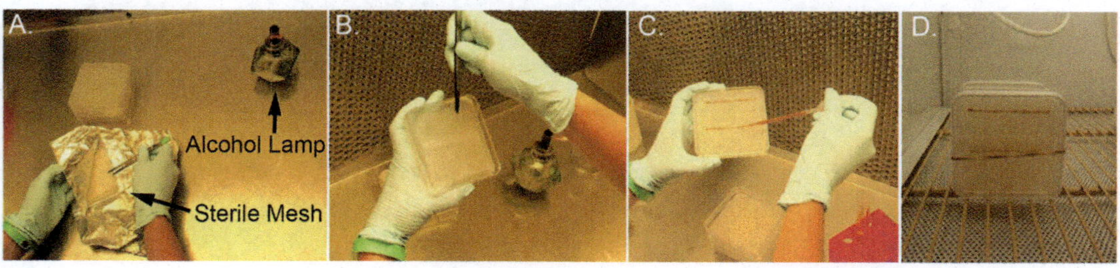

Fig. 1 Plating seeds for FACS analysis. (**a**) Under a laminar flow hood, assemble the necessary materials for seed plating: cut and sterilized mesh, 1× MS plates, an alcohol lamp, and forceps. (**b**) Sterilize forceps in the flame of the alcohol lamp, and place one piece of sterilized mesh onto the surface of each 1× MS plate. (**c**) Use a transfer pipette to plate two rows of sterilized seeds onto the mesh-covered plates. (**d**) Seal the plates with Parafilm and place vertically in a growth cabinet

3.2 Enzymatic Digestion of Cell Walls (Protoplasting)

1. Thaw Solution A and Solution B (or make fresh, *see* **Note 2**).
2. Assemble reagents for protoplasting (Fig. 2). For each sort (where one sort is 4–12 plates):
   (a) Place one 70 μM filter into the petri dish and add 7 mL of Solution B.
   (b) Place the second 70 μM filter into the top of a 50 mL-labeled conical tube.
   (c) Place the 40 μM filter into the top of the second 50 mL-labeled conical tube.
3. Remove Parafilm from the plates and check carefully for contamination (*see* **Note 7**).
4. With a #21 surgical blade, cut roots from four to six plates, and transfer into the petri dish containing the 70 μM filter and Solution B (*see* **Note 8**) (Fig. 2). A fresh blade should be used for every sort.

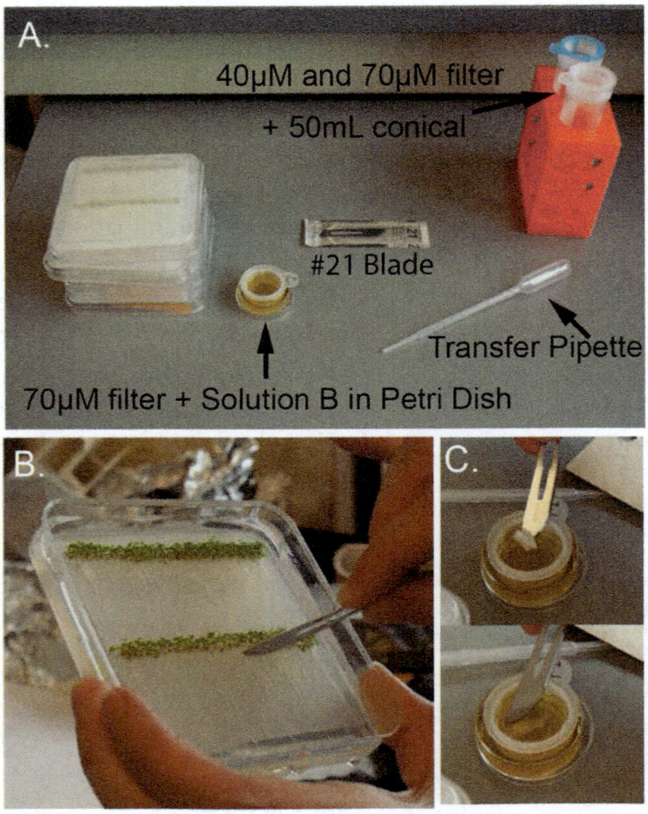

Fig. 2 Cutting roots and setup for protoplasting. (**a**) The material required for protoplasting one sort (here four plates) includes 7 mL of Solution B added to a 70 μM filter in the 35 mm petri dish, #21 blade, transfer pipette, 40 μM (*blue*), and 70 μM filter (*white*) placed in the top of a 50 mL conical tube. (**b**) Cut roots in fragments with the #21 blade. (**c**) Add cut roots into the petri dish containing a 70 μM filter and Solution B

5. Incubate petri dishes on an orbital shaker set to 85 RPM for 1 h. At 20, 50, and 60 min, gently mix the roots in Solution B to provide mechanical assistance for dissociation. Mix by spreading them over the strainer and stirring with a 3.5 mL graduated transfer pipette. Once protoplasts are released, they will pass through the filter into the surrounding solution.
6. Remove the filter with root remnants, and transfer a small amount of the undigested root fragments into a 15 mL conical tube. These undigested fragments will help the protoplasts pellet. Aspirate any solution from the bottom of the filter, and transfer to the same 15 mL conical tube. Discard the filter, and transfer the remaining Solution B + protoplasts into the 15 mL conical tube.
7. Centrifuge the tube(s) at 22 °C, 200 × *g* for 6 min.
8. Remove and discard the supernatant with the transfer pipette, leaving the pellet at the bottom of the tube (approximately 0.5–1 mL of material remaining).
9. With a P1000 micropipette, add 250 mL of Solution A. With the transfer pipette, resuspend the pellet, and transfer to the second 70 μM filter on the 50 mL conical tube.
10. Add an additional 250 mL of Solution A to the 15 mL tube with a P1000 micropipette. Rinse the sides of the tube by pipetting approximately four times. Transfer this volume with the micropipette to the second 70 μM filter on the 50 mL conical tube (*see* **Note 9**).
11. Using a P1000 micropipette, collect liquid from the underside of the 70 μM filter, and transfer to the 40 μM filter on a fresh 50 mL conical tube (*see* **Note 10**).
12. Using a P1000 micropipette, collect liquid from the underside of the 40 μM filter and combining with the rest of the protoplasts in the bottom of the second 50 mL conical tube.
13. At this point the protoplasts are ready for sorting. Depending on the sample holder of the sorting machine, the protoplasts can be kept in the 15 mL tube or transferred to a 5 mL polystyrene round-bottom tube.

3.3 Fluorescence-Activated Cell Sorting

1. Keep the protoplasts at room temperature and transfer to the sorting facility (*see* **Note 3**).
2. Bring tubes for collection with RNA isolation buffer and dry ice to store sorted samples after collection (*see* **Note 11**).
3. To calibrate the sorter, first analyze the non-transgenic protoplasts (*see* **Note 12**).
4. Visualize the forward scatter (FSC) and side scatter (SSC) of the non-transgenic control protoplasts, and gate the population to separate the protoplasts from debris (*see* **Note 13**) (Fig. 3a).

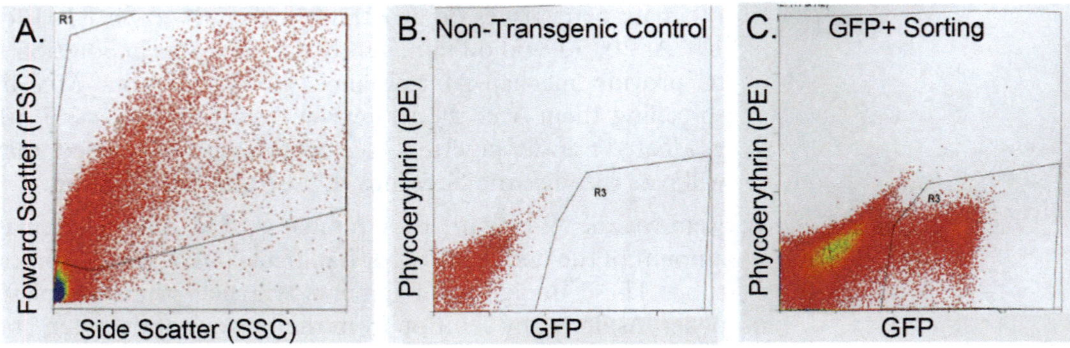

Fig. 3 Representative FACS profiles. (**a**) Example of forward scatter (FSC) and side scatter (SSC) profile from *Arabidopsis* root protoplasts. The region gated in black was selected as the viable portion of cells for subsequent analysis. (**b**) Example plot of a sorting analysis to set gating with a non-transgenic control sample. The spectra of GFP versus Phycoerythrin (or other red spectrum) are used to visualize protoplast autofluorescence. The region gated in black excludes the autofluorescence and aims to gate the GFP-positive population. (**c**) Using the gate established with the control sample, GFP-positive cells are sorted and isolated

5. Gate the sorter for the acquisition of fluorescent-positive protoplasts. This is accomplished by analysis of the autofluorescence from the non-transgenic control and setting a gate that avoids this population (*see* **Note 14**) (Fig. 3b).
6. Flush the sorter with sheath solution (usually phosphate-buffered saline) prior to running each sample to avoid cross-contamination.
7. Apply the gates established with the non-transgenic protoplasts to the first sample of interest, and collect protoplasts that fall within the positive population directly into lysis buffer (*see* **Note 15**) (Fig. 3c).
8. Continue collecting protoplasts until the desired number of positive cells/events is obtained (Table 1).
9. Place collection tube on dry ice and continue to the next sample.

3.4 RNA Isolation and Preparation of Sequencing Libraries

1. Isolate RNA per kit instructions, maintaining RNase-free conditions to avoid RNA degradation (*see* **Note 16**). Include a DNase treatment step to avoid the carry through of contaminating DNA.
2. Check the concentration of the samples using a Qubit fluorometer following the instructions for the instrument (*see* **Note 17**).
3. Check RNA quality by analysis on a microfluidic electrophoresis system such as a Bioanalyzer. The choice of microfluidic chip depends on the concentration of the sample determined by Qubit analysis. Nano kits are for quantification of 25–500 ng/μL, and Pico kits are for quantification of 50–5000 pg/μL of total RNA.

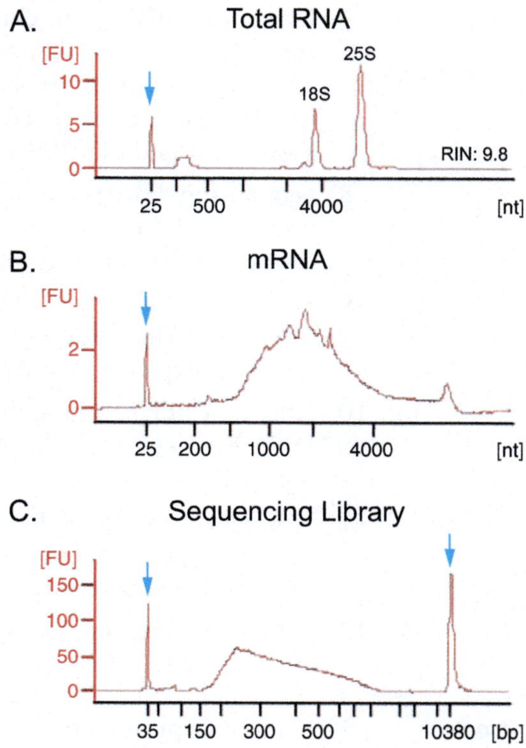

Fig. 4 Bioanalyzer traces from RNAseq library preparation. (**a**) A RNA Nano Bioanalyzer trace for high-quality total RNA. The area under the ribosomal RNA peaks for 18S and 25S are used to calculate a RIN score for RNA integrity. (**b**) A RNA pico Bioanalyzer trace for a poly-A selected sample. (**c**) A DNA High Sensitivity Bioanalyzer trace for a RNAseq library. *Blue arrows* indicate the marker used for Bioanalyzer calibration. Traces were chosen for their representative images, and do not reflect processing of the same sample

4. The Bioanalyzer analysis generates a RNA integrity number (RIN), which is used to estimate the degradation of total RNA samples. RIN is calculated from the ratio of area under the peaks of the two major ribosomal RNAs (18S and 25S in plants). RIN values range from 1 to 10, with higher values indicating less degradation. The RIN cutoff to include samples for subsequent analysis depends on the discretion of the scientist, although generally RIN ≥ 8 will consistently provide high-quality transcriptomic data (Fig. 4a).
5. At this point, the choice of transcriptome analysis depends on the goal of the project, the budget available, and the sequencing machines available for analyses. Here, we will focus on RNAseq analysis with an Illumina HiSeq 2500 and note alternatives where applicable.
6. To analyze the mRNA abundance by RNAseq, first enrich for mRNA by selecting the poly-A fraction (e.g., Dynabeads) or by rRNA depletion (e.g., Plant Ribo-Zero rRNA) (*see* **Note 18**).

7. Check the mRNA enrichment on a Bioanalyzer with a pico chip. Samples should show an absence of the 18S and 25S peaks and a broad distribution of mRNA sizes (Fig. 4b). If rRNA peaks are still detected, samples should be subjected to a second round of mRNA enrichment.
8. Sequencing libraries should be prepared from the mRNA-enriched sample per the manufacturer's instructions (e.g., Illumina TruSeq library kit) (*see* **Note 19**).
9. Follow the instructions for the library preparation protocol. This will include the use of a thermocycler and several size selection steps with the AMPure XP beads.
10. Check the library on the Bioanalyzer using the High Sensitivity DNA Kit to ensure sufficient distribution and quantity of the library (Fig. 4c).
11. Normalize and pool libraries per the instructions of your sequencing facility. Often libraries are multiplexed and pooled to 10 nM concentrations (*see* **Note 20**).
12. Submit libraries for sequencing on an Illumina HiSeq 2500 (*see* **Note 21**).

3.5 Transcriptome Analysis

1. Upon receipt of your sequencing data, conduct a quality control of the reads (*see* **Note 22**). We recommend the FASTX-toolkit for initial analysis and specifically two visualization tools to look at read quality and nucleotide distribution:
   (a) Install the FASTX-toolkit per package instructions.
   (b) Calculate the quality statistics for "your_file" with the fastx_quality_stats argument (*see* **Note 23**).
   (c) `$ fastx_quality_stats -i your_file.fq -o your_file_stats.txt`
   (d) Graph these results for rapid visualization. It is normal for libraries to show a reduction in quality as a function of length (Fig. 5a).
   (e) `$ fastx_quality_boxpot_graph.sh -i your_file_stats.txt -o your_file_quality.png -t "My Library"`
   (f) In addition, you should check the nucleotide distribution across the sequencing reads. You should expect each nucleotide to contribute approximately 25% in an unbiased library (Fig. 5b).
   (g) `$ fastx_nucleotide_distribution_graph.sh -i your_file_stats.txt -o your_file_bp.png -t "My Library"`
   (h) If required you can use the FASTX-toolkit to filter out low-quality reads or remove contaminating sequences (e.g., barcodes or adapters).
2. Map the reads to a reference genome (if available, *see* **Note 24**). We recommend mapping with Tophat and refer you to the detailed instructions provided in ref. [21].

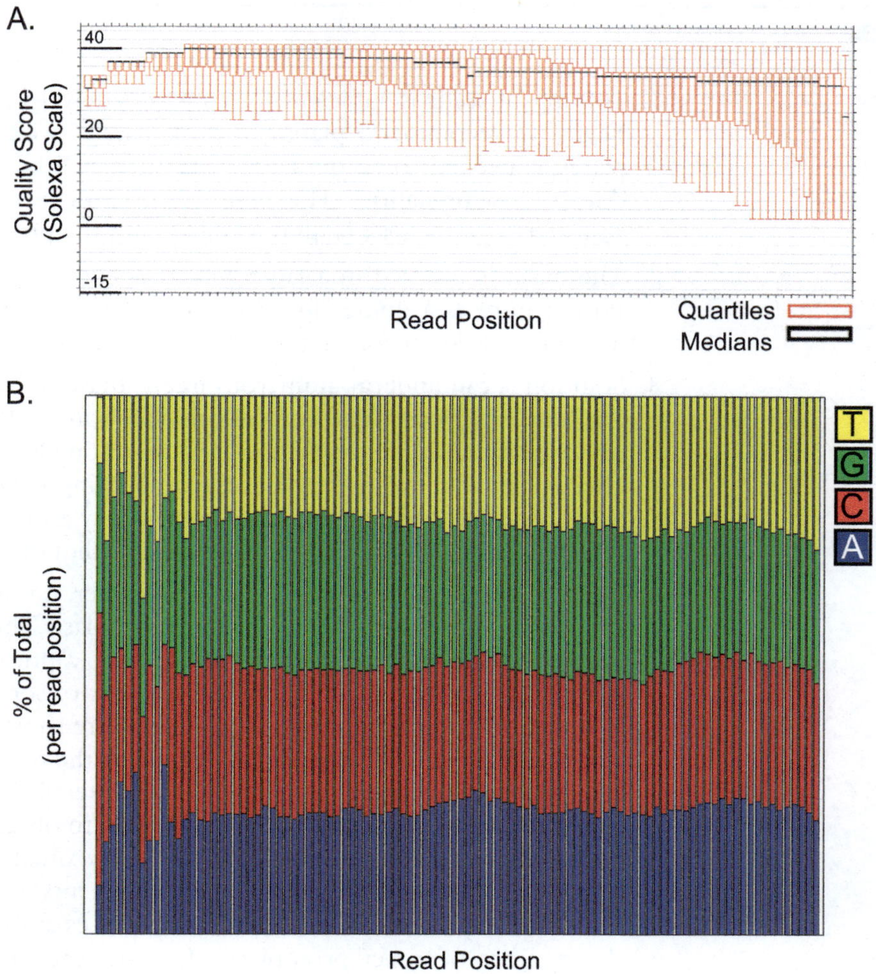

Fig. 5 Quality analysis of RNAseq reads using the FASTX-toolkit. (**a**) Quality as a function of read position. Quality scores are on a Solexa scale with −15 as the lowest and 40 as the highest score. *Black line* indicates mean score, and *red boxes* indicate the quartiles. (**b**) Nucleotide distribution as a function of read position. In an unbiased sample, each nucleotide should contribute approximately 25% at each position

3. After mapping, transcript levels should be normalized for further analysis. To compare the expression levels of two genes within one population, transcripts can be normalized for both the size of the library and the size of the gene (e.g., RPKM or FPKM—reads/fragments per kilobase of transcript per million mapped reads). To compare the expression level of the same gene across different cell types or different conditions, transcripts can be normalized for the size of the library (e.g., CPM—counts per million).
4. Many algorithms exist to examine differential gene expression between RNAseq samples. We refer you to a recent paper that performed an extensive analysis on the number of biological replicates and tools for differential expression [22].

4 Notes

1. 200 mg of seeds is sufficient for approximately 15 plates. To begin we recommend plating approximately twice as many plates as needed to account for plates that will be discarded due to contamination. The number of extra plates can be adjusted once the user is more familiar with the procedure.
2. The mesh is ordered in a large roll. It is cut to size using a Post-It note as a guide, and 30 pieces of mesh are bundled into a foil packet before autoclaving.
3. Solution A can undergo numerous freeze-thaw cycles without any adverse effects. The enzymes in Solution B should not undergo multiple freeze-thaw cycles. Solution B is the most effective if made fresh the day of protoplasting (thaw frozen Solution A and add enzymes); however, it may also be made in advance and frozen at −20 °C in single-use aliquots (7 mL).
4. To establish the correct gating for each marker, it is important to include one sort of non-transgenic protoplast. This enables calibration of the sorter for gating and capture of the fluorophore-positive cells. Limit the number of sorts per day as this protocol is time sensitive, and there are substantial circadian effects on gene expression. Although the total number of sorts will ultimately depend on the marker line (i.e., less abundant populations must be sorted longer to obtain the target number of cells), four sorts will take approximately 1 h. In general, sorting is performed in 1-h increments at the same time each day to minimize circadian influences and maximize the number of intact protoplasts. In our experience, when protoplasts are kept in the sorting solution for extended periods of time, the number of viable cells is reduced.
5. Alternatively, seeds can be sterilized first and then imbibed for 2 days before plating.
6. Plates may also be sealed with Micropore tape; however, plants grow differently in plates sealed with tape versus Parafilm. If any comparisons will be made to existing data, ensure the plates are sealed with the same material.
7. While most contamination is visible by eye, it is important to also smell your plates to identify possible contamination that is not visible. The most frequent indications of plate contamination are "grassy," "earthy," or "sweet" smells. Contaminated plates should be disposed of and not included in the analysis.
8. The root cutting technique depends on the cell population being enriched for. Cutting the roots in multiple sections helps the enzyme access the cell walls and release the protoplasts. However, if the marker marks more differentiated cells, it is best to cut the roots once at the hypocotyl junction and then

1–2 more times in the rest of the root. This is because differentiated cells are larger, and additional cuts will damage the cells and reduce yields. In contrast, if the cells are found within the meristem, then cutting only the root tips is most effective.

9. The volume here can be adjusted based on the anticipated abundance of positive cells. The more diluted the sample, the longer it will take to pass through the sorter and obtain the target number of cells. Between 100 and 350 μL per wash is recommended depending on the marker.

10. Filtering through the 70 μM filter aims to remove the majority of root fragments that were used to facilitate pelleting. While this step is recommended for optimal recovery, it is possible to filter directly through the 40 μM filter. The 40 μM-filtering step is critical to avoid clogging the sorter. Check with your sorting facility to determine if they have specific requirements for sample preparation.

11. The choice of RNA isolation or buffer may depend on the setup of the sorting facility. For example, Trizol should be generally avoided unless the collection can be done under a chemical safety hood to avoid extended exposure to phenol. If necessary protoplasts can be collected in phosphate-buffered saline (PBS), spun down at low speed (i.e., 22 °C, 200 × *g* for 6 min) and frozen. In our experience the yield will drop if protoplasts are not immediately lysed.

12. It is important to be familiar with the specific sorter you are using as each instrument has its own idiosyncrasies. Consult with your sorting core facility for additional details and instrument training.

13. Forward scatter is proportional to the size, and side scatter is proportional to the complexity of the object passing through the sorter. While these ratios can be used to separate different cell types in mammalian systems, in plants this gating is used to enrich for the protoplasts and avoid autofluorescent debris.

14. If a sorted, non-transgenic control sample is required, gate and collect cells based on the FSC and SSC profiles.

15. Sorting of double-positive populations (e.g., mCherry and GFP) is possible, but requires strongly expressed markers to exceed the autofluorescence signal, which appears in the overlap of the red and green spectra (*see* Fig. 3b).

16. If samples were to be processed for RNAseq, it is possible to submit the total RNA samples to a company or core facility for processing. There is a substantial amount of specialized equipment required for RNAseq library preparation, which may not be available to individual laboratories. In addition, most sequencing facilities cannot guarantee the quality of the data unless they process the samples themselves.

17. The use of a fluorometer ensures more accurate quantification of sample concentration than a spectrophotometer such as NanoDrop. In addition, most core facilities recommend replicated fluorometer readings because values can vary as much as 10% or 15% between readings.
18. If non-*Arabidopsis* plants are being utilized, it is important to confirm the rRNA sequences are appropriate for depletion in your system. Small RNAs can also be profiled by enrichment with specific kits (e.g., mirVana miRNA isolation kit, Thermo Fisher Scientific) or through differential molecular weight selection (e.g., [17]).
19. There are several considerations when selecting a library preparation procedure. First is the amount of mRNA for analysis. The TruSeq kit reliably works for ≥10 ng of mRNA; however, other kits (e.g., SMARTer Stranded RNAseq Kit, Clontech) are much better suited for low input. The second consideration is a stranded or unstranded library. Stranded libraries maintain information regarding the origin of the RNA (+ or – strand) and thus provide better mapping of the sequencing reads which is important for poorly annotated genomes. However, stranded library prep kits are often more expensive and may be unnecessary if a high-quality reference genome exists. Alternatively, kit-free and less expensive protocols are also available (e.g., [23]).
20. Pooling multiple libraries into the same lane of a flow cell is an efficient way to obtain data while limiting the cost per sample. The total number of reads per lane varies, and we recommend you to talk to your sequencing facility to estimate their average number of reads. We average between 180 and 200 million reads per lane and have successfully multiplexed up to 12 samples. If transcript discovery is the aim, then increasing sequencing depth can be achieved by decreasing pooling. However, if the aim is to determine differential gene expression, then 10–20 million reads per sample is sufficient for *Arabidopsis*.
21. There are two choices when determining the type of sequencing to obtain: the first is the length of the read (50 bp, 100 bp, etc.) and the second is single-end or paired-end reads. This choice of sequencing type is at the discretion of the scientist.
22. The analysis of the RNAseq data assumes basic familiarity with Unix and the command line. If you are unfamiliar with these tools, consider utilizing web-based platforms such as CyVerse (http://www.cyverse.org/) or Galaxy (https://galaxyproject.org/).
23. Quality scores can be encoded in three different methods: Sanger, Solexa/Illumina <1.3, and Illumina 1.3+.

24. There are several mapping algorithms (e.g., Tophat, BWA, and Bowtie) that rely on the presence of a reference genome. In the absence of a reference genome, de novo transcriptomes can be assembled using the Trinity package (https://github.com/trinityrnaseq/trinityrnaseq/wiki).

Acknowledgments

We gratefully acknowledge members of the Benfey Lab for their comments on this chapter and specifically Colleen Drapek for providing FACS plots, Kevin Lehner for photography, and Masashi Yamada and Louisa Liberman for providing the expected number of cells for different marker lines. Research in the Benfey Lab is supported by funding from the National Institutes of Health, the National Science Foundation, the Howard Hughes Medical Institute, and the Gordon and Betty Moore Foundation.

References

1. Brady SM, Orlando DA, Lee JY, Wang JY, Koch J, Dinneny JR, Mace D, Ohler U, Benfey PN (2007) A high-resolution root spatiotemporal map reveals dominant expression patterns. Science 318:801–806. doi:10.1126/science.1146265
2. Geng Y, Wu R, Wee CW, Xie F, Wei X, Chan PMY, Tham C, Duan L, Dinneny JR (2013) A spatio-temporal understanding of growth regulation during the salt stress response in Arabidopsis. Plant Cell 25:2132–2154. doi:10.1105/tpc.113.112896
3. Shroff R, Vergara F, Muck A, Svatos A, Gershenzon J (2008) Nonuniform distribution of glucosinolates in *Arabidopsis thaliana* leaves has important consequences for plant defense. Proc Natl Acad Sci 105:6196–6201. doi:10.1073/pnas.0711730105
4. Moussaieff A, Rogachev I, Brodsky L, Malitsky S, Toal TW, Belcher H, Yativ M, Brady SM, Benfey PN, Aharoni A (2013) High-resolution metabolic mapping of cell types in plant roots. Proc Natl Acad Sci. doi:10.1073/pnas.1302019110
5. Birnbaum K, Shasha DE, Wang JY, Jung JW, Lambert GM, Galbraith DW, Benfey PN (2003) A gene expression map of the Arabidopsis root. Science 302:1956–1960. doi:10.1126/science.1090022
6. Gifford ML, Dean A, Gutierrez RA, Coruzzi GM, Birnbaum KD (2008) Cell-specific nitrogen responses mediate developmental plasticity. Proc Natl Acad Sci 105:803–808. doi:10.1073/pnas.0709559105
7. Iyer-Pascuzzi AS, Jackson T, Cui H, Petricka JJ, Busch W, Tsukagoshi H, Benfey PN (2011) Cell identity regulators link development and stress responses in the Arabidopsis root. Dev Cell 21:770–782. doi:10.1016/j.devcel.2011.09.009
8. Dinneny JR, Long TA, Wang JY, Jung JW, Mace D, Pointer S, Barron C, Brady SM, Schiefelbein J, Benfey PN (2008) Cell identity mediates the response of Arabidopsis roots to abiotic stress. Science 320:942–945. doi:10.1126/science.1153795
9. Yadav RK, Tavakkoli M, Xie M, Girke T, Reddy GV (2014) A high-resolution gene expression map of the Arabidopsis shoot meristem stem cell niche. Development 141:2735–2744. doi:10.1242/dev.106104
10. Yadav RK, Girke T, Pasala S, Xie M, Reddy GV (2009) Gene expression map of the Arabidopsis shoot apical meristem stem cell niche. Proc Natl Acad Sci 106:4941–4946. doi:10.1073/pnas.0900843106
11. Bailey-Serres J (2013) Microgenomics: genome-scale, cell-specific monitoring of multiple gene regulation tiers. Annu Rev Plant Biol 64:293–325. doi:10.1146/annurev-arplant-050312-120035
12. Pattison RJ, Csukasi F, Zheng Y, Fei Z, van der Knaap E, Catalá C (2015) Comprehensive tissue-specific transcriptome analysis reveals

distinct regulatory programs during early tomato fruit development. Plant Physiol 168:1684–1701. doi:10.1104/pp.15.00287

13. Nakazono M, Qiu F, Borsuk LA, Schnable PS (2003) Laser-capture microdissection, a tool for the global analysis of gene expression in specific plant cell types: identification of genes expressed differentially in epidermal cells or vascular tissues of maize. Plant Cell 15:583–596. doi:10.1105/tpc.008102
14. Nelson T, Tausta SL, Gandotra N, Liu T (2006) Laser microdissection of plant tissue: what you see is what you get. Annu Rev Plant Biol 57:181–201. doi:10.1146/annurev.arplant.56.032604.144138
15. Birnbaum K, Jung JW, Wang JY, Lambert GM, Hirst JA, Galbraith DW, Benfey PN (2005) Cell type-specific expression profiling in plants via cell sorting of protoplasts from fluorescent reporter lines. Nat Methods 2:615–619. doi:10.1038/nmeth0805-615
16. Iyer-Pascuzzi AS, Benfey PN (2010) Methods Mol Biol 655:313–319. doi:10.1007/978-1-60761-765-5_21
17. Breakfield NW, Corcoran DL, Petricka JJ, Shen J, Sae-Seaw J, Rubio-Somoza I, Weigel D, Ohler U, Benfey PN (2012) High-resolution experimental and computational profiling of tissue-specific known and novel miRNAs in Arabidopsis. Genome Res 22:163–176. doi:10.1101/gr.123547.111
18. Li S, Liberman LM, Mukherjee N, Benfey PN, Ohler U (2013) Integrated detection of natural antisense transcripts using strand-specific RNA sequencing data. Genome Res 23:1730–1739. doi:10.1101/gr.149310.112
19. Petricka JJ, Schauer MA, Megraw M, Breakfield NW, Thompson JW, Georgiev S, Soderblom EJ, Ohler U, Moseley MA, Grossniklaus U, Benfey PN (2012) The protein expression landscape of the Arabidopsis root. Proc Natl Acad Sci. doi:10.1073/pnas.1202546109
20. Kawakatsu T, Stuart T, Valdes M, Breakfield N, Schmitz RJ, Nery JR, Urich MA, Han X, Lister R, Benfey PN, Ecker JR (2016) Unique cell-type-specific patterns of DNA methylation in the root meristem. Nat Plants 2:16058. doi:10.1038/nplants.2016.58
21. Trapnell C, Roberts A, Goff L, Pertea G, Kim D, Kelley DR, Pimentel H, Salzberg SL, Rinn JL, Pachter L (2012) Differential gene and transcript expression analysis of RNA-seq experiments with TopHat and cufflinks. Nat Protoc 7:562–578. doi:10.1038/nprot.2012.016
22. Schurch NJ, Schofield P, Gierliński M, Cole C, Sherstnev A, Singh V, Wrobel N, Gharbi K, Simpson GG, Owen-Hughes T, Blaxter M, Barton GJ (2016) How many biological replicates are needed in an RNA-seq experiment and which differential expression tool should you use? RNA 22:839–851. doi:10.1261/rna.053959.115
23. Wang L, Si Y, Dedow LK, Shao Y, Liu P, Brutnell TP (2011) A low-cost library construction protocol and data analysis pipeline for Illumina-based strand-specific multiplex RNA-seq. PLoS One 6:e26426. doi:10.1371/journal.pone.0026426.g008

Chapter 9

Sample Preparation Protocols for Protein Abundance, Acetylome, and Phosphoproteome Profiling of Plant Tissues

Gaoyuan Song, Maxwell R. McReynolds, and Justin W. Walley

Abstract

Peptide mass spectrometry is an invaluable technique to globally quantify the proteome. Central to proteome profiling are efficient methods to extract proteins, digest proteins into peptides, and enrich for posttranslationally modified peptides prior to mass spectrometry. In this chapter, we describe methods to extract proteins, process them into peptides, and optionally enrich for phospho- and acetyl-peptides prior to analysis by mass spectrometry.

Key words Plant proteomics, Protein extraction, Phosphoproteome, Acetylome, Mass spectrometry

1 Introduction

The field of proteomics encompasses a large area of study intent on characterizing essential biological molecules, proteins. These key macromolecules perform a vast array of tasks within living organisms serving as organizers of cellular function, product manufacturers, and biological signaling constituents [1]. The unique amino acid sequences and posttranslational modifications that make up proteins allow for great diversity among molecules relating to their shapes, functional capabilities, and regulation [1]. The term *proteomics* was first used in the 1990s as an analogy to the study of the genome or *genomics*. The proteome refers to the complete set of produced or modified proteins within an organism that varies temporally and spatially in response to cellular development and environmental conditions [2].

Liquid chromatography coupled with mass spectrometry (LC-MS/MS) is a powerful platform that is employed for deep and quantitative proteomic analyses [3]. In this setup, high-performance liquid chromatography (HPLC) is utilized to fractionate and deliver peptides to the mass spectrometer (MS). Although mass

Wolfgang Busch (ed.), *Plant Genomics: Methods and Protocols*, Methods in Molecular Biology, vol. 1610,
DOI 10.1007/978-1-4939-7003-2_9, © Springer Science+Business Media LLC 2017

spectrometers differ in some regards, there are three common components that are typically utilized for proteome profiling experiments. The first is the ionization source, which is involved in generating gas-phase charged species. For global proteomics, ionization of the sample is typically achieved by electrospray ionization (ESI) [4]. The charged ions are then separated by their mass-to-charge ratio (m/z) using a variety of mass analyzers including time of flight (TOF) [5], quadrupole mass filters [6], and/or ion traps (e.g., an orbitrap [7, 8]). Finally, a detector is used to convert the impact of the ion into an electronic signal that is read and reported as a spectrum. In a typical workflow (Fig. 1), the generated spectra are matched to peptides using the "target-decoy" database search method [9].

Sample preparation is crucial to obtaining high-quality resolution of proteins in a proteome analysis, and thus a wide variety of protocols have been developed. Sodium dodecyl sulfate (SDS) detergent is a frequent choice for total extraction and solubilization of proteins, but it can impede enzymatic digestion and will

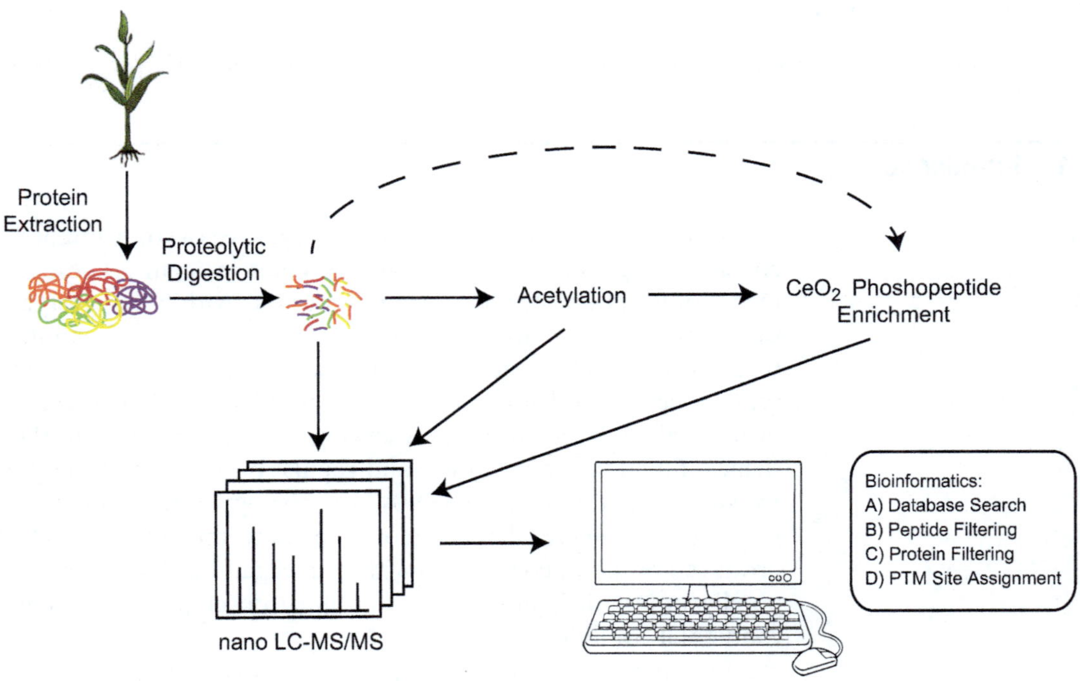

Fig. 1 Proteome profiling workflow. In a typical workflow, proteins are extracted and then digested into peptides. The digested peptides can directly be analyzed by LC-MS/MS to quantify total protein abundance. Additionally, peptides containing PTMs such as acetylation or phosphorylation can be enriched to enable PTM detection and quantification by LC-MS/MS. The isolated peptides are then introduced to the mass spectrometer using nano-flow liquid chromatography, and tandem mass spectra are generated. Finally, the spectra are searched against a reference proteome database to match spectra to peptides, which are used for protein/PTM identification and quantification

also overwhelm spectra data due to its high ionizability [10]. Thus, avoidance or removal of SDS is a must for efficacious MS analysis. A number of protocols use various combinations of solvent washes (methanol and/or acetone), TCA/acetone precipitation, methanol/chloroform precipitation, and/or phenol extraction in conjunction with SDS [11–14]. We often employ a protocol that utilizes a series of methanol and acetone steps, to remove metabolites and precipitate proteins, which are followed by addition of dilute (0.1%) SDS. After proteins are digested into peptides, the SDS is acid precipitated, and residual SDS is further removed by mixed-mode reverse-phase/cation exchange Cleanup (MCX) [15–19]. Alternatively, chaotropic reagents such as urea or guanidine hydrochloride can be used instead of SDS to extract and solubilize proteins prior to digestion [20, 21]. Finally, a widely used method for SDS removal is the filter-aided sample preparation (FASP) method. This method retains high molecular weight substances, including proteins, on the filter unit and allows the impurities (SDS and metabolites) to pass through the filter. Proteins are then digested into peptides on the filter, which thereby enables the peptides to pass through the filter for collection [22]. Below we describe a modified FASP method that substitutes guanidine hydrochloride for SDS in the FASP workflow, which we find results in increased protein IDs and reduced SDS artifacts.

Posttranslational modification (PTM) increases functional diversity of the proteome through covalent addition of functional groups, proteolytic cleavage, and/or protein degradation. Phosphorylation, ubiquitination, and acetylation are critical modifications that affect regulatory activity, localization, and interaction of the protein with other molecules [23]. Phosphoproteomic studies require enrichment of phosphopeptides prior to MS analysis since phosphorylated proteins can rarely be captured in "shotgun" MS analysis due to their low relative abundance. Metal oxide affinity enrichment is a common approach that uses metal ions (TiO_2, CeO_2, Fe^{3+}) to bind the negatively charged phosphate for enrichment prior to MS analysis. Acetylation of the ε-amino group of lysine also plays a major role in regulation of target proteins. Much like other PTMs, acetylation often occurs at low stoichiometry, with only a small fraction of the protein population carrying the PTM, making identification of modified proteins difficult. To improve analysis, acetylated peptides are enriched with acetyl-lysine-specific antibodies [24]. Detection of endogenous ubiquitination sites by MS has also benefited by availability of antibodies that recognize the Lys-ε-Gly-Gly remnants of trypsin-digested proteins with ubiquitinated lysine side chains [25, 26]. Finally, PTM enrichments can be done sequentially in order to reduce the total amount of peptides necessary and to enable direct comparison of PTMs in the same sample (Fig. 1). For example, the approach termed serial enrichments of PTMs (SEPTM) uses the flow-through of a primary PTM enrichment step to enrich a second PTM, whose flow-through then is used for a third PTM enrichment step. Critically, the authors demonstrated that only 0.3% of posttranslationally modified peptides carried multiple different types of modifications (e.g., phospho- plus acetyl-sites) [27].

Below we describe a workflow for protein extraction and digestion into peptides as well as optional steps for enrichment of phospho- and acetyl-peptides prior to LC-MS/MS (Fig. 1).

2 Materials

2.1 Protein Extraction

1. 0.2 mM Na_3VO_4 (phosphatase inhibitor) in 100% methanol (MeOH).
2. 50× phosphatase inhibitor mix: 125 mM NAF, 12.5 mM $NaVO_4$, 12.5 mM $NaPyroPO_4$, and 12.5 mM glycerol-P.
3. 200× sodium butyrate stock: 2 M sodium butyrate (histone deacetylase inhibitor, HDACi).
4. 200× nicotinamide stock: 2 M nicotanamide (HDACi).
5. Gnd lysis solution: 6 M guanidine hydrochloride, 100 mM Tris–HCl pH 8.0, 10 mM tris(2-carboxyethyl)phosphine (TCEP) (*see* **Note 1**).
6. 2.4 mm metal beads (e.g., The Homogenizer Company™, Cat # 19–640).
7. Acetone (100 and 80%).
8. Heated shaking block (e.g., ThermoFisher Scientific Digital Shaking Drybath).

2.2 Filter Aided Sample Preparation (FASP)

1. UA solution: 8 M urea in 0.1 M Tris–HCl pH 8.0 (*see* **Note 2**).
2. IAA solution: 0.05 M iodoacetamide (IAA) in UA solution.
3. Trypsin, 1 μg/μl in water.
4. Lys-C, 0.1 μg/μl in water.
5. ABC solution: 0.05 M NH_4HCO_3.
6. TCEP stock: 100 mM TCEP in 1 M Tris–HCl pH 7.5.
7. Amicon Ultracel—30 K centrifugal filters, 4 ml; Cat # UFC803008; Max 5 mg protein.
8. pH indicator paper.

2.3 Mixed-Mode Reverse-Phase/Cation Exchange Cleanup (MCX) (See Note 3)

1. MCX column (Oasis® MCX Vac Cartridge; Cat # 186000776) (*see* **Note 4**).
2. 100% MeOH.
3. Water.
4. 80% acetonitrile (ACN).

5. 500 mM NH_4HCO_3 in 45% isopropanol (IPA).
6. Vacuum manifold (e.g. Visiprep™ SPE Vacuum Manifold; Cat # 57030-U).
7. Speedvac system (e.g. ThermoFisher Scientific ISS110 SpeedVac).
8. pH indicator paper.

2.4 Acetylated Lysine Enrichment

1. Agarose-conjugated anti-acetyl lysine antibody (ImmuneChem-ICP0388).
2. 0.2 μM size exclusion centrifugal column (Microsep MCPM02C68).
3. 50 mM Tris–HCl pH 7.5.
4. 1 M Tris–HCl pH 7.5.
5. 0.1% trifluoroacetic acid (TFA).
6. pH indicator paper.

2.5 Phosphopeptide Enrichment

1. 50% trifluoroacetic acid (TFA).
2. 0.1% formic acid (FA).
3. 10% FA.
4. 1% colloidal CeO_2 (Sigma 643,009).
5. Elution buffer: 10% FA, 100 mM citric acid.
6. pH indicator paper.

3 Methods

3.1 Protein Extraction

1. Grind plant tissue in liquid N_2 into fine powder using a ceramic mortar and pestle for 15 min.
2. Weight 250 mg leaf tissue into 50 ml centrifuge tube.
3. Add 50 ml MeOH/0.2 mM Na_3VO_4 (pre-chilled to −20 °C) to leaf tissue.
4. Add 250 μl 200× sodium butyrate and 250 μl 200× nicotinamide and vortex.
5. Keep the tube at −20 °C for 30 min to overnight.
6. Spin at 4000 × *g* for 20 min at 4 °C; discard supernatant.
7. Add 1 ml of Gnd lysis solution to pellet, pipette up and down several times, and transfer to a 2 ml tube with two metal beads (*see* **Note 5**).
8. Vortex ~1 min to resuspend the pellet.
9. Incubate at 95 °C for 5 min while shaking in heated shaking block (*see* **Note 6**).

10. Put the tube on ice for 15 min.
11. Sonicate 1.5 min (*see* **Note 7**).
12. Incubate at 95 °C for 5 min while shaking in heated shaking block (*see* **Note 6**).
13. Clarify sample by centrifugation at 20,000 × *g* for 10 min.
14. Add supernatant to new tube and mix with 1× volume water, then add 4× volume acetone (pre-chilled at −20 °C), mix well, and store at −20 °C overnight to precipitate the protein.
15. Spin at 4000 × *g* for 15 min at 4 °C and discard supernatant.
16. Add 10 ml 80% acetone (precooled at −20 °C), sonicate on ice to resuspend the pellet, and spin at 4000 × *g* for 15 min at 4 °C.
17. Repeat **step 15** once.
18. Discard the supernatant, air-dry to remove residual acetone for 5 min, resuspend the pellet with 4 ml UA solution, and measure the protein concentration using the BCA assay [28].

3.2 FASP

1. Add proteins (max 5 mg protein) resuspended in UA solution to filter unit. Bring final volume of UA to 4 ml, and then centrifuge at 4000 × *g* for 20–40 min (*see* **Note 8**).
2. Add 4 ml UA and centrifuge at 4000 × *g* for 20–40 min.
3. Add 4 ml of UA w/2 mM TCEP to the filter unit and centrifuge at 4000 × *g* for 20–40 min.
4. Discard the flow-through from the collection tube.
5. Add 2 ml IAA solution, mix, and incubate without mixing at room temperature for 30 min in the dark.
6. Centrifuge the filter unit at 4000 × *g* for 20–40 min.
7. Add 2 ml of UA to the filter unit and centrifuge at 4000 × *g* for 20–40 min. Repeat this step once.
8. Add 2 ml of ABC to the filter unit and centrifuge at 4000 × *g* for 20–40 min. Repeat this step once.
9. Add 2 ml ABC with trypsin (enzyme to protein ratio 1:100) and mix.
10. Incubate the units at 37 °C overnight.
11. Add 1/5 original amount of trypsin (1 μg/μl) and equal volume of Lys-C (0.1 μg/μl). Incubate 2–4 h at 37 °C.
12. Transfer the filter unit to a new collection tube.
13. Centrifuge the filter units at 4000 × *g* for 20–40 min.
14. Add 1 ml ABC and centrifuge the filter unit at 4000 × *g* for 20–40 min.
15. Acidify samples to pH 3 with 100% formic acid (measure pH using indicator paper).

16. Centrifuge at 21,000 × *g* for 20 min.
17. Transfer supernatant to a fresh tube for further processing.

3.3 Mixed-Mode Reverse-Phase/Cation Exchange Cleanup (MCX)

1. Setup a MCX column on the vacuum manifold, and place a 15 ml centrifuge tube in vacuum chamber to collect the flow-through.
2. Rinse the MCX column with 5 ml MeOH followed by 5 ml water; repeat with another 5 ml water.
3. Replace the 15 ml centrifuge tube with a clean 15 ml tube. Load the digested sample at a flow rate of less than 1 ml per minute (*see* **Note 9**).
4. Replace collection tube with a 15 ml centrifuge tube. Wash MCX column with 5 ml water and repeat once.
5. Wash MCX column with 5 ml 80% ACN and repeat once.
6. Replace 15 ml vials for clean 15 ml vials. Elute peptides with 4 ml of 45%IPA/500 mM NH_4HCO_3.
7. Speedvac the peptide solution until almost dry.
8. Add 0.1% FA to a protein concentration of ~2 mg/ml, and vortex to dissolve well.
9. Centrifuged at 21,000 × *g* for 20 min at room temperature. Transfer supernatant to a clean tube (*see* **Note 10**).

3.4 Acetyl-Lysine Enrichment

1. Bring 10 mg of MCX processed peptides to 4 ml final volume using H_2O.
2. Add 250 μl 1 M Tris–HCl pH 7.5.
3. Verify that pH is ~7.5 using pH strips; adjust as necessary.
4. Add 2 mg of anti-acetyl lysine antibody to a 0.2 μm spin filter unit.
5. Add 4 ml of 50 mM Tris–HCl pH 7.5 to the antibodies and centrifuge at 4100 × *g* (*see* **Note 11**).
6. Repeat antibody preparation washes for a total of three times.
7. Resuspend antibodies in the solution from **step 3**, which contains 10 mg of peptides. Transfer the antibody/peptide mix to a 15 ml centrifuge tube.
8. Rotate at 4 °C for 1 h.
9. Transfer antibody/peptide mix back onto the 0.2 μm filter unit and centrifuge at 4100 × *g* until the non-bound peptides have passed through (*see* **Note 12**).
10. Add 4 ml of 50 mM Tris–HCl pH 7.5 to the antibodies and centrifuge at 4100 × *g* (*see* **Note 11**).
11. Repeat these 50 mM Tris–HCl washes for a total of three times.

12. Add 1.5 ml of 0.1% TFA and incubate for 5 min at room temperature.
13. Centrifuge at 4100 × *g* to elute the bound peptides.
14. Repeat elution with a second addition of 1.5 ml of 0.1% TFA (*see* **Note 13**).
15. Merge the two elutions.
16. The acetyl-enriched peptides can now be desalted using standard C18 reverse-phase methods and analyzed by LC-MS/MS.

3.5 Phosphopeptide Enrichment

1. Add 50% stock TFA to 2 mg of digested peptides to a final concentration of 3–5% TFA to reach a pH of 1 (*see* **Note 14**).
2. Centrifuge 5 min at 21,000 × *g* to clear solution. Transfer supernatant to clean tube.
3. To supernatant, add 20 μl 1% colloidal CeO_2 (*see* **Note 15**).
4. Vortex 1 min.
5. Centrifuge at 500 × *g* for 2 min.

   Optional: repeat enrichment **steps 3–5** on the supernatant with additional 20 μl CeO_2.
6. Remove supernatant (non-phosphopeptides).
7. Wash CeO_2 pellet with 1 ml 5% TFA and vortex 5 s.
8. Centrifuge at 500 × *g* for 2 min and then discard the supernatant.
9. Add 40 μl of 200 mM $(NH_4)_2HPO_4$/2 M $NH_3.H_2O$/10 mM EDTA.
10. Check that the pH is >9.5 using pH strips and then vortex 15 min.
11. Add 40 μl 10% FA/100 mM citric acid for a final concentration of ~5% FA; pH should be 3–4. CeO_2 pellet should precipitate and phosphopeptides remain in solution.
12. Centrifuge at 16,000 × *g* for 1 min. Transfer supernatant containing phosphopeptides to a clean microfuge tube.
13. The phospho-enriched peptides can now be desalted using standard C18 reverse-phase methods and analyzed by LC-MS/MS.

4 Notes

1. A 100 mM (10×) TCEP stock solution prepared in 1 M Tris–HCl, pH 7.5 can be stored at 4 °C for several weeks.
2. UA solution should be prepared fresh.

3. MCX is an optional step to remove residual contaminants.
4. Up to 10 mg peptides can be loaded and purified by each column; other size columns are available for processing smaller amounts of peptides.
5. The amount of Gnd lysis solution used in this step may need to be optimized for the specific tissue of interest for optimal protein extraction.
6. It is necessary to shake vigorously (e.g., >1000 rpm on a ThermoFisher Scientific Digital Shaking Drybath) in this step for a high yield of protein.
7. This shears DNA and reduces the viscosity of the sample. Time and power level will need to be optimized for the specific sonicator/probe combination.
8. The centrifuge time in each step may need to be adjusted for different samples. It is not necessary to completely remove the liquid above the filter; up to 10% of the volume can remain above the filter.
9. Optional: check the flow-through for peptides using BCA assay [28] to verify that the column is not overloaded.
10. At this point ~100 μg (depending on LC-MS/MS setup) of the sample can be desalted using standard C18 reverse-phase chromatography and analyzed by LC-MS/MS to quantify total protein abundance. If performing phospho- or acetyl-enrichment proceed without C18 desalting.
11. Adjust spin time to almost completely pass the solution through the filter unit. ~100–200 μl can remain on the antibodies.
12. Retain the flow-through peptides. These peptides can be used for another round of enrichment, saved for LC-MS/MS (total/non-modified proteome) or used for enrichment of additional PTMs (ubiquitin or phosphorylation).
13. Optional: the antibody-conjugated beads can be washed two times with 50 mM Tris–HCl pH 7.5 and then used for a second round of enrichment of the same sample (i.e., the original flow-through) to increase recovery of acetylated peptides.
14. The amount of peptides used for phospho-enrichment can be adjusted based on the desired depth of coverage as well as the amount of peptides available.
15. The amount of CeO_2 used in this step should be adjusted for different species/tissue types. A good starting point is a ratio of 1:10 CeO_2:protein (wt:vol) for most tissues (e.g., roots or reproductive tissues) and a 2:10 ratio for green plant tissues.

References

1. Wilkins MR, Pasquali C, Appel RD et al (1996) From proteins to proteomes: large scale protein identification by two-dimensional electrophoresis and amino acid analysis. Nat Biotechnol 14:61–65
2. Graves PR, Haystead TA (2002) Molecular biologist's guide to proteomics. Microbiol Mol Biol Rev 66:39–63. doi:10.1128/MMBR.66.1.39
3. Glish GL, Vachet RW (2003) The basics of mass spectrometry in the twenty-first century. Nat Rev Drug Discov 2:140–150. doi:10.1038/nrd1011
4. Ho CS, Lam CWK, Chan MHM et al (2003) Electrospray ionisation mass spectrometry: principles and clinical applications. Clin Biochem 24:3–12
5. Tanaka K, Waki H, Ido Y et al (1988) Protein and polymer analyses up to m/z 100 000 by laser ionization time-of-flight mass spectrometry. Rapid Commun Mass Spectrom 2:3
6. Syed SU, Maher S, Taylor S (2013) Quadrupole mass filter operation under the influence of magnetic field. J Mass Spectrom 48:1325–1339. doi:10.1002/jms.3293
7. Hu Q, Noll RJ, Li H et al (2005) The Orbitrap: a new mass spectrometer. J Mass Spectrom 40:430–443. doi:10.1002/jms.856
8. Makarov A (2000) Electrostatic axially harmonic orbital trapping: a high-performance technique of mass analysis. Anal Chem 72:1156–1162. doi:10.1021/ac991131p
9. Elias JE, Gygi SP (2010) Target-decoy search strategy for mass spectrometry-based proteomics. Methods Mol Biol 604:55–71
10. Yeung Y, Nieves E, Angeletti R, Stanley ER (2008) Removal of detergent from protein digests for mass spectrometry analysis. Anal Biochem 382:135–137. doi:10.1016/j.ab.2008.07.034.Removal
11. Wang W, Scali M, Vignani R et al (2003) Protein extraction for two-dimensional electrophoresis from olive leaf, a plant tissue containing high levels of interfering compounds. Electrophoresis 24:2369–2375. doi:10.1002/elps.200305500
12. Saravanan RS, Rose JKC (2004) A critical evaluation of sample extraction techniques for enhanced proteomic analysis of recalcitrant plant tissues. Proteomics 4:2522–2532. doi:10.1002/pmic.200300789
13. Isaacson T, Damasceno CMB, Saravanan RS et al (2006) Sample extraction techniques for enhanced proteomic analysis of plant tissues. Nat Protoc 1:769–774. doi:10.1038/nprot.2006.102
14. Wang W, Vignani R, Scali M, Cresti M (2006) A universal and rapid protocol for protein extraction from recalcitrant plant tissues for proteomic analysis. Electrophoresis 27:2782–2786. doi:10.1002/elps.200500722
15. Walley JW, Shen Z, Sartor R et al (2013) Reconstruction of protein networks from an atlas of maize seed proteotypes. Proc Natl Acad Sci U S A 110:E4808–E4817. doi:10.1073/pnas.1319113110
16. Castellana NE, Shen Z, He Y et al (2013) An automated proteogenomic method utilizes mass spectrometry to reveal novel genes in Zea mays. Mol Cell Proteomics 13(1):157–167. doi:10.1074/mcp.M113.031260
17. Walley J, Xiao Y, Wang J-Z et al (2015) Plastid-produced interorgannellar stress signal MEcPP potentiates induction of the unfolded protein response in endoplasmic reticulum. Proc Natl Acad Sci U S A 112:6212–6217. doi:10.1073/pnas.1504828112
18. Marcon C, Malik WA, Walley JW et al (2015) A high-resolution tissue-specific proteome and phosphoproteome atlas of maize primary roots reveals functional gradients along the root axes. Plant Physiol 168:233–246. doi:10.1104/pp.15.00138
19. Walley JW, Sartor RC, Shen Z et al (2016) Integration of omic networks in a developmental atlas of maize. Science 353(6301):814–818. doi:10.1126/science.aag1125
20. Glatter T, Ahrné E, Schmidt A (2015) Comparison of different sample preparation protocols reveals lysis buffer-specific extraction biases in gram-negative bacteria and human cells. J Proteome Res 14(11):4472–4485. doi:10.1021/acs.jproteome.5b00654
21. Kulak NA, Pichler G, Paron I et al (2014) Minimal, encapsulated proteomic-sample processing applied to copy-number estimation in eukaryotic cells. Nat Methods 11:319–324. doi:10.1038/nmeth.2834
22. Wiśniewski JR, Zougman A, Nagaraj N, Mann M (2009) Universal sample preparation method for proteome analysis. Nat Methods 6:359–362. doi:10.1038/nmeth.1322
23. Jensen ON (2004) Modification-specific proteomics: characterization of post-translational modifications by mass spectrometry. Curr Opin Chem Biol 8:33–41. doi:10.1016/j.cbpa.2003.12.009
24. Choudhary C, Kumar C, Gnad F et al (2009) Lysine acetylation targets protein complexes and co-regulates major cellular functions. Science 325(5942):834–840. doi:10.1126/science.1175371

25. Udeshi ND, Mertins P, Svinkina T, Carr SA (2013) Large-scale identification of ubiquitination sites by mass spectrometry. Nat Protoc 8:1950–1960. doi:10.1038/nprot.2013.120
26. Kim W, Bennett EJ, Huttlin EL et al (2011) Systematic and quantitative assessment of the ubiquitin-modified proteome. Mol Cell 44: 325–340. doi:10.1016/j.molcel.2011.08.025
27. Mertins P, Qiao JW, Patel J et al (2013) Integrated proteomic analysis of post-translational modifications by serial enrichment. Nat Methods 10:634–637. doi:10.1038/nmeth.2518
28. Smith PK, Krohn RI, Hermanson GT et al (1985) Measurement of protein using bicinchoninic acid. Anal Biochem 150(1):76–85. doi:10.1016/0003-2697(85)90442-7

Chapter 10

Automated High-Throughput Root Phenotyping of *Arabidopsis thaliana* Under Nutrient Deficiency Conditions

Santosh B. Satbhai, Christian Göschl, and Wolfgang Busch

Abstract

The central question of genetics is how a genotype determines the phenotype of an organism. Genetic mapping approaches are a key for finding answers to this question. In particular, genome-wide association (GWA) studies have been rapidly adopted to study the architecture of complex quantitative traits. This was only possible due to the improvement of high-throughput and low-cost phenotyping methodologies. In this chapter we provide a detailed protocol for obtaining root trait data from the model species *Arabidopsis thaliana* using the semiautomated, high-throughput phenotyping pipeline BRAT (Busch-lab Root Analysis Toolchain) for early root growth under the stress condition of iron deficiency. Extracted root trait data can be directly used to perform GWA mapping using the freely accessible web application GWAPP to identify marker polymorphisms associated with the phenotype of interest.

Key words High-throughput root phenotyping, Root development, Iron deficiency stress, Abiotic stress, *Arabidopsis thaliana*

1 Introduction

Plant roots are important to plants for a wide variety of processes, including water and nutrient uptake, as well as for anchoring and mechanical support. Importantly, plant root growth is highly plastic and responds to changes in the surrounding environment through developmental alterations of root system architecture (RSA) [1, 2]. Thereby, small changes in the root growth rate, root growth direction, and branching density can eventually have a strong influence on RSA. These fundamental traits are modulated based on environmental cues, one of which is nutrient abundance. One of the most limiting nutrients for plants as well as animals is iron, and low Fe availability drastically limits crop yields in many parts of the world.

Understanding the genetic basis of how plants synchronize their growth with nutrient availability has important consequences

Wolfgang Busch (ed.), *Plant Genomics: Methods and Protocols*, Methods in Molecular Biology, vol. 1610,
DOI 10.1007/978-1-4939-7003-2_10,

for our understanding of plant growth. Genome-wide association mapping is a suitable method to study this. In particular, this is easy to do in *Arabidopsis thaliana*, as a large number of accessions (naturally occurring strains) that have been collected over the past few decades from all over the world are available, together with high-resolution data of their genomes [3, 4]. Importantly these accessions show a wide range of genetic and trait variation. Different accessions successfully cope with environmental challenges in local habitats. These adaptations are manifested in changes in the genome. This has already been successfully exploited to find genes involved in developmental processes such as flowering time as well as plant architecture and morphology, growth, primary and secondary metabolism, and responses to biotic and abiotic stresses [5–9].

In this chapter, we present a protocol for performing high-throughput root phenotyping under Fe deficiency conditions using *Arabidopsis thaliana* accessions that is based on an advanced version of the previously published BRAT software [10, 11]. BRAT V2 uses updated algorithms for image processing. This not only results in better plate and plant detection, and therefore in a significantly reduced number of false positives, but also speeds up the quality control step in most cases. Detected plates are also corrected for possible tilting, which results in a higher accuracy of the measured root angle traits.

While we focus on an Fe deficiency condition for screening in this chapter, this can be applied to many other abiotic stress conditions (such as other nutrient deficiencies, salt stress, low pH stress, high or low temperature stresses, and heavy metal stresses). Moreover, the high-throughput root phenotyping protocol provided here can be applied to other resources, such as large mutant populations or large collections of recombinant inbred lines (RILs).

2 Materials

2.1 Seed Sterilization

1. 1.5 mL microcentrifuge tubes.
2. Seeds of *Arabidopsis thaliana* accessions.
3. Tube rack for 1.5 mL microcentrifuge tubes.
4. 250 mL glass beaker.
5. 10% sodium hypochlorite.
6. 15 mL polypropylene conical centrifuge tube.
7. 37% hydrochloric acid.
8. Autoclaved distilled water.
9. Polycarbonate lockable airtight box, e.g., 36.8 × 18.4 × 17.1 cm, Nalgene 7135-0001 (*see* Fig. 1).

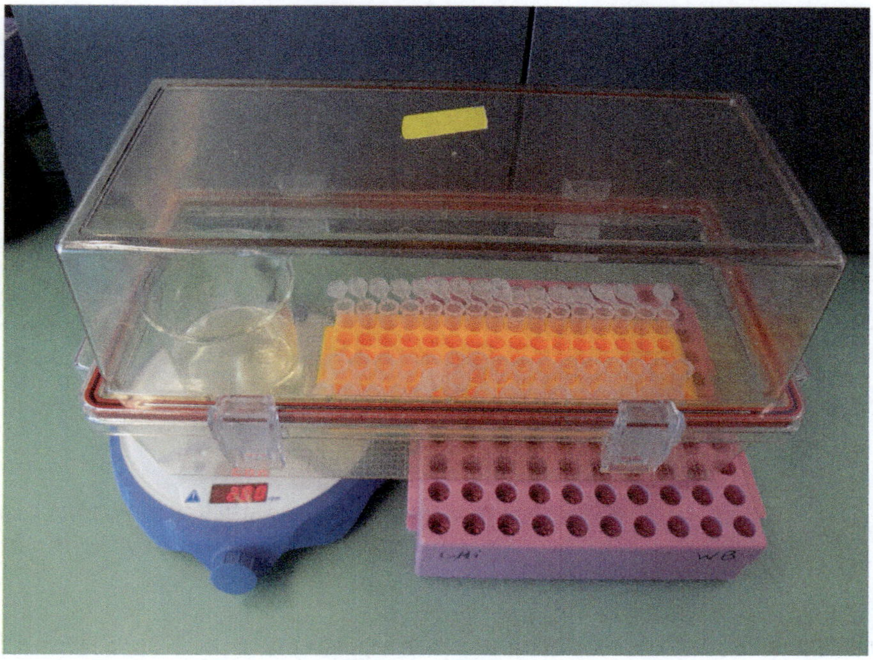

Fig. 1 Airtight box setup for gas sterilization of *Arabidopsis thaliana* seeds

2.2 Plant Growth Medium and Other Supplies

1. Petri plates, 12 × 12 × 1.7 cm.
2. 1 M KOH.
3. Ferrozine, a strong iron chelator [3-(2-pyridyl)-5,6-diphenyl-1,2,4-triazinesulfonate, Sigma Aldrich].
4. -Fe plant growth medium: 1× Murashige and Skoog [4] mineral salts, 1% (w/v) sucrose, 0.8% (w/v) plant agar, 300 μM ferrozine, pH 5.7 (*see* **Notes 1** and **2**).
5. Pipetman and tips (2–200 μL).
6. Micropore tape.
7. Percival incubator or plant growth room with a setting of constant temperature of 22 °C with long-day lighting conditions (16 h light and 8 h dark).
8. Support rack to hold plates vertically (*see* Fig. 2b).

2.3 Image Acquisition

1. One or more (up to eight) conventional flatbed scanners (*see* **Note 3**).
2. Desktop computer with Linux operating system and the open source scanning framework "SANE" installed (*see* **Note 4**).
3. "Multi-Scan" scanner operating software tool (download link: https://gitlab.com/buschlab/Multi-Scan-Binaries/tree/master).

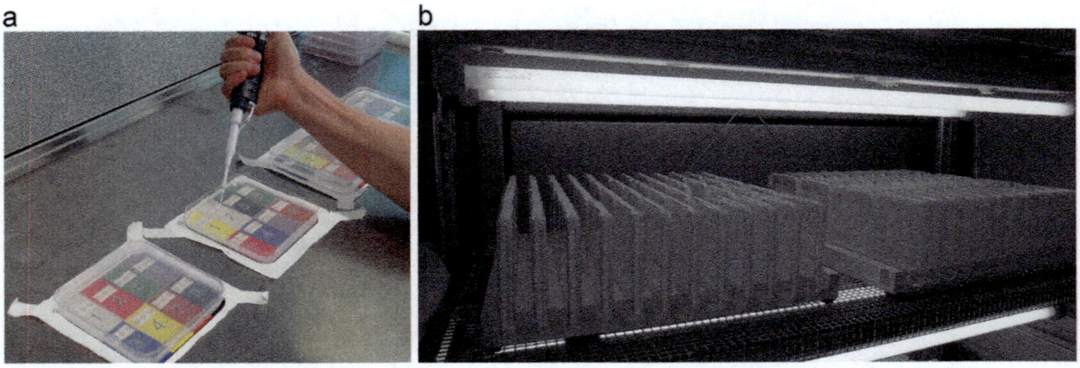

Fig. 2 (**a**) Plating seeds on media plates. (**b**) Plates on custom-made plate racks in growth room

4. Scanner support frames for consistent positions of the plates (https://gitlab.com/buschlab/Multi-Scan-Binaries/tree/master) [10].

2.4 Trait Extraction

1. Workstation or desktop computer (*see* **Note 5**) with Fiji software (a distribution of ImageJ; http://fiji.sc/Downloads).
2. BRAT plug-ins (https://gitlab.com/buschlab/BratV2-Arabidopsis-Binaries/tree/master).

3 Methods

3.1 Growing *Arabidopsis thaliana* Seedlings

3.1.1 Surface Sterilization and Stratification of Seeds

1. Place dry *Arabidopsis thaliana* seeds into open 1.5 mL microcentrifuge tubes (about 15–20 seeds in each tube, *see* **Note 6**) on a tube rack into an airtight lockable box along with a beaker containing a magnetic stirrer bar and 130 mL of 10% sodium hypochlorite.
2. Prepare 3.5 mL of 37% hydrochloric acid in a conical centrifuge tube.
3. Add HCl rapidly into the beaker to generate chlorine gas (*see* **Note 7**).
4. Quickly close the lid of the box, and secure clamps to make it airtight.
5. Place the airtight box on a magnetic stirrer, and leave the chemicals stirring continuously for 1 h of sterilization time (*see* Fig. 1).
6. Unclamp the box, and slightly tilt the gassing box lid to vent it in a chemical hood for 30 min.
7. Transfer the box with seeds to a laminar flow hood, and add 200 μL sterile water in each tube; close the individual seed tubes gently (*see* **Note 6**).

8. Imbibe seeds in water and stratify in the dark at 4 °C for 3 days. Alternatively, surface-sterilized seeds can be stored in the tubes at room temperature or used immediately for plating if stratification of seeds on the plates is planned.

3.1.2 Seed Plating

1. In sterile laminar hood, place the seeding layout grid (Fig. 2a, *see* **Note 8**) below a petri plate. We highly recommend using permutated block design (*see* Fig. 4) to account for positional effects within and between petri plates.
2. Using a 2–10 μL pipette, transfer a single seed onto the agar surface to a position specified by the seed layout grid.
3. Repeat until all positions on the plate have been occupied.
4. Seal the completed plates with micropore ventilating tape in order to ensure sterile conditions during growth.
5. Place the plates on a support rack.
6. Transfer the racks into a plant growth chamber providing the desired controlled conditions (*see* Fig. 2b; light cycle, light intensity, temperature, humidity). Seed germination should be visible within 2–3 days.

3.2 Image Acquisition

1. Start image acquisition 3 days after plating the seeds. Transfer the rack with petri plates to the image acquisition room (*see* **Note 9**). Optional: We recommend starting image acquisition on the day of plating seeds (termed the day 0 scan). A day 0 image enables better plant detection and reduces the false positive rate.
2. Place unopened petri plates with plants onto the horizontally oriented scanners (*see* **Note 10**) in the positions specified by the support frames (Fig. 3).
3. Start the BRAT image acquisition tool to control all scanners at the same time. Specify in the main window which scanners should be used (check box "Use:") and if images of one or both plate positions should be acquired (check box "Row 1" for bottom and "Row 2" for upper scanning position) (*see* Fig. 5).
4. Select the folder path to save the image data by clicking on the folder selection button (labeled with "...") and specifying the experiment prefix (*see* **Note 11**).
5. Click on "Scan" to initiate scanning. The images are stored as 1200 dpi 24-bit RGB TIFF files and automatically named based on the initial user input into the image acquisition tool. For better quality images, the scanning should be performed in the dark with the scanner lid open. If a dark room is not available you can use a cardboard box to cover the entire scanner.

Fig. 3 Multi-scanner system for image data acquisition

a

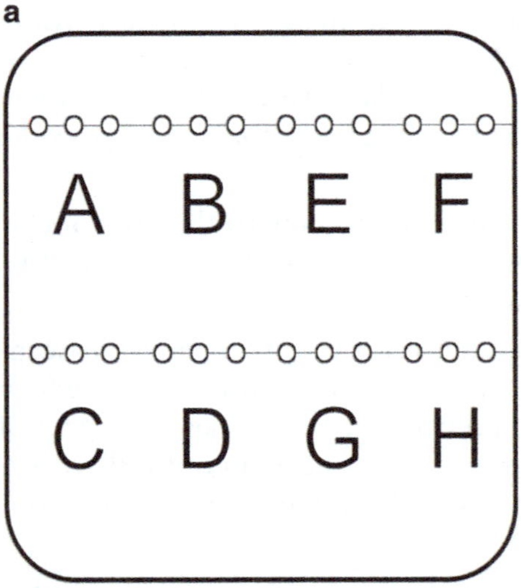

b

Plate number	Sections 1	2	3	4	5	6	7	8
001	A A A	B B B	C C C	D D D	E E E	F F F	G G G	H H H
002	D D D	A A A	B B B	C C C	H H H	E E E	F F F	G G G
003	C C C	D D D	A A A	B B B	G G G	H H H	E E E	F F F
004	B B B	C C C	D D D	A A A	F F F	G G G	H H H	E E E

Fig. 4 Grid layout template and positional permutations for plating. (**a**) Grid is placed under the plate for accurate placement of the seeds on the agar medium surface. Multiple genotypes can be used per plate (e.g., eight with three replicates on this template). (**b**) Permutated block design was used to account for positional effects within and between petri dishes. A schematic representation of the permutations for three replicates of each of eight accessions on four plates is shown

6. After scanning, return the rack with plates to the growth chamber. We scan the plates every 24 h to track root growth for several days.

Fig. 5 Screenshot of the BRAT image acquisition tool for simultaneous control of the scanners

3.3 Trait Extraction

Measuring various plant traits from a large number of images is often a time-consuming task. To increase throughput and precision, the BRAT software was developed [11]. The updated BRAT V2 performs many of the necessary tasks automatically. BRAT provides three different tools responsible for image segmentation, quality control, and trait evaluation. The tools are implemented as plug-ins for the popular open-source image analysis platform Fiji [12]. Due to the availability of Fiji for all major operating systems (Windows, MacOS, Linux), the BRAT tools can be used on any of these without modification. If the tools are run on desktop computers, they provide graphical user interfaces that make it easy for the user to configure and start the different tasks. The image segmentation tool can also be run from the command line, which enables its use on a compute cluster to reduce computation times. BRAT currently extracts 16 traits that are relevant for plant root growth and development.

1. BRAT image segmentation (plug-in BRAT V2): Select the base directory by clicking the "Browse" button or typing the correct path in the "Base Directory" field. The base directory is the folder where your plate images are found. All newly created files will be saved in a folder structure below this directory (*see* Fig. 6).
2. Specify the file extension of your images in the field: "File Extension" (*see* **Note 12**). All images in the base directory with the given file extension will be taken into account for analysis.
3. If your images were acquired using scanners, you should check "flip horizontal" (*see* **Note 13**).

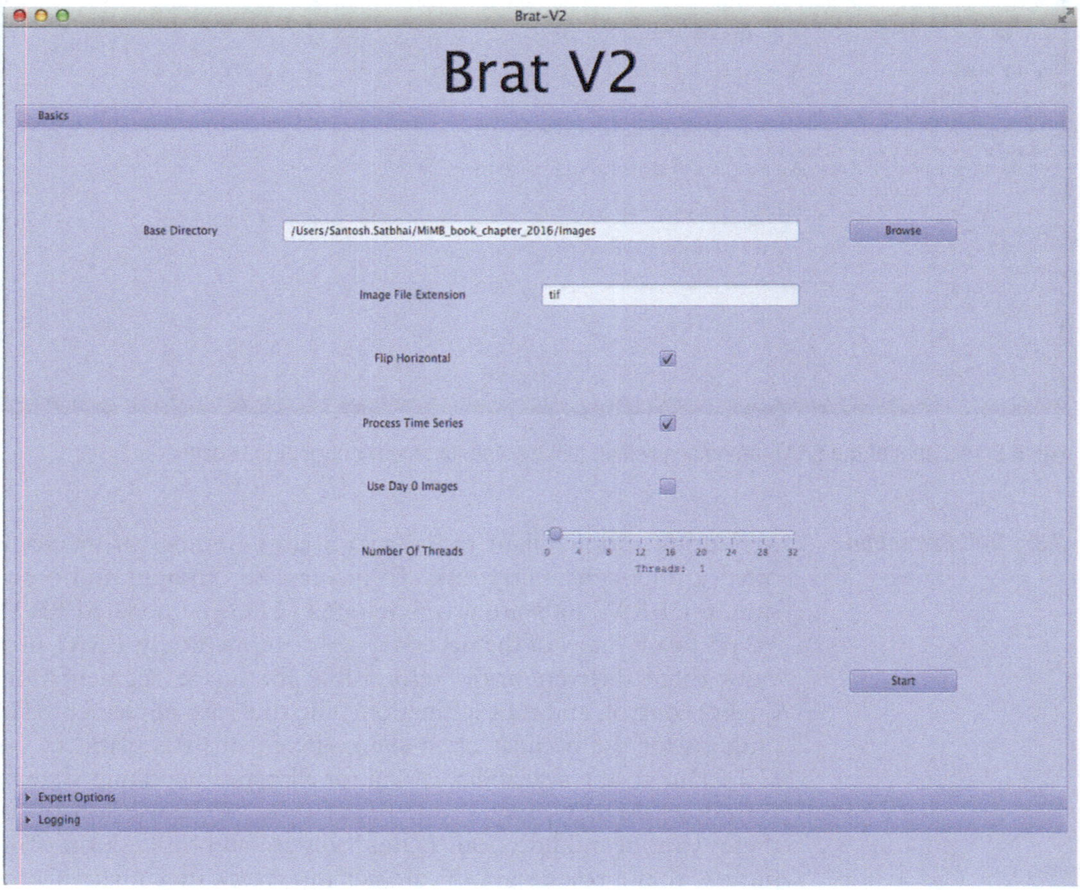

Fig. 6 Screenshot of the BRAT V2 segmentation plug-in

4. Select "process time series" if you have images of the same plate acquired at different time points. If "process time series" is selected, the same plates at different times will be recognized automatically as a time series, and the images will be aligned with the root-hypocotyl boundary point fixed across the time series (*see* **Notes 11** and **14**).
5. Select "use day 0 images" if you have images acquired on day 0 (i.e., images taken on the day of plating seeds).
6. Adjust the number of threads that are used in parallel for computation based on the resources available in your computer (*see* **Note 5**).
7. Multiple "expert options" are available (*see* Fig. 7), however we recommend using default settings.
8. Click the "Start" button to initiate the segmentation process. The progress will be shown in the "Logging" field (*see* Fig. 8).

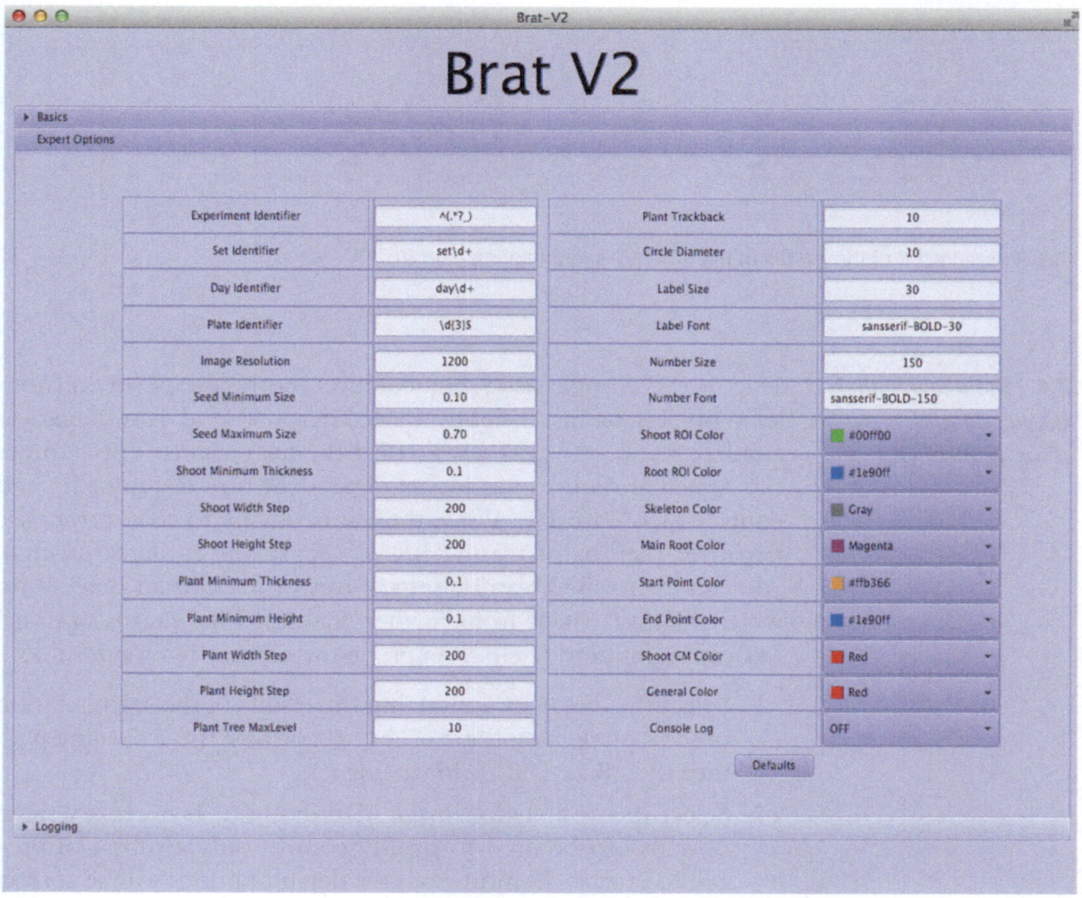

Fig. 7 Screenshot of the expert options available in the BRAT V2 segmentation plug-in

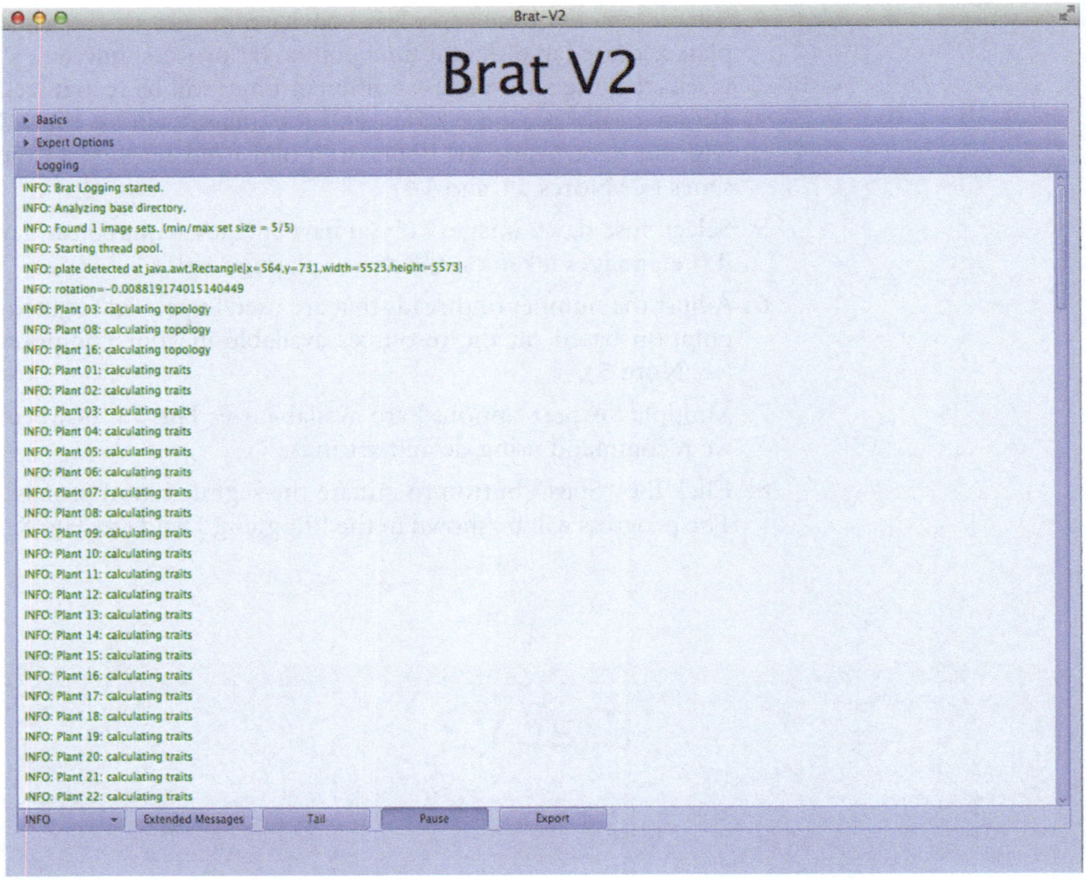

Fig. 8 Screenshot of the log file in the BRAT V2 segmentation plug-in

3.4 Optional Manual Quality Control (Plug-In BRAT QC)

Due to the unsupervised segmentation, both false positives (non-root objects or insufficiently detected roots) and false negatives (undetected root objects) will affect the final results. The number of segmentation errors is highly dependent on image quality and contrast (*see* **Note 10**). While there is no ability to correct for false negatives (i.e., by increasing BRAT's sensitivity to detect plants), you can manually discard false positives or insufficient detections, which removes them from further analysis. This step is optional but can significantly help to improve the accuracy of the results.

1. Select the directory containing the results of the segmentation process by clicking the "…" button or type the according path into the "Base Dir" field (*see* Fig. 9).
2. Select the options for image identification. If necessary these options can be used to quality control only a subset of processed images. In most cases the default options will be correct (*see* **Note 15**).

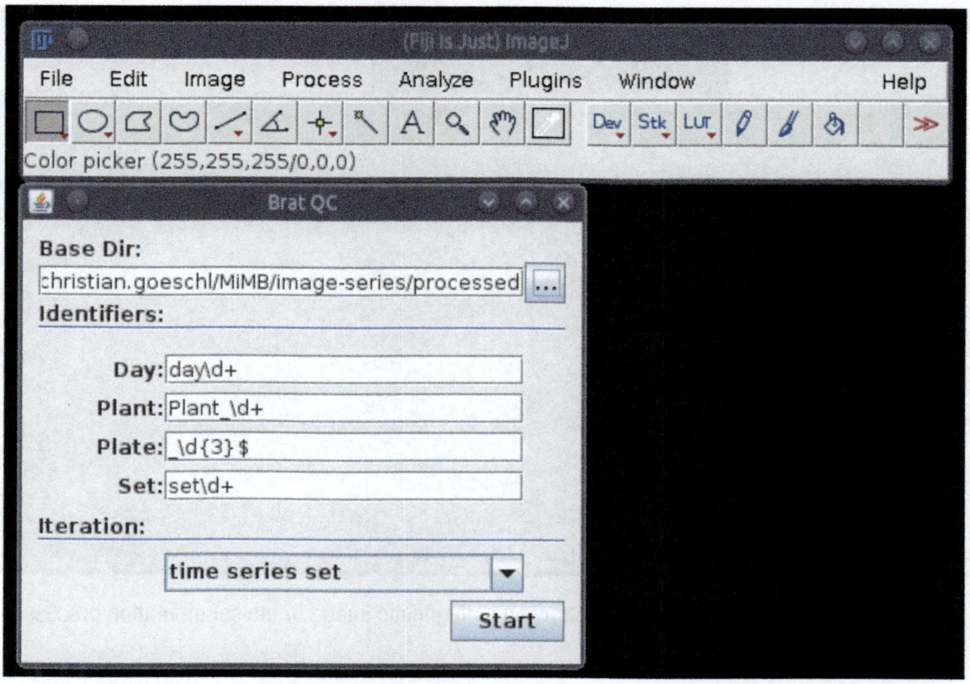

Fig. 9 Screenshot of the BRAT QC plug-in

3. Select iteration method. If "time series set" is selected, the images will be reordered so that all time points of a plant will be shown in direct order. Selecting "independent set" causes no ordering of images.
4. Click "Start" to initiate the quality control process. The BRAT QC plug-in will guide you through all diagnostic images of the segmentation process. Use the mouse to move and resize the image to fit your monitor resolution. The position and size will be preserved for the whole process. The left side of the diagnostic image shows an overview of the whole plate and the position of the detection marked in red. On the right side you will see the detection details. The detected shoot parts (leaves) will be marked in green. An orange circle marks the shoot-root transition, and a blue circle marks the root tip. The segmented path is shown by magenta color line. On the very right position of the diagnostic image, the detection is shown without any overlay (*see* Fig. 10).
5. To mark detection as correct or to correct the position of a root, input the correct position number (integer) into the field "Plant Nr." The plug-in will try to guess the correct plant number, so in many cases it is enough to just click ok (or press ENTER).
6. If you replace the plant number by some arbitrary character (e.g., "x"), the detection will be discarded and not be included in further analysis.

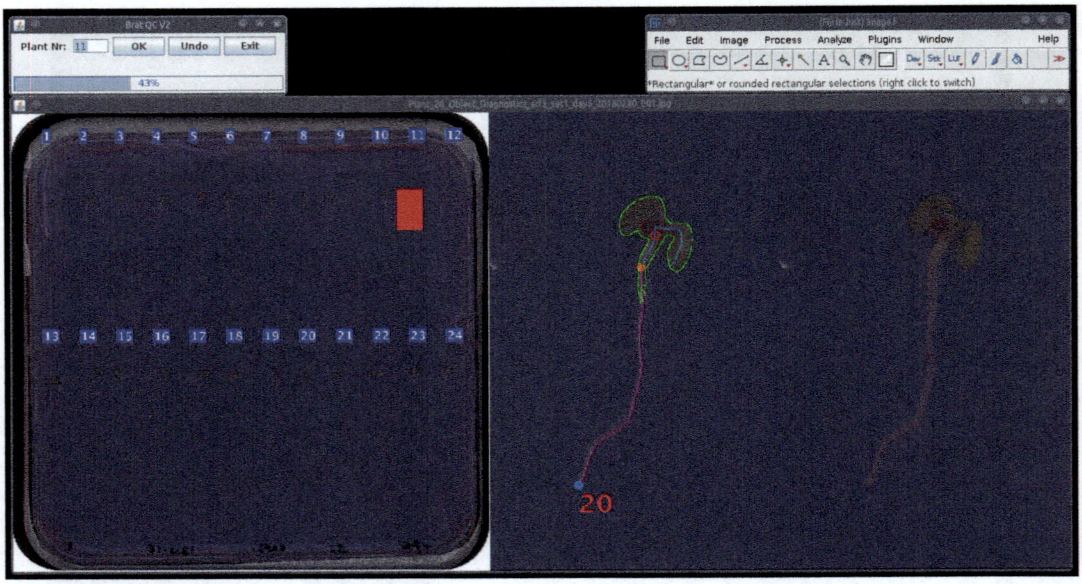

Fig. 10 Screenshot of the BRAT QC plug-in showing a diagnostic image of the segmentation process

7. To go back to a previous image click "Undo" (or press CTRL-Z on your keyboard).
8. Click "Exit" (or ESC on your keyboard) if you want to leave the application. If you restart the application with the same base directory, the quality control will be resumed from the current state. The progress will be saved automatically every minute or whenever you press the "Exit" button, to a newly created file named "brat-qc-results.txt" in the previously specified base directory.

3.5 Trait Evaluation

The final step in the BRAT pipeline is the evaluation (BRAT Eval plug-in) of measured traits. In this step, the traits and newly created descriptive statistics are written to output files. The graphical interface of the tool provides three tabs for diverse options. Options need to be set in each tab.

1. On the first tab labeled "Base Dir," select the path to the directory containing the results from the segmentation step by clicking the "..." button or typing the correct path into the according text field. The needed files all start with "Object_Measurements*.txt." If you follow the recommended file name nomenclature, the default values for the identifiers should be unchanged, otherwise you have the possibility to specify how to identify your files within these fields (*see* Fig. 11).
2. Choose whether you wish to convert the internal length units (pixels) to some physical units. As default, the conversion from pixels to millimeters is selected. Make sure that the given

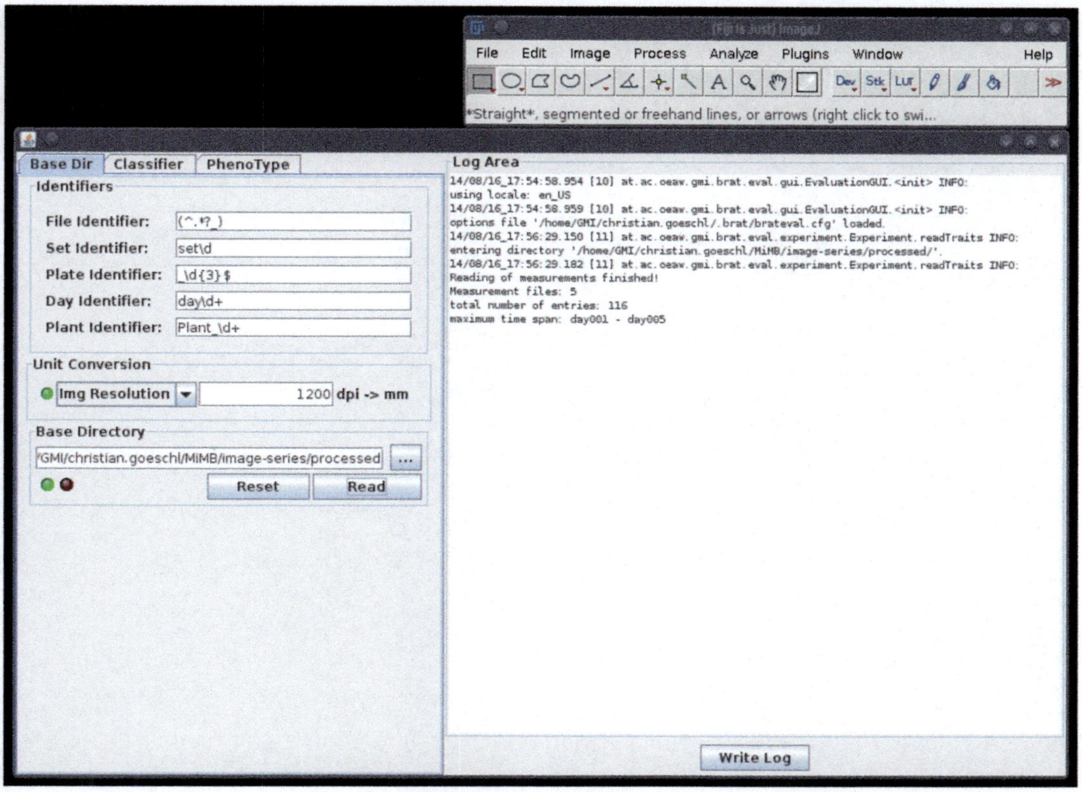

Fig. 11 Screenshot of the BRAT evaluation plug-in

resolution (1200 dpi by default) matches that of your images. Other conversion options include the multiplication with an arbitrary factor (option: "custom factor") or no conversion and can be selected from the corresponding drop-down menu.

3. To start inputting the specified result files from the segmentation step, click on "Read" at the bottom of the tab.
4. Move to the "Classifier" tab (*see* Fig. 12).
5. If you used manual quality control (Subheading 3.4), select the file created by the BRAT QC plug-in (normally a file called "brat-qc-results.txt" which can be found in the directory containing the segmentation results) by clicking on the "..." button. Click the "Read" button nearby to load the quality control results. Make sure that the numbers in the fields "Columns" and "Rows" match the layout used for placing the seeds. You already specified the correct plant numbers in the quality control step, so it is safe to disregard the other "Plate Layout" options.
6. If only a subset of segmentation results was manually classified, load the "User classification" file (as described in **step 5** of Subheading 3.5) and then click the "Auto" button in the "Plate Layout" section.

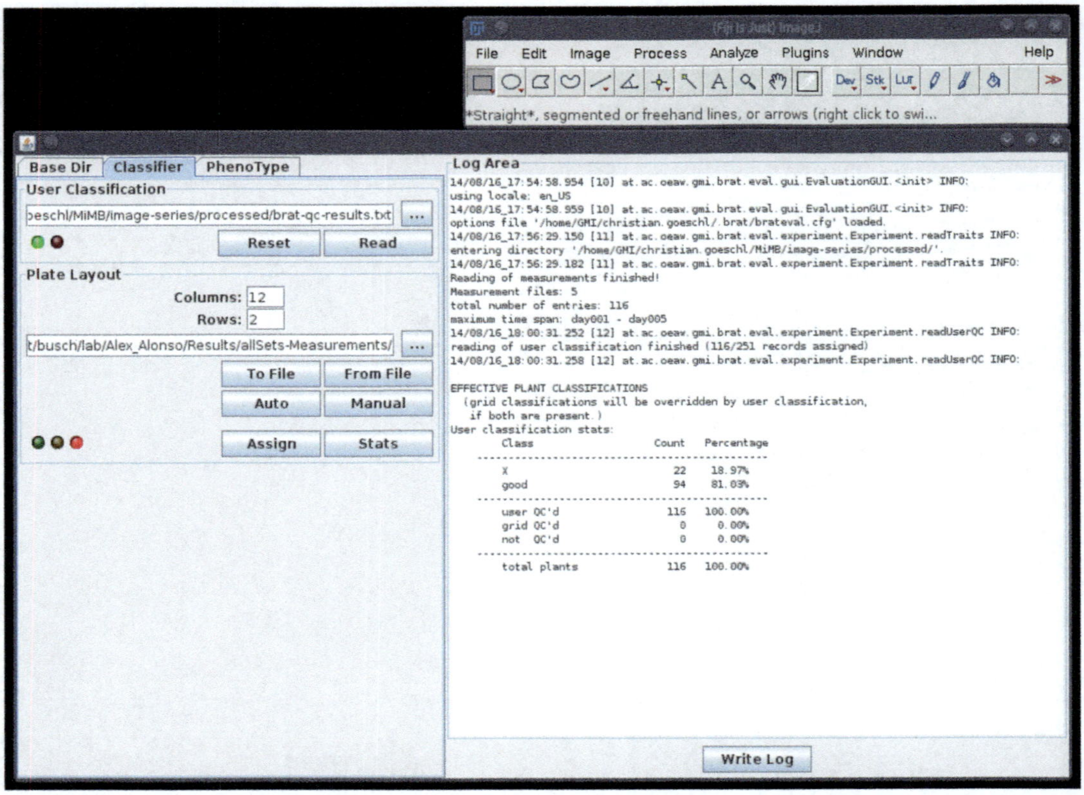

Fig. 12 Screenshot of the second step of the BRAT evaluation plug-in

7. To provide the plate layout coordinates manually (only necessary if no manual quality control was done), click "Manual" in the "Plate Layout" section and select one of the images you used as input for the segmentation plug-in. In the case of a scanned image, flip horizontally (use the secondary mouse button, i.e., right click on a two-button mouse or press control while clicking on a one-button Apple mouse). Once the image is shown, click on the expected start point for each root. Coordinates specified appear in a table (*see* Fig. 13). The algorithm automatically moves to the next root entry row. To modify a root's start coordinate, click on the respective row entry. If you want to save a newly created plate layout, specify the wanted file path and press "To File." You can later reload this layout by using the "From File" button.

8. To load the genotype layout data, move to the third tab, specify the path to the text file in the field "Accession layout" (use the folder selection button "..." to select the file), and click on "Read." Specify the path to the "Output directory" or select the folder with the folder selection button (labeled with "..."). To label the result files, specify the "Prefix." Click on "Write" to save the results (*see* Fig. 14).

Fig. 13 Screenshot of the window for manual specification of the plate layout and the table of coordinates

3.6 Genome-Wide Association Mapping

Trait data extracted from BRAT software is fully compatible to run GWA analysis without any modifications and can be directly used to perform GWA mapping using the freely accessible web application GWAPP, http://gwapp.gmi.oeaw.ac.at/ [13]. Please *see* [11] for a detailed explanation of how to run GWA mapping using GWAPP.

4 Notes

1. Iron-deficient media (Fe medium) is prepared by adding filter-sterilized ferrozine to autoclaved medium before pouring the media into the plates.
2. Avoid water droplets due to condensation on the plate lid by leaving the plates to cool down for a defined period (~60 min) with the lids slightly open.

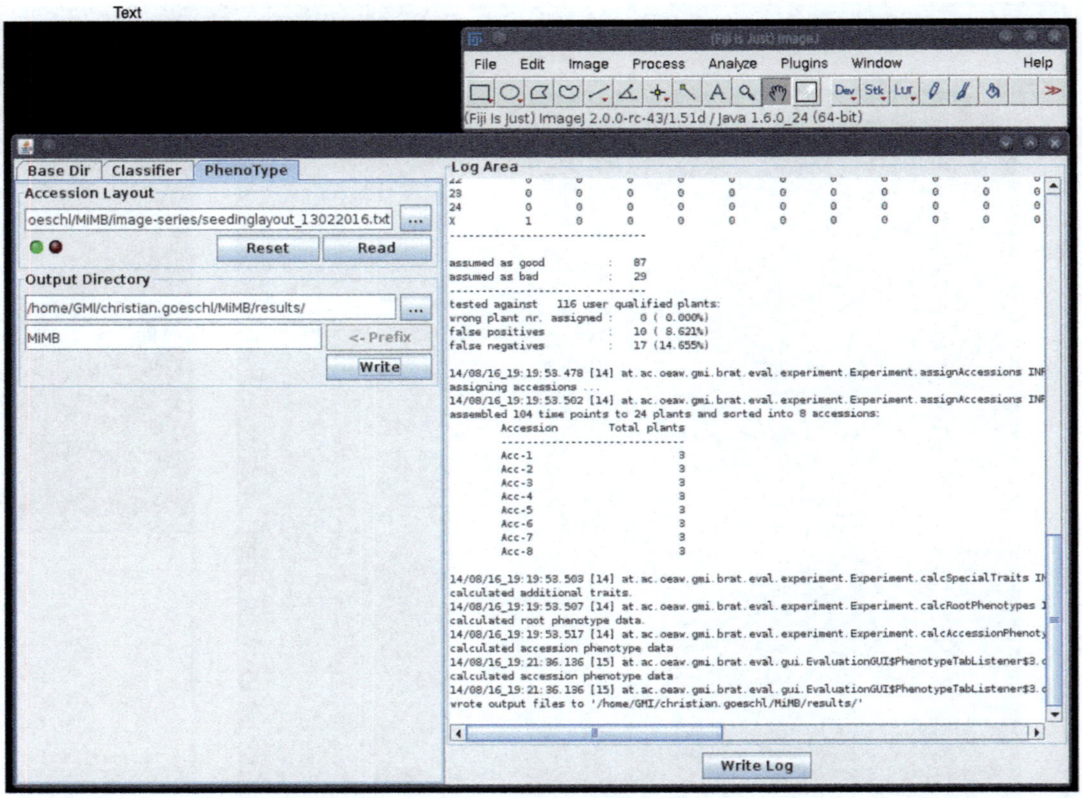

Fig. 14 Screenshot of the third step of the BRAT evaluation plug-in

3. For scanning the plates, we recommend an image resolution of 1200 dpi. The scanners used need to fulfill this specification (e.g., Epson Perfection V600, Catalogue Number EPSV600). Multiple scanners can be used to scan different plates in parallel thereby reducing the total image acquisition time. The "Multi-Scan" software tool supports up to eight scanners. If alternative methods are used for image acquisition, the image data should be of 1200 dpi or higher resolution for best results.

4. Any desktop computer should be fine as long as it is able to run Linux (we tested on Ubuntu 14.04 (https://www.ubuntu.com)) and provides enough USB ports to connect the wanted number of scanners. The SANE scanning framework (https://sane-project.org) needs to be installed as a prerequisite for Multi-Scan.

5. Image processing is a compute-intensive task. The most important computer feature for a successful run of the image segmentation tool is the RAM. Processing one of the 1200 dpi images requires about 1 GB of RAM. For example, 5 GB of memory is required to process a time series consisting of five images. To assure that all available RAM is used, you also might need to

change Fiji's memory allocation by adjusting the value at Edit → Options → Memory & Threads and restarting Fiji thereafter. By default Fiji uses two thirds of the available memory. For large screens, the utilization of a compute cluster is advisable. BRAT can be run on a computer cluster if Fiji and Java are available. Depending on the cluster scheduling system, we can provide cluster submission scripts. Please, contact us if you are interested in running BRAT on a compute cluster.

6. Avoid wearing gloves when handling dry seeds for sterilization because static electricity can displace dry seeds.
7. Chlorine is a toxic gas and also a strong oxidizing agent. Wear gloves, lab coat, and, safety goggles and work in a well-ventilated chemical hood.
8. To prevent artifacts during image processing: (a) do not disturb the media surface with a pipette tip while plating the seeds, as this often causes scratches, (b) plate the seeds evenly to avoid seedlings touching each other or the plate borders, and (c) avoid heavy condensation droplets under the plate lid by replacing the lid if necessary.
9. The image acquisition room should have the same, or a slightly higher, temperature to avoid condensation of droplets.
10. For better image quality, always clean the scanner surface and bottom lid of petri plates with 70% ethanol. We also use the scanners in a dark room and leave the scanner lids open. This increases the contrast in the images and reduces reflections on the lids of the petri plates.
11. The recommended naming pattern is "ArbitraryExperiment Name_setX_dayY." Parameter "set" allows you to divide your experiment into different sets, where "X" is an arbitrary integer number. The time of a scan can be specified by parameter "day," where "Y" is again an arbitrary integer and should denote the current day of your experiment (counting from its initiation or the first day of scanning). The program will automatically append the date of the scan in YYYYMMDD format. The plate (or image) order number will be automatically calculated and appended as well. The final filename will have the following format: "ArbitraryExperimentName_setX_dayY_YYYYMMDD_NNN". Do not use an underscore "_" as part of the experiment name.
12. BRAT accepts any of the popular image formats as input (.TIFF, .bmp, .gif, .png). The images have to be in 24-bit RGB format with a resolution preferably 1200 dpi or higher. Although BRAT could, in theory, support other image formats supported by Fiji, we strongly discourage using "lossy" image compression formats (such as .jpg) or a low image resolution, as these can impact the efficiency and accuracy of the phenotyping pipeline.

13. Images acquired on scanners are scanned from the bottom (through agar) and therefore the layout appears in a mirrored orientation.
14. To process the time series, the filenames have to match the recommended naming pattern (*see* **Note 11**). Without the recommended naming pattern, each image will be treated as a separate plate.
15. The identifier fields allow the specification of regular expressions, which are used to identify the needed files or images (*see* e. g., http://www.vogella.com/tutorials/JavaRegular Expressions/article.html for more information on regular expressions).

Acknowledgment

We wish to acknowledge Bonnie Wohlrab, Takehiko Ogura, Monica Meijon, and Radka Slovak for development and optimization of the wet laboratory methods used in this article. We are grateful to the members of the Busch lab for critically reading this protocol and for Matt Watson for manuscript editing. This work was supported by funds from the Austrian Academy of Sciences through the Gregor Mendel Institute and an FWF stand-alone project (P27163-B22).

References

1. Forde B, Lorenzo H (2001) The nutritional control of root development. Plant and Soil 232(1–2):51–68. doi:10.1023/A:1010329902165
2. Lynch J (1995) Root architecture and plant productivity. Plant Physiol 109(1):7–13
3. Genomes Consortium. Electronic address mngoaa, Genomes C (2016) 1,135 genomes reveal the global pattern of polymorphism in *Arabidopsis thaliana*. Cell 166(2):481–491. doi:10.1016/j.cell.2016.05.063
4. Horton MW, Hancock AM, Huang YS, Toomajian C, Atwell S, Auton A, Muliyati NW, Platt A, Sperone FG, Vilhjalmsson BJ, Nordborg M, Borevitz JO, Bergelson J (2012) Genome-wide patterns of genetic variation in worldwide *Arabidopsis thaliana* accessions from the RegMap panel. Nat Genet 44(2):212–216. doi:10.1038/ng.1042
5. Doyle MR, Bizzell CM, Keller MR, Michaels SD, Song J, Noh YS, Amasino RM (2005) HUA2 is required for the expression of floral repressors in *Arabidopsis thaliana*. Plant J 41(3):376–385. doi:10.1111/j.1365-313X.2004.02300.x
6. Loudet O, Saliba-Colombani V, Camilleri C, Calenge F, Gaudon V, Koprivova A, North KA, Kopriva S, Daniel-Vedele F (2007) Natural variation for sulfate content in *Arabidopsis thaliana* is highly controlled by APR2. Nat Genet 39(7):896–900. doi:10.1038/ng2050
7. Mouchel CF, Briggs GC, Hardtke CS (2004) Natural genetic variation in Arabidopsis identifies BREVIS RADIX, a novel regulator of cell proliferation and elongation in the root. Genes Dev 18(6):700–714. doi:10.1101/gad.1187704
8. Shindo C, Aranzana MJ, Lister C, Baxter C, Nicholls C, Nordborg M, Dean C (2005) Role of FRIGIDA and FLOWERING LOCUS C in determining variation in flowering time of Arabidopsis. Plant Physiol 138(2):1163–1173. doi:10.1104/pp.105.061309
9. Svistoonoff S, Creff A, Reymond M, Sigoillot-Claude C, Ricaud L, Blanchet A, Nussaume L, Desnos T (2007) Root tip contact with low-phosphate media reprograms plant root architecture. Nat Genet 39(6):792–796. doi:10.1038/ng2041

10. Slovak R, Goschl C, Seren U, Busch W (2015) Genome-wide association mapping in plants exemplified for root growth in *Arabidopsis thaliana*. Methods Mol Biol 1284:343–357. doi:10.1007/978-1-4939-2444-8_17
11. Slovak R, Goschl C, Su X, Shimotani K, Shiina T, Busch W (2014) A scalable open-source pipeline for large-scale root phenotyping of Arabidopsis. Plant Cell 26(6):2390–2403. doi:10.1105/tpc.114.124032
12. Schindelin J, Arganda-Carreras I, Frise E, Kaynig V, Longair M, Pietzsch T, Preibisch S, Rueden C, Saalfeld S, Schmid B, Tinevez JY, White DJ, Hartenstein V, Eliceiri K, Tomancak P, Cardona A (2012) Fiji: an open-source platform for biological-image analysis. Nat Methods 9(7):676–682. doi:10.1038/nmeth.2019
13. Seren U, Vilhjalmsson BJ, Horton MW, Meng D, Forai P, Huang YS, Long Q, Segura V, Nordborg M (2012) GWAPP: a web application for genome-wide association mapping in Arabidopsis. Plant Cell 24(12):4793–4805. doi:10.1105/tpc.112.108068

Chapter 11

Large-Scale Phenotyping of Root Traits in the Model Legume *Lotus japonicus*

Marco Giovannetti, Anna Małolepszy, Christian Göschl, and Wolfgang Busch

Abstract

Plants are sessile organisms that can tune their body architecture to the environment. This is very pronounced in their root system. In particular, nutrient availability strongly influences the architecture of the root system; depending on the abundance of specific nutrients, root growth rates and lateral root number are modulated. The extent of these effects is important for plant adaptation and has a major impact on plant fitness. However, the assessment of quantitative effects on a scale large enough for identifying genes and variants using quantitative genetics is difficult, and well-developed methods have been largely restricted to the model species *Arabidopsis thaliana*. In this chapter, we present a protocol for high-throughput phenotyping of early root traits in the model legume plant *Lotus japonicus*. This species allows for the study of important root-associated traits that are not present in Arabidopsis, such as symbioses with nitrogen-fixing *Rhizobia* and arbuscular mycorrhizal fungi. The methods described in this chapter can be used in the context of reverse and forward genetics approaches to dissect the genetic basis of root growth in legumes.

Key words Large-scale root phenotyping, *Lotus japonicus*, Quantitative genetics, Root development

1 Introduction

Legumes are an important plant family for humans because of their ability to fix atmospheric nitrogen, the main limiting factor for plant growth. *Lotus japonicus* is a diploid legume plant, which was proposed as a model since the early 1990s due to its small genome (~480 Mbp), its ability to produce many seeds within 3–4 months, and ease of *Agrobacterium* transformation [1]. Legumes are essential models for endosymbiosis, being able to interact with both *Rhizobia* and arbuscular mycorrhizal (AM) fungi in a mutualistic manner. In the intervening decades, using the genetic resources and tools available, research on *L. japonicus* has massively contributed to our understanding of the genetic bases of plant endosymbiosis [2–5]. *Rhizobia* symbiosis is believed to have evolved from AM symbiosis, and a portion of the genetic program controlling the establishment and functioning of

Wolfgang Busch (ed.), *Plant Genomics: Methods and Protocols*, Methods in Molecular Biology, vol. 1610,
DOI 10.1007/978-1-4939-7003-2_11, © Springer Science+Business Media LLC 2017

both endosymbionts is overlapping. This has allowed for a fruitful exchange of knowledge between the two fields of study. The two endosymbionts are known to induce a deep reprogramming in plant development and physiology at both root and systemic level. Although they share core common-symbiosis pathway genes, they lead to completely different outcomes. In particular, arbuscular mycorrhizal fungi are the *default* microbial machinery used by the majority of terrestrial plants to absorb phosphate from the surrounding soil [6]. On the other hand, symbiosis with nitrogen-fixing bacteria constituted a key evolutionary step in the acquisition of the abundant atmospheric nitrogen, triggering the formation of a de novo structure within the roots called the nodule [7, 8]. In both cases, plants reward the symbionts with carbon compounds obtained by photosynthesis, and the nutrient exchange between the two partners is one of the key regulators. Since both the availability of nutrients in the soil, such as phosphorous and nitrogen, and the interaction with the symbiont modulate the root architecture, understanding how legumes adjust their root system upon different cues could be useful to unravel as-yet unknown adaptive strategies and responses of plant root growth to the changing environment.

One possible strategy for studying genetics in legume root development and interaction with symbionts is to conduct large-scale phenotyping that quantitatively assesses root traits in the context of forward genetic approaches such as mutant screens, linkage or association mapping, as well as reverse genetics approaches.

The self-fertile species *L. japonicus* is well suited for such approaches because it can be maintained as inbred lines and the same genotypes can therefore be phenotyped multiple times. Moreover, its limited size allows for plants to be grown in 12 × 12 cm plates for several weeks, facilitating high-throughput root imaging techniques similar to those previously implemented for *Arabidopsis* [9, 10]. In addition, the availability of a large collection of non-transgenic *LORE1* retrotransposon insertional mutants [11–13] and the possibility of efficient root-specific transformation techniques allows for further characterization of genes of interest.

In this chapter, we present an efficient and low-cost method to conduct large-scale phenotyping of root growth and development traits in *L. japonicus* in different nutrient conditions. This approach is based on the established BRAT method in *Arabidopsis* [9].

2 Materials

2.1 Culturing Plants

1. *Lotus japonicus* seeds, approx. 50 seeds per genotype.
2. 2.0 ml microcentrifuge or 5 ml tubes.
3. Sand paper.
4. Tube rack for 1.5 ml microcentrifuge tubes.

5. Sodium hypochlorite.
6. Triton-X.
7. Round Petri dishes, 9 cm diameter.
8. Squared Petri dishes, 12 × 12 × 1.7 cm.
9. Aluminum foil.
10. Pipette and tips.
11. Sterile forceps.
12. Non-woven ventilating tape.
13. Custom-made support rack to hold plants vertically (as described in [6]).
14. Plant growth medium (*see* **Note 1**): 1.5 mM $Ca(NO_3)_2 \times 4H_20$, 1 mM KNO_3, 0.75 mM KH_2PO_4, 0.75 mM $MgSO_4 \times 7H_2O$, 0.1 mM Fe-EDTA, 10 μM $MnCl_2 \times 4H_2O$, 50 μM H_3BO_3, 1.75 μM $ZnCl_2$, 0.5 μM $CuCl_2$, 0.8 μM Na_2MoO_4, 1 μM KI, 0.1 μM $CoCl_2 \times 6H_20$, MES hydrate 0.8 g/L, pH 5.7 with KOH, agarose 8 g/L.

2.2 Inoculation with Rhizobium

1. *Mesorhizobium loti* R7A expressing DsRed fluorescence marker [14].
2. *Rhizobia* growth medium: 2 g/l mannitol, 0.4 g/l yeast extract, 0.1 g/l NaCl, 0.5 g/l K_2HPO_4, 0.2 g/l $MgSO_4 \times 7H_2O$, pH 6.8.

2.3 Image Data Acquisition and Analysis

Image acquisition is the same as described in [7]. The main difference in the pipeline is caused by the tendency of Lotus roots to develop a greener color upon exposure to light. A major change from Arabidopsis BRAT to Lotus BRAT is therefore a user-guided selection of the root-shoot interface.

1. One or more conventional flatbed scanners capable of 1200 dpi image data resolution, e.g., Epson V600 Perfection, catalog number EPSV600. Multiple scanners allow for an increased speed of image acquisition [10].
2. Desktop computer for operating scanners and image segmentation analysis (*see* **Note 2**). Fiji (a distribution of ImageJ: http://fiji.sc [15]) and Lotus BRAT plugins (https://gitlab.com/buschlab/Brat-Lotus-Binaries) should be installed (*see* **Note 3**).
3. Scanner support frames for consistent positions of the plates (https://gitlab.com/buschlab/Multi-Scan-Binaries/tree/master) [7].

3 Methods

3.1 Culturing Plants

1. Scarify *L. japonicus* seeds by placing around 50 seeds into a mortar, and scrape them vigorously with a squared 2 × 2 cm piece of fine microgrit sandpaper (P280–P320) for 1–2 min or until the seeds show signs of the treatment, such as small white dots. Since Lotus genotypes can be highly heterogeneous, sulfuric acid sterilization can damage some of them, impairing further trait quantification (*see* **Note 4**).
2. Surface sterilize seeds by placing scarified seeds into open 2/5 ml tubes (about 50 seeds in each tube) on a tube rack and treating them with 2/5 ml of sterilizing solution (0.5% sodium hypochlorite and 0.1% Triton-X solution). Shake the vials for 18 min at room temperature, remove the sterilizing solution under the laminar flow hood, and extensively rinse the seeds five to six times with sterile water.
3. Soak and shake the seeds in sterile water at room temperature for 3 h or until they are imbibed and swollen. Alternatively, seeds can be incubated at 4 °C overnight (*see* **Note 5**). In a laminar flow hood, spread the sterilized seeds on wet filter paper in a Petri dish and seal the plates with Parafilm (Fig. 1a).
4. Wrap the plates in aluminum foil, and place them for 3 days at 21 °C upside down or in a vertical position to allow the root to grow straight (Fig. 1b).

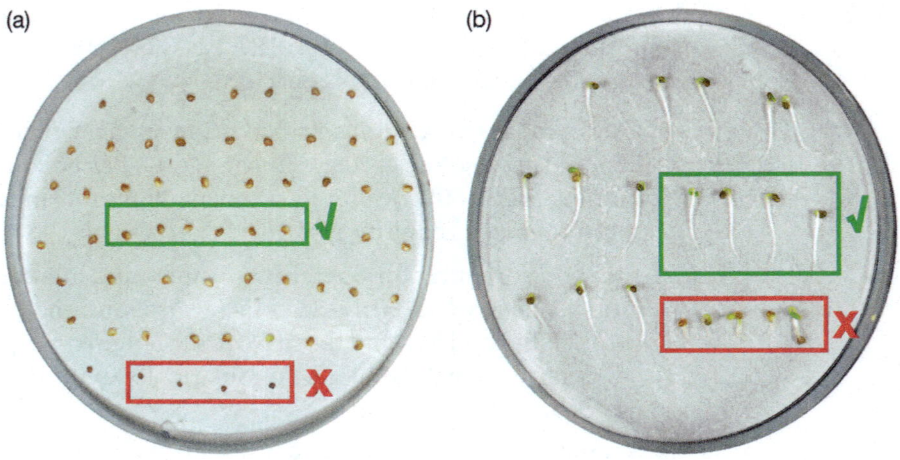

Fig. 1 Germination of Lotus plants. (**a**) Lotus seeds germinated and placed on wet filter paper. The green rectangle indicates nicely imbibed, good-quality seeds, while the red rectangle indicates low-quality, not imbibed seeds, which will probably not germinate and might lead to fungal contamination (*see* **Note 4**). (**b**) Three-day-old seedlings grown in a vertical position at 21 °C covered with foil. Exemplary seedlings, which are suitable for further assays, are marked with a green rectangle, while seedlings which are not suitable are marked with a red rectangle

5. In a laminar flow hood, place a seed layout grid (Fig. 2) under the squared plate containing 60 ml of the desired plant growth medium.
6. Using sterile forceps, carefully remove the seed coat from each seedling (*see* **Note 6**) and transfer onto the surface of plant growth medium in the positions specified by the seed layout grid (Fig. 3) (*see* **Notes 7** and **8**). Place eight plants from two different genotypes on each plate. Avoid abnormal seedlings with short and/or swollen rootlets that generally result in seedling growth arrest (a curved radicle or radicle meristem entrapped within the seed coat often leads to radicle breakage).
7. Optional: If conducting nodulation-related experiments, inoculate plates with compatible *Rhizobia*. Grow bacterial culture at 28 °C for 2–3 days in Yeast Mannitol Broth and dilute to $OD_{600} = 0.02$. Spot 30 μl of diluted culture on each Lotus root.
8. Seal the complete plates with non-woven ventilating tape impermeable to bacteria. Alternatively, ¾ of the plate can be closed with Parafilm and the remaining ¼ with ventilating tape to allow sufficient gas exchange (*see* **Note 9**).
9. Place the plates in the vertical position in the plateholder at 21 °C (16 h day and 8 h night).
10. Scan plants daily for the desired period as described in Subheading 3.2.

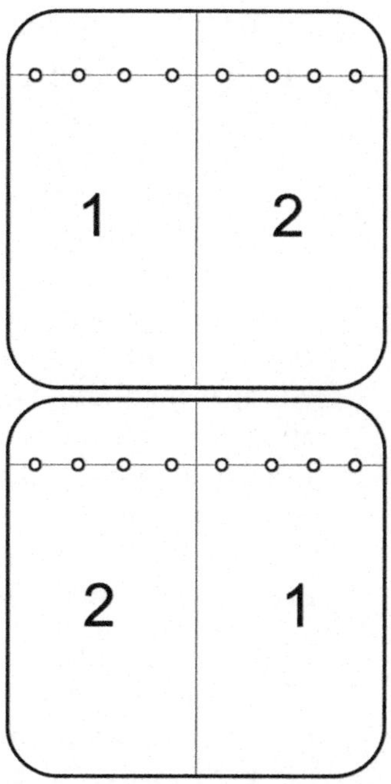

Fig. 2 Grid layout template for *Lotus japonicus*. Four plants of two different genotypes are grown on each plate

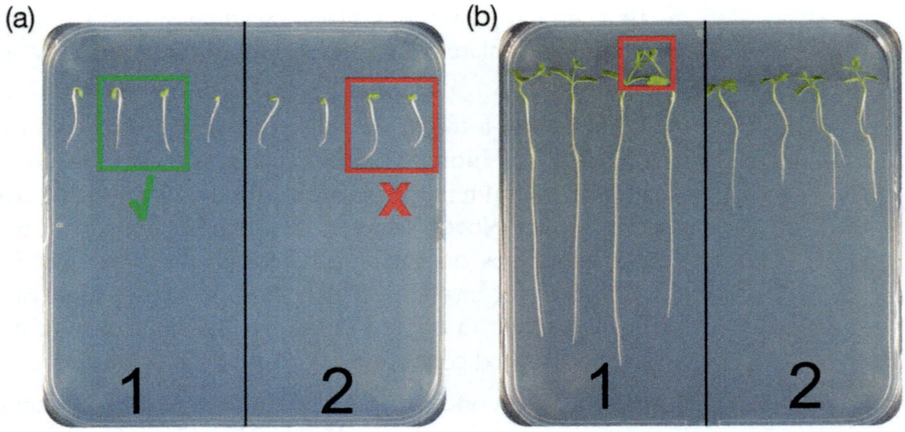

Fig. 3 Example of Lotus seedling layout on the plate. (**a**) Four 3-day-old plants from each genotype are placed on the plate. The green rectangle indicates plants at the right stage for further analysis, while the red one shows plants with strong root curling which should be excluded from analysis, in order not to alter the normal root growth, introducing bias. (**b**) Plants grown for 10 days. The red rectangle indicates overlapping plants, which might disturb the image segmentation analysis

3.2 Image Data Acquisition

1. Place the plates onto scanners within the support frames.
2. Start the BRAT image acquisition tool on the scanner control computer. You can mark which scanner and rows you want to use for image acquisition by clicking on the check boxes "Use" and "Row."
3. Select the folder in which the images will be saved by clicking on "..." and then specifying the name of the folder.
4. Initiate scanning by clicking on the "scan" button. All images are saved as 1200 dpi 8-bit RGB TIFF files and are automatically labeled with the chosen prefix and consecutive numbers. Optimal scanning results are achieved by scanning with open scanner lids in a dark room.

3.3 Lotus Root Trait Extraction

The BRAT plugins automatically detect plant roots from images of Petri dishes containing multiple plants growing on a transparent agar medium. The Lotus BRAT plugins consist of four tools allowing for marking the root/shoot boundary, image segmentation, quality control, and trait evaluation. All of them are open-source plugins that can be run on most operating systems where Java 1.8 or higher and Fiji have been installed. Currently, BRAT extracts 16 traits relevant for plant root growth, development, and physiology.

1. To initiate marking the root/shot boundary, use the BRAT Click Interface (plugin BRAT), and on the first tab, specify the path to the "Base directory" or click on the folder selection button (labeled with "...") to select the folder that contains the scanned images.

2. Manually select the beginning of each root with the cursor. This will create a specific .txt file for each image containing the coordinates of the beginning of each root (Fig. 4). After selecting the root start for the eight plants on a plate, the Root of Interest (Roi) Table opened by Fiji will be automatically filled with x and y coordinates. The image is automatically flipped by the plugin but a right click of the mouse allows for it to be manually flipped.
3. Click "Done" to process the next image contained in the selected folder. Continue with **step 2** until all image files are processed.
4. Perform BRAT image segmentation (plugin BRAT V2): the interface is composed of three different tabs called Basics, Expert Options, and Logging. In the first tab (Fig. 5a), indicate the path for the Base Directory containing the plate image files and the .txt files with root coordinates, previously created. In the field "File Extension," specify the image extension of your files (*see* **Note 10**). If you used a scanner to take pictures of plates, you should check "Flip Horizontal." If you are quantifying root traits from different time point, check also "Process Time Series." Only check "Use Day 0 Images," if the first image of the time series was acquired before germination (day 0) and plants were not transferred to the plates. Finally adjust the number of threads used in parallel, based on the performance of your computer (*see* **Note 3**). Finally, click the "Start" button to initiate the segmentation process. The "Logging"

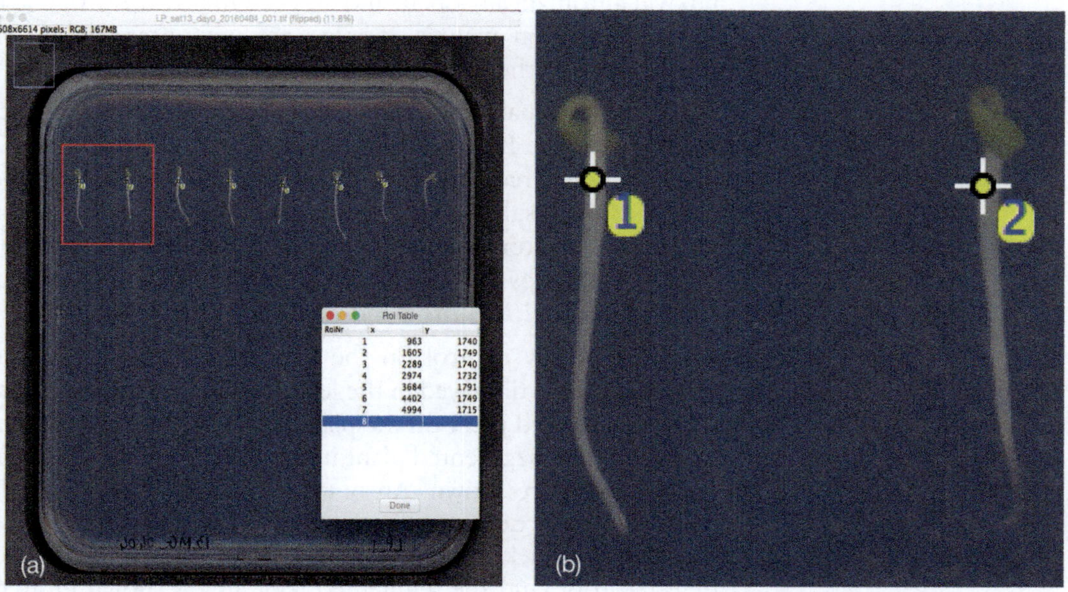

Fig. 4 Click-assisted version of Lotus BRAT. (**a**) The numbers indicate the coordinates of root/shoot border. (**b**) Close-up of plants indicated by the red rectangle in (**a**)

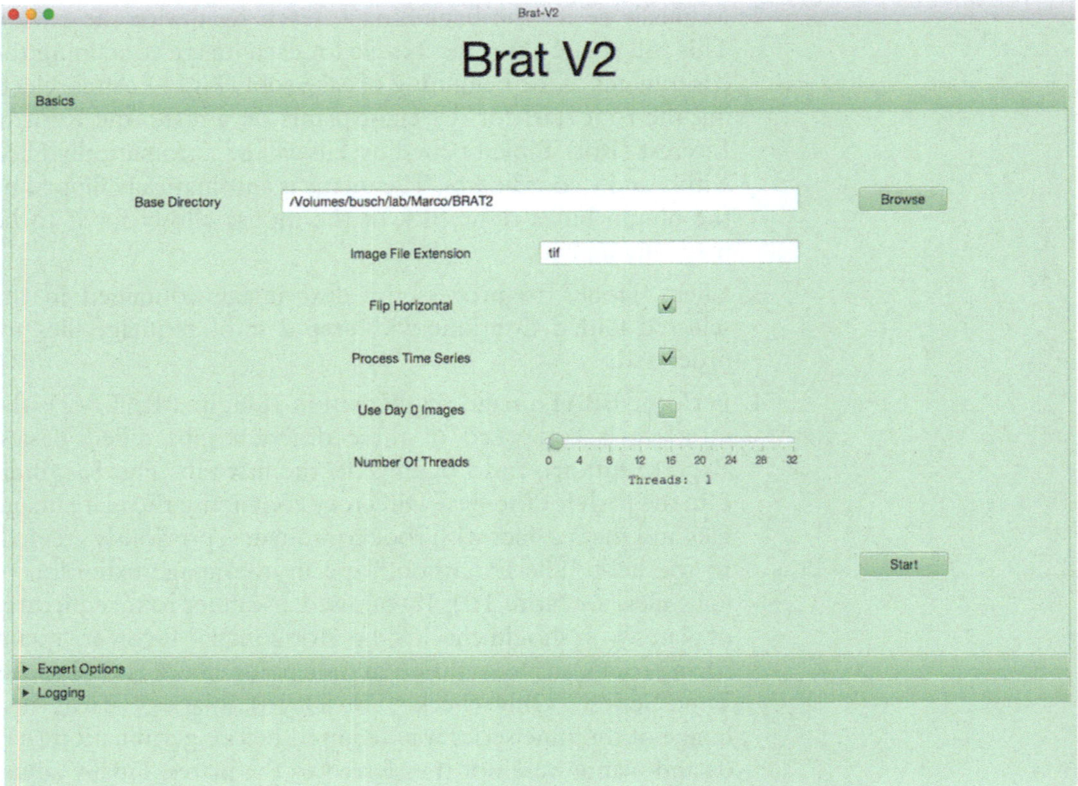

Fig. 5 BRAT V2 segmentation analysis interface. The "Basics" tab is shown where the user specifies the main parameters as well as the folder containing the images that are to be processed

tab will automatically open, and you can follow the progress of the segmentation analysis. If errors are encountered, they will be displayed in the "Logging" tab.

5. Performing a quality control check of the segmented images, by initiating the BRAT Quality Control (plugin BRAT QC2) in Fiji, is highly recommended. Once started, specify the directory with the results of the image segmentation from **step 4**, as well as the directory containing the root/shoot junction points obtained in **steps 1–3** in the "Base Dir" field. Then click on the folder selection button and then on "Start."

6. Conduct quality control on the diagnostic images displayed. Each red square displayed in the left panel indicates the position of the segmented plant on the plate. In the right panel, magnifications of the segmented plant in its original context and after segmentation are shown. An orange circle corresponds to the root/shoot junction coordinates that were created by the user in the BRAT Click Interface, a blue circle marks the end of each detected root, and the segmented root area is shown in green (Fig. 6). To confirm the correct segmentation of the plant root, confirm the "Plant Nr" by clicking on "OK" or pressing the

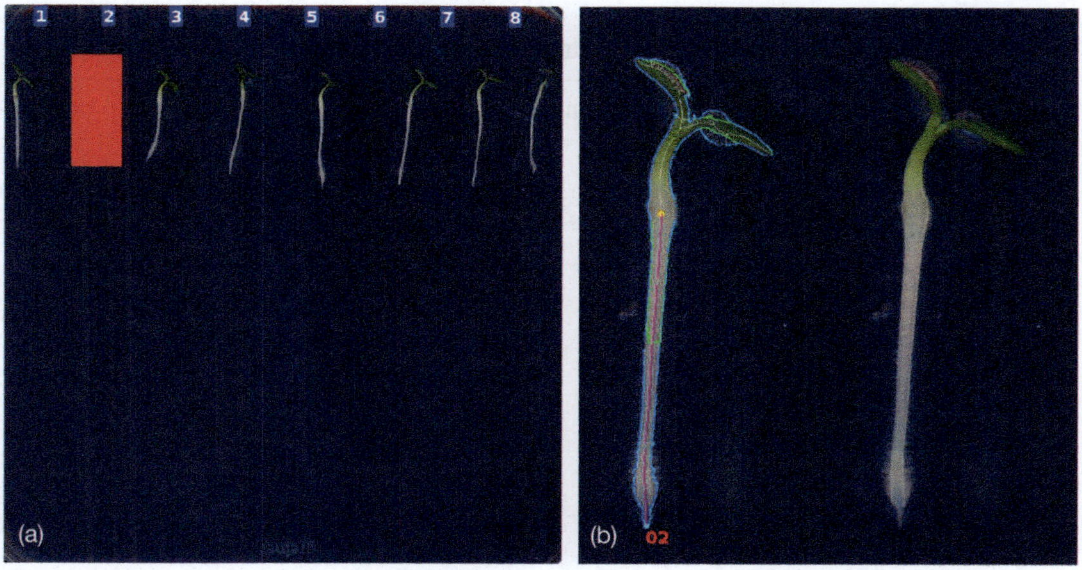

Fig. 6 The BRAT quality control interface. (**a**) The image of the whole plate with eight plants. (**b**) The magnified image of Lotus seedlings where the root and shoots are marked. The orange dot represents the beginning of the root based on the user's selection

ENTER key. To change the plant numbering, input a correct integer value for the desired position. To reject flag an incorrect segmentation, type in any letter or leave an empty space before pressing ENTER. The "Undo" button returns to the previous image.

7. Repeat **step 6** for all images. By pressing the ESC key or clicking "Exit," you will quit the plugin. Every classification is automatically saved, and quality control can be resumed at another time in case the whole folder was not processed. The progress bar shows the percentage of images already processed.

3.4 Lotus Root Trait Evaluation

Finally, to assign the genotype/treatment to each plant root and create the output file with every trait value and a summary of descriptive statistics of each genotype, the BRAT Evaluation plugin is needed. Here, the user will fill three different tabs, indicating the location of the data from previous steps and the plating layout file.

1. To initiate the BRAT evaluation, on the first tab—Base Dir—specify the path for the processed folder that was created by the image segmentation analysis in Subheading 3.3, **step 4** by clicking the "..." button. If the files can be read, the green circle on left should light up after clicking the "Read" button. If the red circle lights up, the segmentation file is not correct.
2. Select the "Classifier" tab and specify the path to the BRAT QC2 output file (User Classification) and click "Read" to load and check the data. In the Plate Layout space, you should

Table 1
Plate layout template

set_nr.	plate_nr.	plate_id	acc_id	row	column
3	1	1	50	1	1
3	1	1	51	1	2
3	2	2	51	1	1
3	2	2	50	1	2
3	3	3	52	1	1
3	3	3	53	1	2
3	4	4	53	1	1
3	4	4	52	1	2

The plate layout allows the BRAT Evaluation plugin to assign to each plant the specific treatment/genotype name. In particular, the .txt has to contain set number (denoting the experimental batch), plate number, plate ID, accession ID, row, and column. Two columns are considered, one for each group of four plants

manually select how many rows and columns of plants are set on each plate (in our case eight columns and one row).

3. Select the third tab, "Phenotype," to load the genotype/treatment layout data by specifying the path to the relative text file in the field "Accession Layout" (*see* Table 1 as a template). The last should be a tab-separated .txt file containing five columns specifying set number, plate number, plate ID, genotype/treatment, row, and column (as exemplified in Table 1).

   If a suitable accession layout file can be read, the circle will light up green after clicking the "Read" button. Finally, in the section "Output Directory," specify where to write the output files by clicking on the "..." button. You can also set a specific label per each output file, by filling the "<– Prefix" field.

4. Click on "Write" to create the output files. These will contain three different .txt files with individual and summary trait values and descriptive statistic values (median, average, and standard deviation) for each genotype/treatment as indicated by the "Accession Layout."

4 Notes

1. This is a modified version of Long Ashton nutrient solution that allows a physiological balance between elements and reproducible root growth. If preferred, ¼ B&D medium can be used instead [17]. We didn't observe differences in the nodulation kinetics between these two media. The use of agarose

prevents the effect of trace nutrients often present in other agar sources. A cheaper alternative could be to wash commercial agar. For Rhizobium inoculation, it is recommended to supplement growth medium with 1.4% agarose instead of 0.8% to avoid the root growing into the medium, which decreases the nodulation efficiency.

2. Any common desktop computer should fulfill the needed specifications for controlling the scanners so long as it provides enough USB-ports to connect the desired number of scanning devices (using an external USB hub is not recommended). If the images should be stored locally, a hard disk of appropriate size is needed. We use the open-source scanning framework "SANE" (http://www.sane-project.org/) for communication with the scan units. A Linux distribution is obligatory as the operating system in this case. Any distribution should be fine but our software was tested on Ubuntu 14.04 (https://www.ubuntu.com/desktop).
3. The segmentation tool of the BRAT pipeline is provided as a plugin for Fiji (ImageJ). This gives the possibility to run it on any computer which is able to execute Fiji and fulfills the minimal memory requirements. (There are installation packages for Linux, macOS, and Windows available at http://fiji.sc). With the recommended image resolution of 1200 dpi, each image processed in a single run uses about 1GB of RAM. For example, if you want to process a time series composed of five images (i.e., five time points), your workstation should provide at least 5GB of free RAM. If you want to analyze many time series, a compute cluster is advisable, which would significantly reduce processing time by running many jobs in parallel.
4. The maternal plants for seed amplification should be grown side by side at the same time to ensure that no batch/maternal effects confound the genetic effects of early root growth phenotypes during phenotyping.
5. In order to avoid thrip and aphid contamination, treatment with 0.075% (v/v) Conserve is advisable (Dow AgroScience, Denmark).
6. Whereas bacterial contaminants are easily eliminated by bleach treatment, fungal contamination of some seed batches can survive. This is probably due to the presence of spores within the seed tegument. Removing the teguments just after germination helps to reduce the occurrence of such contamination (and also ensures a more synchronous expansion of cotyledons). Additionally, before seed germination, manually select high-quality seeds by avoiding dark brown/black seeds which might be sources of fungal contamination.

7. The positioning of Lotus seedling is crucial to the further root development. Initially, plant roots need to be parallel to the plate to prevent the formation of adventitious roots from the hypocotyl.
8. It is advisable to avoid any overlap between plants' leaves that could impair the image segmentation and further trait acquisitions (Fig. 3b).
9. It is important to allow proper gas exchange in the plate, as closing it completely with Parafilm might lead to ethylene accumulation, which has an inhibitory effect on nodulation kinetics and affects root growth.
10. Most common image formats (.TIFF, .bmp, .gif, .png) are accepted as input files. The images are required to be in 24-bit RGB format with a minimal resolution of 1200 dpi. It is not recommended to use image compression formats (such as .jpg) or a low image resolution as this could impact the phenotyping accuracy.

Acknowledgments

We are grateful to Stig Anderson (Aarhus University) for materials, help, and collaboration. We thank Matt Watson for manuscript editing. This work was supported by funds from the Austrian Academy of Sciences through the Gregor Mendel Institute.

References

1. Handberg K, Stougaard J (1992) Lotus Japonicus, an autogamous, diploid legume species for classical and molecular genetics. Plant J 2:487–496. doi:10.1111/j.1365-313X.1992.00487.x
2. Madsen EB, Madsen LH, Radutoiu S et al (2003) A receptor kinase gene of the LysM type is involved in legume perception of rhizobial signals. Nature 425:637–640. doi:10.1038/nature02045
3. Radutoiu S, Madsen LH, Madsen EB et al (2003) Plant recognition of symbiotic bacteria requires two LysM receptor-like kinases. Nature 425:585–592. doi:10.1038/nature02039
4. Oldroyd GED (2013) Speak, friend, and enter: signalling systems that promote beneficial symbiotic associations in plants. Nat Rev Microbiol 11:252–263. doi:10.1038/nrmicro2990
5. Gutjahr C, Parniske M (2013) Cell and developmental biology of arbuscular Mycorrhiza Symbiosis. Annu Rev Cell Dev Biol 29:593–617. doi:10.1146/annurev-cellbio-101512-122413
6. Smith SE, Smith FA (2011) Roles of arbuscular mycorrhizas in plant nutrition and growth: new paradigms from cellular to ecosystem scales. Annu Rev Plant Biol 62:227–250. doi:10.1146/annurev-arplant-042110-103846
7. Lerouge P, Roche P, Faucher C et al (1990) Symbiotic host-specificity of rhizobium meliloti is determined by a sulphated and acylated glucosamine oligosaccharide signal. Nature 344:781–784. doi:10.1038/344781a0
8. Schauser L, Roussis A, Stiller J, Stougaard J (1999) A plant regulator controlling development of symbiotic root nodules. Nature 402:191–195. doi:10.1038/46058
9. Slovak R, Göschl C, Su X et al (2014) A scalable open-source pipeline for large-scale root phenotyping of Arabidopsis. Plant Cell 26:2390–2403. doi:10.1105/tpc.114.124032
10. Slovak R, Göschl C, Seren Ü, Busch W (2015) Genome-wide association mapping in plants

exemplified for root growth in *Arabidopsis thaliana*. In: Alonso JM, Stepanova AN (eds) Plant functional genomics. Springer, New York, pp 343–357

11. Fukai E, Soyano T, Umehara Y et al (2012) Establishment of a Lotus Japonicus gene tagging population using the exon-targeting endogenous retrotransposon LORE1. Plant J 69:720–730. doi:10.1111/j.1365-313X.2011.04826.x
12. Urbański DF, Małolepszy A, Stougaard J, Andersen SU (2012) Genome-wide LORE1 retrotransposon mutagenesis and high-throughput insertion detection in Lotus Japonicus. Plant J 69:731–741. doi:10.1111/j.1365-313X.2011.04827.x
13. Małolepszy A, Mun T, Sandal N et al (2016) The LORE1 insertion mutant resource. Plant J. doi:10.1111/tpj.13243
14. Kelly S, Sullivan J, Ronson C et al (2014) Genome sequence of the Lotus spp. microsymbiont Mesorhizobium loti strain R7A. Stand Genomic Sci 9:6. doi:10.1186/1944-3277-9-6
15. Schindelin J, Arganda-Carreras I, Frise E et al (2012) Fiji: an open-source platform for biological-image analysis. Nat Methods 9:676–682. doi:10.1038/nmeth.2019
16. Seren Ü, Vilhjálmsson BJ, Horton MW et al (2012) GWAPP: a web application for genome-wide association mapping in Arabidopsis. Plant Cell 24:4793–4805. doi:10.1105/tpc.112.108068
17. Broughton WJ, Dilworth MJ. Control of leghaemoglobin synthesis in snake beans. Biochem J. 1971;125:1075–1080

Chapter 12

Long-Term Confocal Imaging of *Arabidopsis thaliana* Roots for Simultaneous Quantification of Root Growth and Fluorescent Signals

Delyana Stoeva, Christian Göschl, Bruce Corliss, and Wolfgang Busch

Abstract

Observing cellular and molecular processes in living organisms is key for understanding many important biological processes. Confocal microscopy is excellently suited for this as it enables the observation of molecules and cells in tissue layers of living organisms in three dimensions over time. However, in continuously growing organs, such as plant roots, observations over extended time spans become difficult as the specimen quickly grows out the field of view. Here, we provide a protocol that allows for the acquisition of confocal microscope time-lapse images of root tips spanning many hours, as the growing root tip is tracked and the microscopy is automatized to change the position of the stage. Importantly, due to its specific setup, this protocol allows for observing the effects of chemical stimuli or for creating specific growth conditions by precisely defining the growth medium during imaging. The protocol is suitable for observing multiple fluorophores, thereby moving beyond the level of individual genes. It is also simple enough to conduct larger numbers of these assays. Here we exemplify our method by describing the observation of root growth and GFP intensity in root tips under iron depletion conditions.

Key words Confocal microscopy, Root growth, Tracking, Automated imaging, Cell biology

1 Introduction

Biological processes happen in time and space. It is therefore crucial to measure both where these processes happen and when. For instance, a notable proportion of genes of multicellular organisms are not expressed in all tissues at all times. To study the spatiotemporal aspects of gene and protein expression, it is necessary to observe them in intact organs, since single-cell cultures cannot reproduce the three-dimensional properties of a living organism. A well-established method to observe gene and protein activity in multilayer tissues is confocal microscopy in conjunction with fluorescent proteins that are either driven by endogenous gene promoters or fused to the protein of interest. In particular, the root of *Arabidopsis thaliana* has proven to be a useful system for live

Wolfgang Busch (ed.), *Plant Genomics: Methods and Protocols*, Methods in Molecular Biology, vol. 1610,
DOI 10.1007/978-1-4939-7003-2_12, © Springer Science+Business Media LLC 2017

imaging, as it is relatively thin and optically transparent [1]. Moreover, its cell types are arranged in concentric cylinders around the central vascular tissue displaying radial symmetry, and there is no cell migration as cells divide only in the tip of the root and maintain their relative position later on during elongation and differentiation. However, a major constraint is that a living root is constantly growing and cells leave the field of view within minutes or hours (depending on the magnification). Here we provide a protocol for automatically imaging growing root tips using a confocal microscope under defined conditions. In particular, we describe the observation of a fluorescent reporter gene under control and reduced iron levels. Importantly, this method can be utilized using different growth conditions or chemical stimuli.

2 Materials

2.1 Plant Growth Media [3]

1. 100× MS stock solution: 2.062 M NH_4NO_3, 1.879 M KNO_3, 125 mM KPO_4H_2, 0.5 mM KI, 150.1 mM $(MgSO_4)7H_20$, 10 mM $(MnSO_4)H_20$, 2.99 mM $(ZnSO_4)7H_20$, 10.027 mM H_3BO_3, 10 mM $(Na_2EDTA_2)H_20$, 0.01 mM $(CuSO_4)5H_20$, 0.103 mM $(Na_2MoO_4)2H_20$, 0.0105 mM $(CoCl_2)6H_20$, NH_4NO_3. KNO_3 requires heat to dissolve. Start with ~350 mL deionized water and bring up to 0.5 L. Then pass through a sterile filter.
2. 100× $CaCl_2$ stock solution: 299 mM $CaCl_2$ in deionized water. Sterile filtered.
3. MES buffer: 235 mM MES.
4. 1 M potassium hydroxide.
5. Agar.
6. 30× sucrose solution: 876 mM sucrose. Sterile filtered.
7. Iron–EDTA solution: 0.1 M iron stock solution—ethylenediaminetetraacetic acid iron(III) sodium salt in deionized water. Sterile filtered.
8. Ferrozine stock solution: 77 mM ferrozine–3-(2-pyridyl)-5,6-diphenyl-1,2,4-triazine-p,p′-disulfonic acid monosodium salt hydrate in deionized water. Sterile filtered.
9. Pipetman, 5 mL and 25 mL glass pipettes.
10. Plastic Pasteur pipette.
11. 1000 μL micropipette and 1000 μL tips.
12. 2 L plastic beaker.
13. 2× 1 L glass bottles.
14. pH meter.
15. Metal spoon.
16. Magnetic stir bars and magnetic stirrer.

17. Autoclave.
18. Water bath.
19. Leveling table and level.
20. Square petri dishes, 12 × 12 × 1.7 cm in their original plastic bags.
21. Sealing tape.
22. 500 mL glass bottles.
23. 5 L plastic beaker.
24. Sucrose.
25. 4 M sodium hydroxide solution.

2.2 Gas Sterilization of Seeds [4]

1. *Arabidopsis thaliana* seeds of a line expressing a fluorescent protein.
2. 2 mL microcentrifuge tubes.
3. Tube rack for 2 mL microcentrifuge tubes.
4. Permanent marker and transparent tape.
5. 15 mL plastic falcon tube.
6. Polycarbonate lockable airtight box, e.g., 36.8 × 18.4 × 17.1 cm, Nalgene.
7. 250 mL glass beaker.
8. Stir bar and magnetic stirrer.
9. 10% sodium hypochloride.
10. 37% hydrochloric acid.
11. Pipetman and tips (0.5–10 μL).
12. Adhesive ventilating tape.
13. Support rack to hold square petri dishes vertically.

2.3 Ferrozine Infiltration

1. 5 mL microcentrifuge tubes.
2. Plastic forceps.
3. 5 mL glass pipette.
4. 20 μL micropipette with appropriate tips.
5. 77 mM ferrozine stock solution.
6. Liquid ½ MS medium: 50 g sucrose, 11.05 g MS salts in 5 L deionized water. pH adjusted to 5.6–5.7 using 4 M sodium hydroxide solution. Autoclave in 500 mL bottles for 15 min, 125 °C. Store at 4 °C.
7. Desiccator attached to a vacuum peristaltic pump.

2.4 Confocal Imaging

1. Microscopy chambers: Chambered 1.0. borosilicate coverglass system (Lab Tech/Nunc).
2. Plastic forceps.

3. Propidium iodide solution: 0.15 mM PI solution in deionized water.
4. Microscope: LSM 700, Zeiss.
5. Zeiss Zen 2010.
6. Zen controller: https://code.google.com/archive/p/zen-controller/.
7. Fiji [2]: https://fiji.sc/.
8. Confocal_Tracking Tool plug-in: https://gitlab.com/buschlab/Confocal-Tracking.
9. Confocal Stitching plug-in: https://gitlab.com/buschlab/Confocal_Tracking.

3 Methods

3.1 Plant growth medium preparation

1. Mix 21 mL of 100× MS stock solution with 21 mL of 100× $CaCl_2$ stock solution and add 1958 mL deionized water to obtain a 2 L final volume.
2. Transfer 1931.34 mL of the MS solution to a 2 L plastic beaker and add 20 mL of MES buffer solution to maintain the pH value of the 1× MS base solution after pH adjustment.
3. Adjust pH to 5.75 using 1 M potassium hydroxide (*see* **Note 1**).
4. Weigh 10 g agar into two 1 L glass bottles and add in one bottle 955.65 mL 1× MS base solution for control medium and in the other bottle 952.65 mL 1× MS base solution for ferrozine medium. Add stir bars to bottles.
5. Mix using magnetic stirrer, so that the agar does not clump on the bottom.
6. Autoclave for 15 min, 125 °C. Take the bottle out after the temperature drops to 70 °C and put in into a water bath until it cools down to 55–60 °C.
7. In a sterile hood, place the bottle onto a magnetic stirrer with temperature set to 55–60 °C.
8. Add 33.33 mL of the 30× sucrose solution to medium.
9. Add 1 mL of sterile iron–EDTA solution for control plates or 4 mL of sterile 77 mM ferrozine solution for iron deficiency plates. After adding iron–EDTA or ferrozine, place the bottle in the water bath for at least 30 min to allow efficient distribution of the chemical in the liquid.
10. Use a leveling table with a level to create an even surface in the hood.
11. Distribute 53 mL of medium to each square petri dish and leave the plates with lids half-open for 2 h, or overnight with closed lids, to allow the medium to polymerize.

12. Put the plates in their original plastic bag, seal it with tape, and store with the lids facing down at 4 °C. The ferrozine containing plates must be stored protected from light.

3.2 Seed Gas Sterilization

1. For seed sterilization, put dry *A. thaliana* seeds in a 2 mL (*see* **Note 2**) microcentrifuge tube on a tube rack in the polycarbonate lockable airtight box. Label the tubes with a permanent marker and seal with a piece of transparent tape to protect the writing. Place a 250 mL beaker containing a magnetic stir bar and 130 mL of 10% sodium hypochlorite next to the rack in the box.
2. Aliquot 3.5 mL 37% hydrochloric acid into the 15 mL falcon tube.
3. Place the box on a magnetic stirrer in a chemical hood.
4. Slowly add the hydrochloric acid to the sodium hypochlorite and quickly close the box with the lid.
5. Turn on the stirrer at 200 rpm.
6. Incubate for 1 h (*see* **Note 3**).
7. Keep in the chemical hood and open the lid slightly to let the gas vent for 30 min.
8. Under sterile conditions, add 500 μL sterile deionized water to each tube.
9. Vortex the tubes and spin down shortly.
10. Wrap the tubes in aluminum foil and stratify at 4 °C for a minimum of 2 days.
11. Under a sterile hood and using a pipette, distribute 24 stratified seeds in a regular pattern in two rows on the MS agar plates.
12. Close plates with ventilating tape and place them vertically on a support rack in a long day (16 h light, 21 °C) plant growth chamber for 4–7 days.

3.3 Ferrozine Infiltration

1. Pipette 4 mL 1/2MS medium into a 5 mL tube and add 16 μL of the ferrozine solution. Use this as ferrozine treatment.
2. Pipette 4 mL 1/2MS medium into a 5 mL tube. Use this as control treatment.
3. Using forceps, transfer the 4–7-day-old seedlings into the liquid so that they are fully submerged or floating on the surface of the solution.
4. Place the tube into a desiccator attached to a vacuum peristaltic pump and turn on the vacuum for a maximum of 10 min.

3.4 Confocal Imaging: Sample Preparation

These steps illustrate how seedlings are transferred to a growth chamber (Fig. 1) that allows them to grow for long time periods while being imaged. In this growth chamber, chemical stimuli can

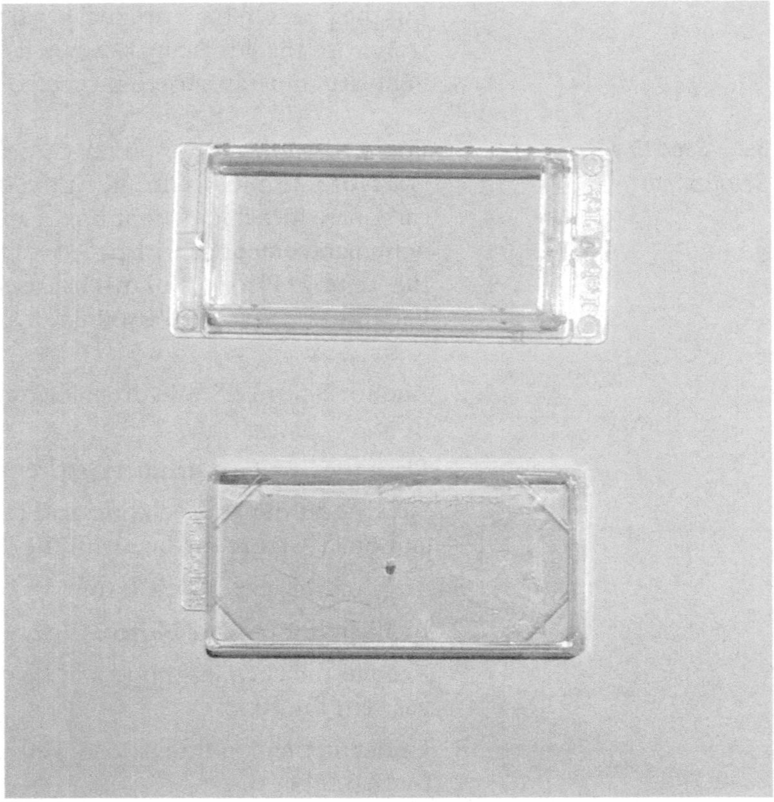

Fig. 1 Empty chamber and its lid

be performed, e.g., by the addition of hormones or chemicals or by changing the nutrient composition that the roots encounter. As fluorescence, cellular properties, and root growth rate can be measured simultaneously over time, this setup also allows for correlations between gene expression and cellular and root growth rate.

1. Slowly press the lid from the microscopy chamber against a fresh plate for either 1× MS medium (control treatment) or ferrozine medium (iron deficiency treatment) (Fig. 2a).
2. Turn the plate upside-down so that the microscopy chamber lid with the gel still attached points down. Carefully move the lid first up and down and then to the left or right until the gel detaches from the plate surface (Fig. 2b).
3. Carefully pull the lid of the microscopy chamber out of the agar plate, until the gel block falls into it (Fig. 2c). For the testing of two conditions simultaneously, e.g., ferrozine treatment and control, a layer of ferrozine containing gel and a layer of a control iron-containing gel is cut in half with a scalpel after being placed in the lid. Then, each block is separately transferred to the chamber.

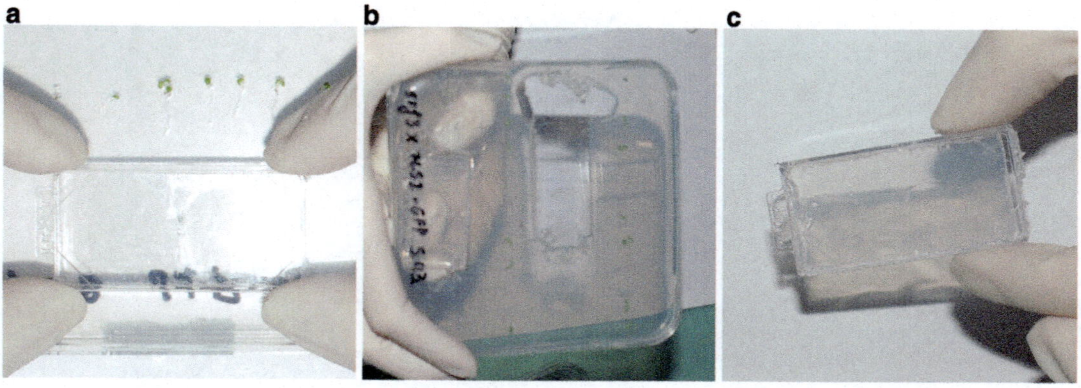

Fig. 2 Cutting out a gel block for the chamber. (**a**) Placement of the microscopy chamber to cut out gel block. (**b**) The original plate after cutting out the gel block. (**c**) Cut out gel block in the microscopy chamber

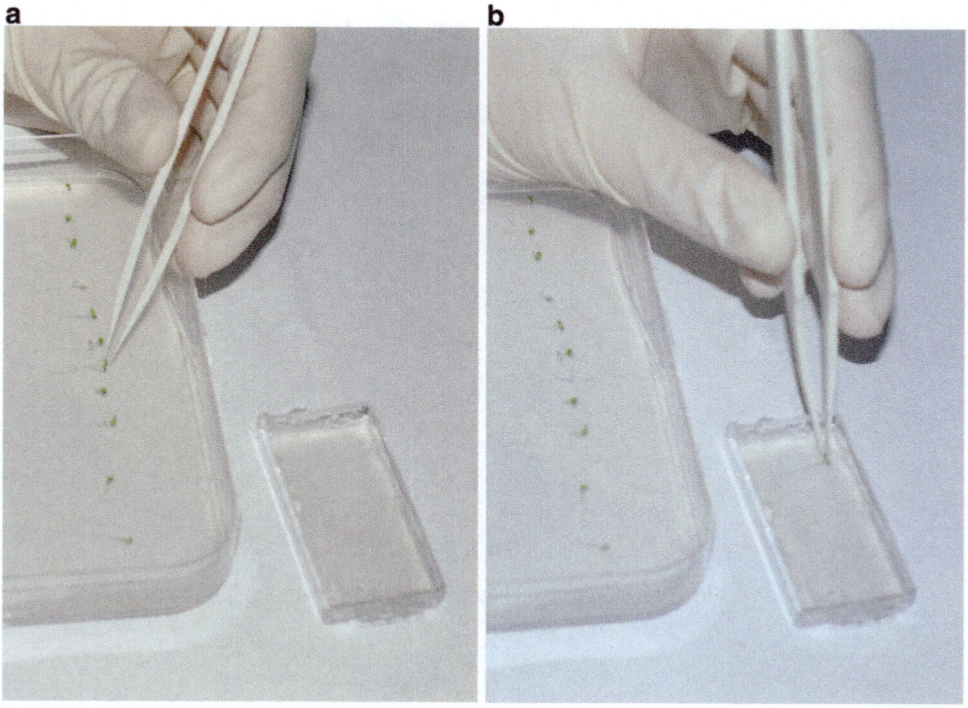

Fig. 3 Placing seedlings on the gel block. (**a**) Grasping of a seedling for transfer onto the gel block. (**b**) Placement of seedlings on the gel block

4. Using forceps, transfer seedlings (Fig. 3a) infiltrated in control or ferrozine containing liquid ½ MS to the control or ferrozine gel block, respectively, in the lid of the microscopy chamber in a way that allows the root to still grow (*see* **Note 4**) and with sufficient distance between the individual plants (Fig. 3b).
5. Place the confocal chamber, facing downward, onto the lid. Press slowly until the edges of the confocal chamber sink down into the gel.

Fig. 4 Transferring the gel block with seedlings into the chamber

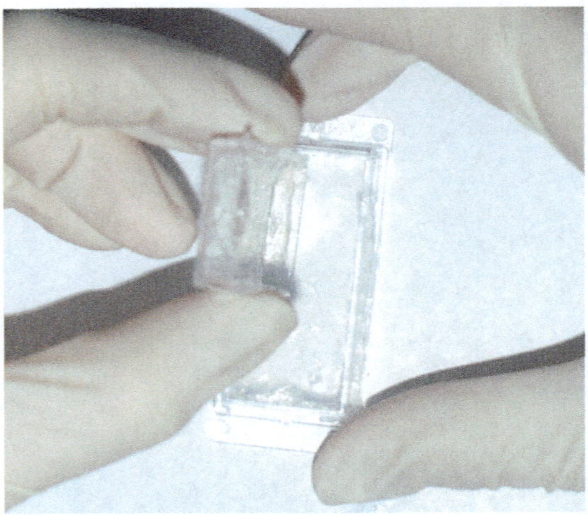

Fig. 5 Moving the gel block with seedlings to the bottom of the chamber

6. Now turn the chamber with the lid assembled, so that the lid is facing up. Slowly remove the lid—the gel should now detach and slide into the chamber (Fig. 4) (*see* **Note 5**).
7. Use the short smooth side of the lid to gently press the gel down and thereby remove the remaining air between the bottom side of the chamber and the gel (Fig. 5). For imaging under ferrozine treatment and control conditions, the gel layers are pressed down one after another in the same chamber (Fig. 6).
8. Put the lid onto the chamber to cover it and place it on the microscope.

Fig. 6 Placing seedlings in ferrozine and control treatments in the same chamber

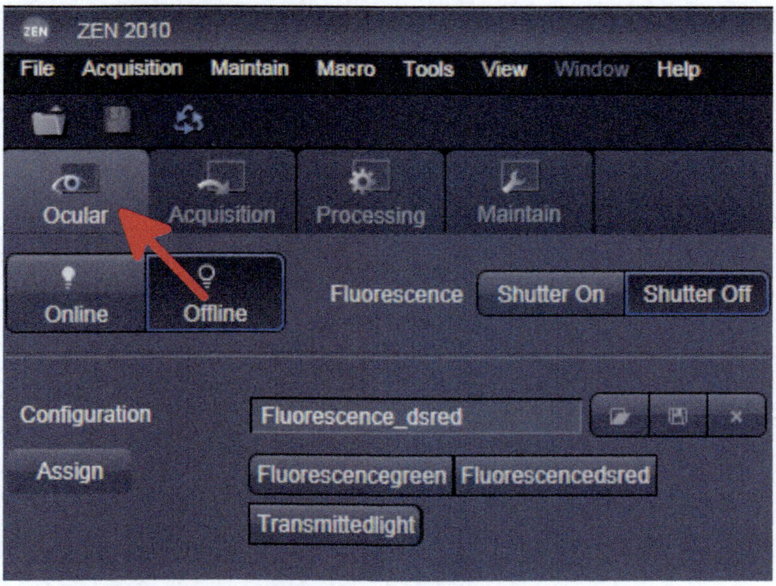

Fig. 7 Switching to "Ocular" mode

3.5 Image Acquisition

1. Open the "Zen" software on the microscope-controlling computer and select "Start system" in the first dialogue.
2. Switch to "Ocular" mode and find the first root (Fig. 7).
3. In "Acquisition" mode click on the "open" symbol to load a preexisting setting (Fig. 8). If no setting has been saved, a new setting must be configured. This setting should contain at least one channel that can be used for tracking. If the fluorescent signal is expressed throughout the root, this channel suffices for tracking. If not, a separate PI channel should be acquired.

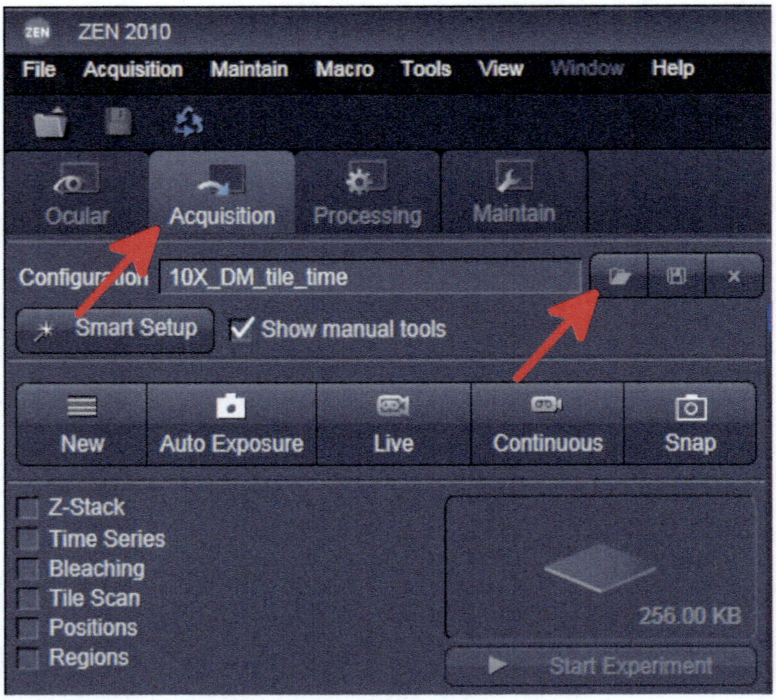

Fig. 8 Switching to "Acquisition" mode

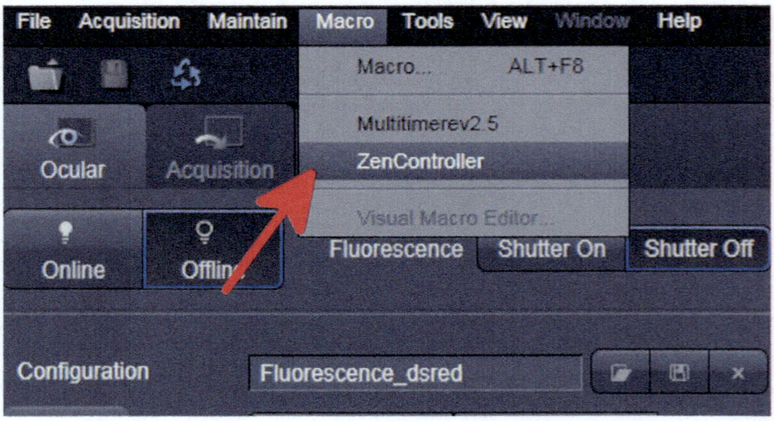

Fig. 9 Opening "Zen controller"

4. Run "Zen Controller" from the Tools—Macro menu in the ZEN software (Fig. 9).
5. Run "Fiji" on the microscope-controlling computer.
6. In "Fiji," choose "Confocal Tracking" from the "Plug-ins" menu (Fig. 10).
7. Press "Start" at Zen Controller (Fig. 11) on the microscope-controlling computer.
8. In the "General" tab of the "Confocal tracking" Fiji plug-in, select the "Tcp" button, create an output folder in which the

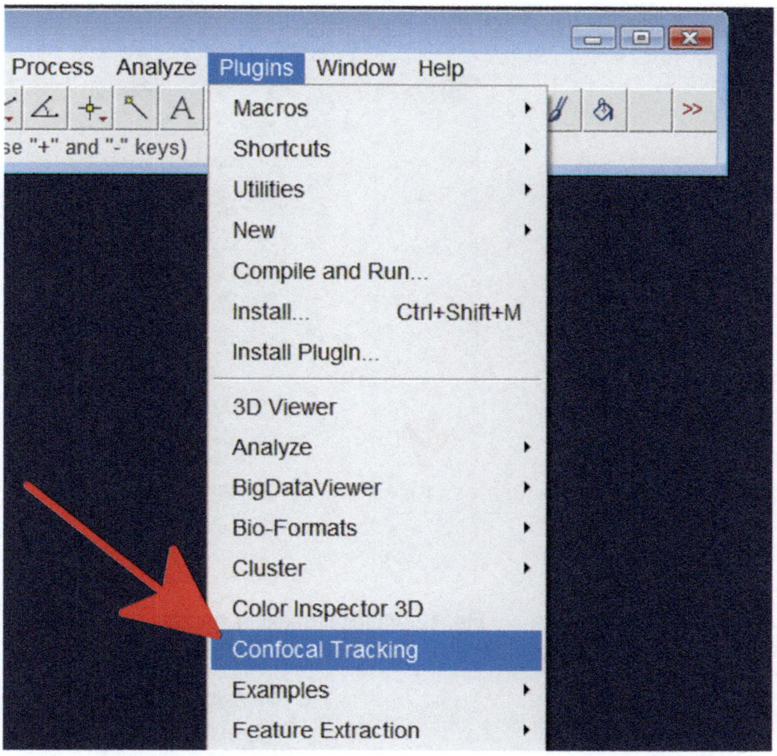

Fig. 10 Starting "Confocal tracking" plug-in

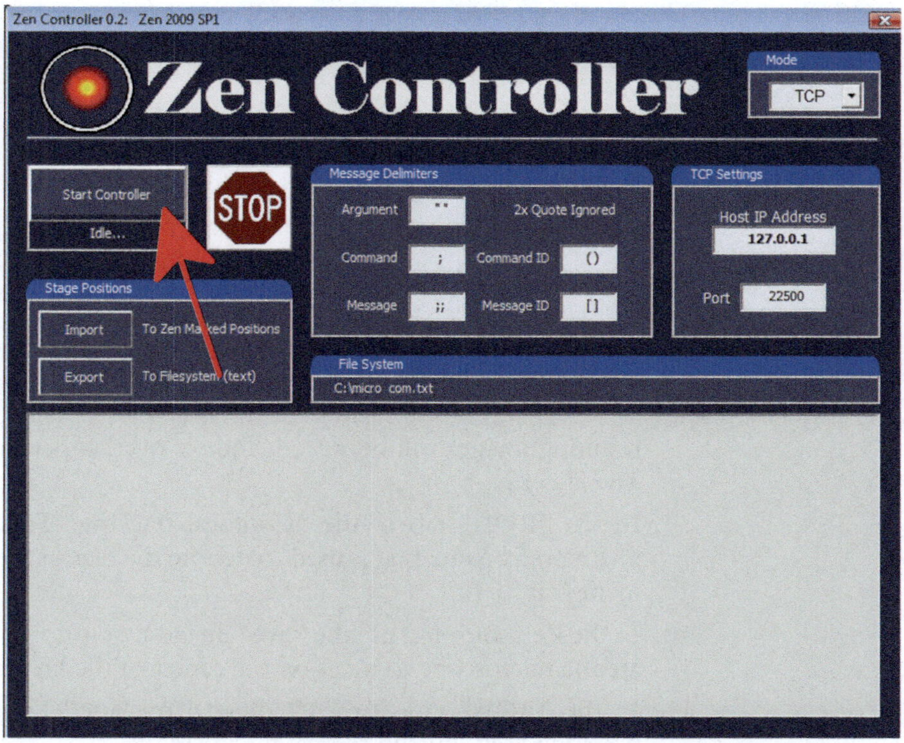

Fig. 11 Starting Zen controller

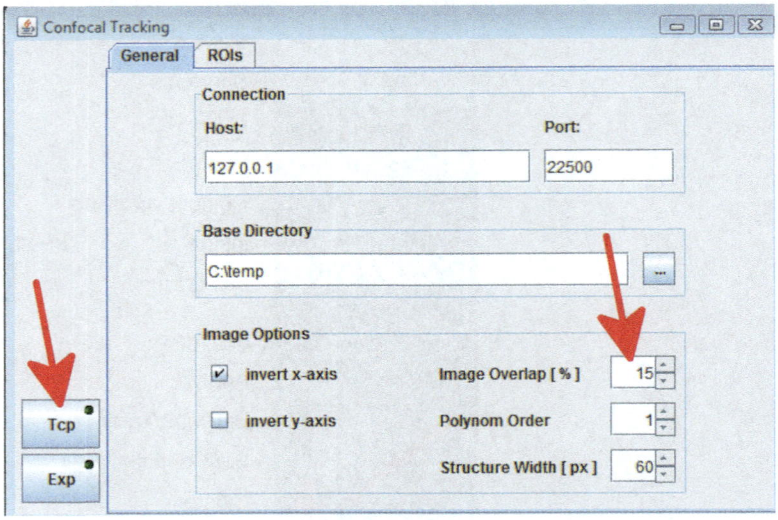

Fig. 12 Confocal tracking, general set up

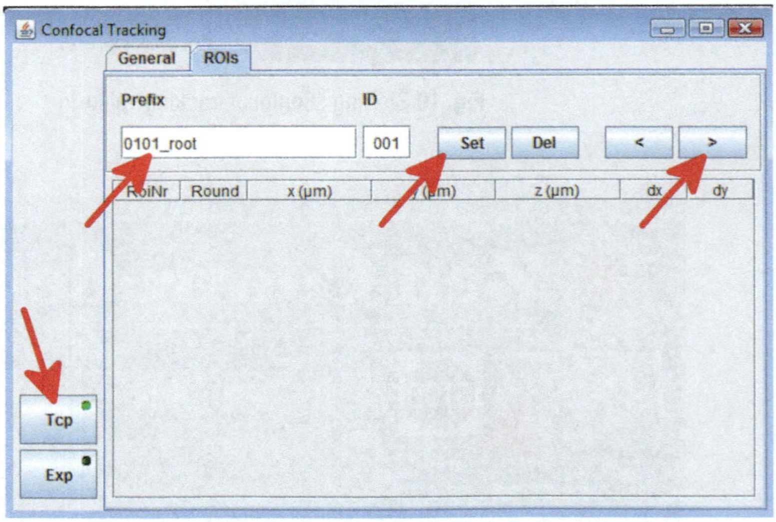

Fig. 13 Setting up positions for imaging

resulting images will be stored, and set overlap percentage to **15** (Fig. 12).

9. In the "ROIs" tab of the "Confocal tracking" Fiji plug-in, give a prefix name that is used to denote the files of the experiment (Fig. 13).
10. In the Zen software, use the "live" image acquisition and operate the microscope to focus on the center of the first root tip.
11. In the "ROIs" tab of the "Confocal tracking" Fiji plug-in, press the "set" button to save the position.
12. Move to the center of the next root tip using Zen and the microscope.

13. In the "ROIs" tab of the "Confocal tracking" Fiji plug-in, press the arrow pointing to the right and then press "set" button to save the position of the current root.
14. Repeat **steps 12** and **13** until all positions have been defined and then press the "Exp" button to start the experiment.
15. For stopping the experiment, press the "Exp" button and wait until the acquisition of the last image has finished. Save the automatically generated log.txt file into the output folder. The images will be saved in the "Output" folder in a TIFF format.
16. To assemble the time-lapse images from the individual images, start the "Confocal stitching" plug-in in Fiji (Fig. 14).

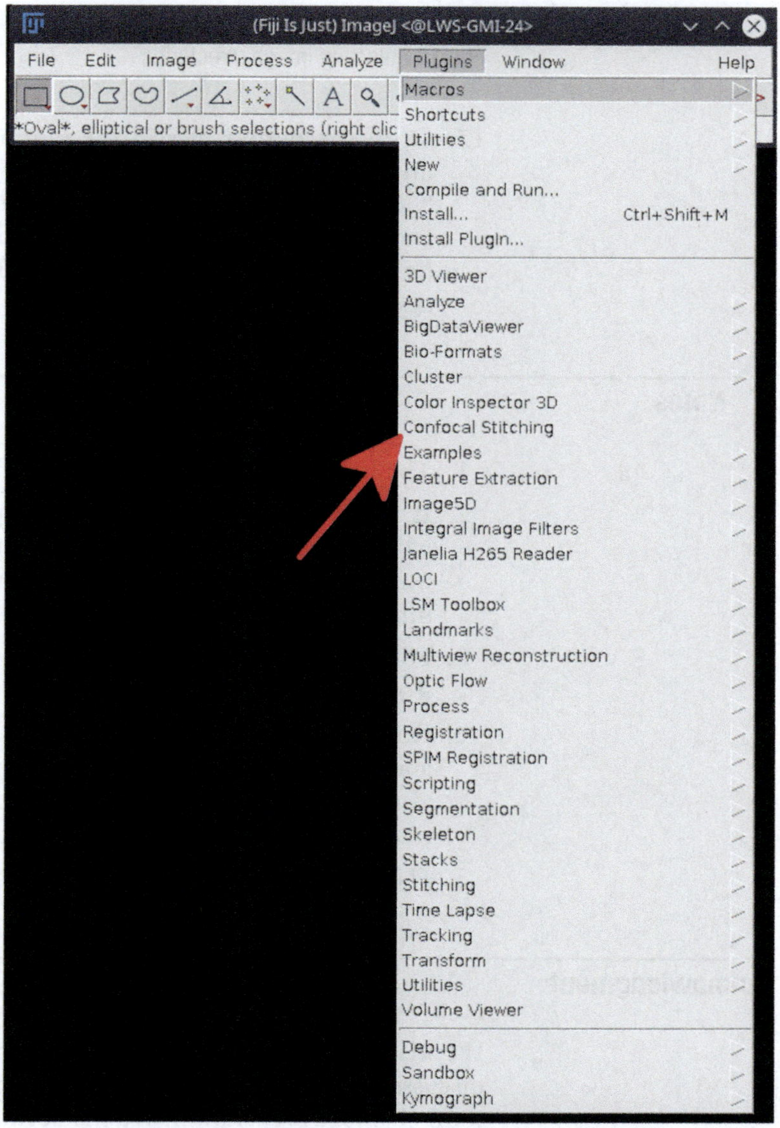

Fig. 14 Starting the "Confocal stitching" plug-in

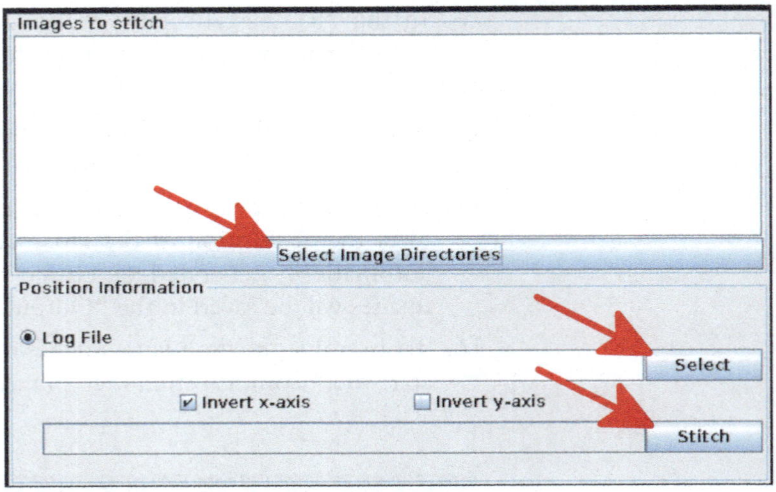

Fig. 15 Selecting image directories

17. Open the output files in the plug-in by clicking on "Select image directories" and the log.txt file by clicking on the "Select" button. Press "Stitch" to stitch the images (Fig. 15) and generate a "movie" from the stitched images, which will be saved in the output folder in TIFF format and can be opened with Fiji for further analysis.

4 Notes

1. The pH value can decrease by 0.5 after autoclaving.
2. It is important that all seeds are exposed to the gas. That is why 2 mL tubes are better that 1.5 mL, since the round bottom allows better exposure. For the same reason, the seed number must not exceed 300 seeds.
3. For *Arabidopsis thaliana* seeds, the sterilization time can be from several hours to overnight.
4. Roots can be stained for 30 s with 0.15 mM (0.1 mg/mL) PI and washed in tap water before transfer to the gel for better visualization of root cells. However, this will not improve tracking efficiency if the signal of interest is too weak.
5. Plastic forceps can be used as additional aid for detaching the gel.

Acknowledgment

We wish to thank all Busch lab members for support and helpful discussions, especially Ceren Tabak for taking pictures of the confocal chamber preparation. We thank Matt Watson for manuscript editing. This work was supported by funds from the Austrian Academy of Sciences through the Gregor Mendel Institute and an FWF stand-alone project (P27163-B22).

References

1. Busch W et al (2012) A microfluidic device and computational platform for high-throughput live imaging of gene expression. Nat Methods 9(11):1101–1106
2. Schindelin J et al (2012) Fiji: an open-source platform for biological image-analysis. Nat Methods 9(7):676–682
3. Murashige T, Skoog F (1962) A revised medium for rapid growth and bioassays with tobacco tissue cultures. Physiol Plant 15:473–449
4. Miché L, Balandreau J (2001) Effects of rice seed surface sterilization with hypochlorite on inoculated Burkholderia vietnamiensis. Appl Environ Microbiol 67(7):3046–3052

Part III

Molecular Bases of Phenotypes

Chapter 13

Identification of Protein–DNA Interactions Using Enhanced Yeast One-Hybrid Assays and a Semiautomated Approach

Allison Gaudinier, Michelle Tang, Anne-Maarit Bågman, and Siobhan M. Brady

Abstract

Yeast one-hybrid assays are an in vitro gene-centered approach to map transcription factor–DNA interactions. Here we describe this method and adaptations to screen for interactions between plant transcriptional regulators and their targets. Of particular note, the use of yeast one-hybrid assays fills in an important gap in available methodologies. When one is interested in a specific biological process of interest, the yeast one-hybrid assay is the only method that allows researchers to identify upstream regulators of the biological process of interest. This technique can be also used to further validate physical protein–DNA interactions or as a hypothesis-generating tool. In this method, promoters or DNA regions of interest are cloned and transformed into yeast and tested for interaction against a collection of transcription factors (TFs). Yeast one-hybrid screens are adaptable to the question the researcher is asking and the tools and components available. In this chapter we will describe large-scale and high-throughput Y1H screening; however, this can easily be scaled down for smaller studies.

Key words Yeast one-hybrid, High-throughput, Transcription factor, Promoter, Transcriptional regulation, Transcriptional regulatory network

1 Introduction

In a yeast one-hybrid (Y1H) assay, a single hybrid protein (a transcription factor fused with a yeast transcriptional activation domain) is screened for a physical interaction with DNA, typically a known regulatory region. The readout for this interaction is the activation of a reporter gene to which the regulatory region is fused. Thus, the two critical components needed for a Y1H screen are a collection of transcription factors (TFs) (preys) and DNA sequences of interest (baits), usually promoters or enhancers. In the system that we will describe, a TF collection is individually transformed into a haploid **α** yeast strain, and baits are individually transformed into a haploid **a** yeast strain. Mating these strains together will introduce both the bait and prey into a single yeast

Wolfgang Busch (ed.), *Plant Genomics: Methods and Protocols*, Methods in Molecular Biology, vol. 1610,
DOI 10.1007/978-1-4939-7003-2_13,

diploid colony, and performing reporter assays to characterize the readout will determine if a TF interacts with the regulatory region of interest. The yeast one-hybrid protocol was first described in [1–3] and the enhanced yeast one-hybrid assay in [4]. The protocols described below have used these systems as their basis and have been adapted to screen Arabidopsis transcription factors with regulatory regions [5].

In designing these assays, the collection of bait and/or prey strains as well as transformation methods can be modified and parallelized, depending on what is already available (Figs. 1 and 2). Moreover, these steps do not have to be carried out linearly in order to optimize screen efficiency (Fig. 3).

Here, we organize the protocol into sections describing each of the major components of Y1H, including recommendations for how to scale the protocol to become high-throughput as well as a description of a realistic workload for one person to carry out the mating screen given a single robot.

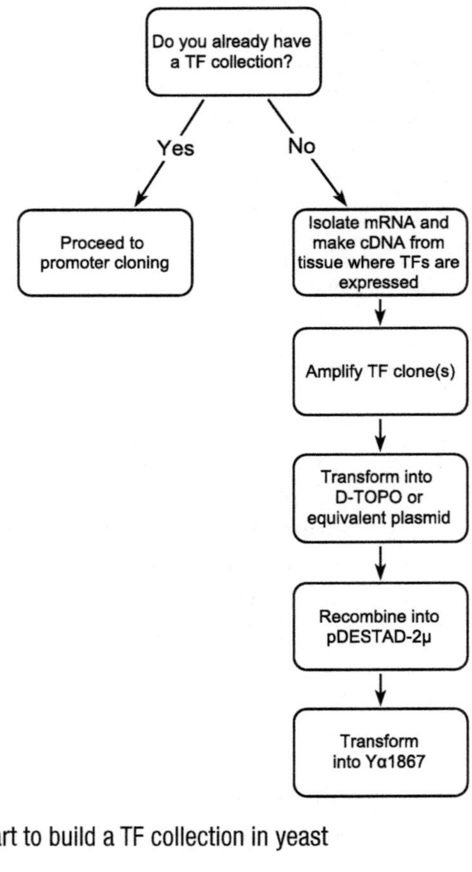

Fig. 1 Flow chart to build a TF collection in yeast

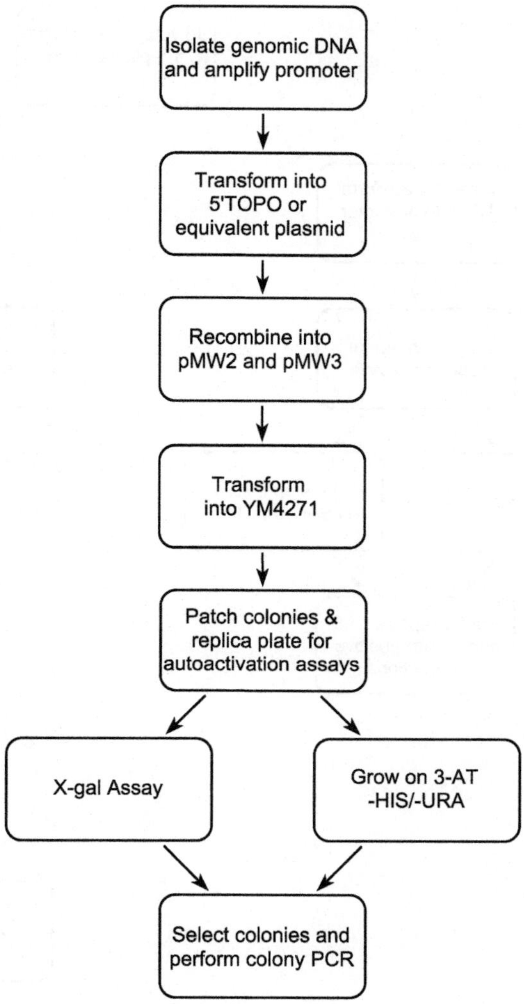

Fig. 2 Flow chart for promoter cloning and yeast transformation

2 Materials

2.1 Transcription Factor and Bait Promoter Cloning

1. Small round plates: 100 mm × 15 mm.
2. Large round plates: 150 mm × 15 mm.
3. LB broth: 10 g peptone, 5 g yeast extract, 5 g NaCl in 1 L of dH_2O.
4. LB agar: 10 g peptone, 5 g yeast extract, 5 g NaCl, 12 g agar in 1 L of dH_2O.
5. Ampicillin [Amp 100]. Stock solution concentration: 100 mg/1 mL H_2O. Working concentration: 100 μg/mL.
6. Chloramphenicol [Chlor 34]. Stock solution concentration: 34 mg/1 mL in 100% ethanol. Working concentration: 34 μg/mL.

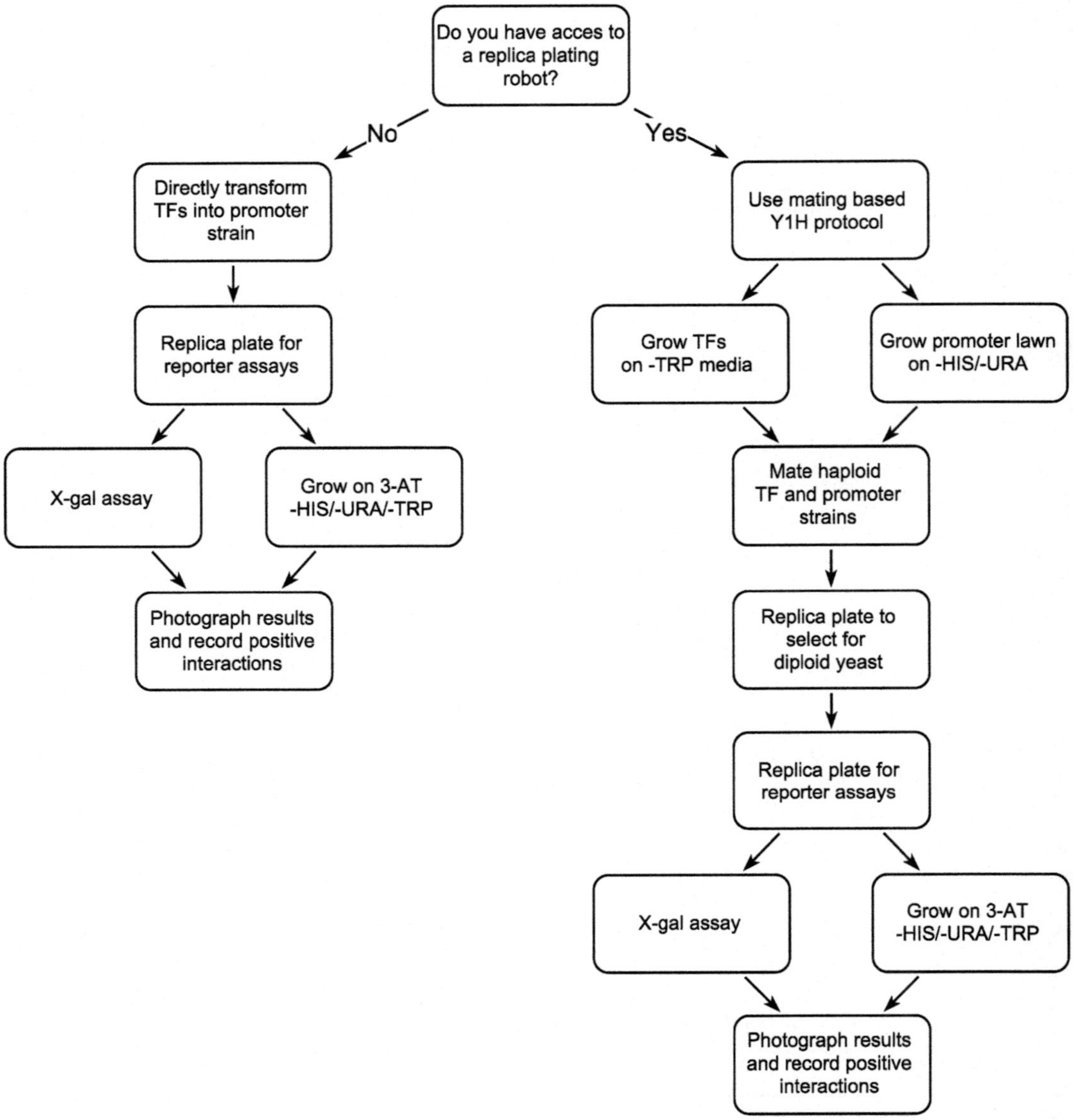

Fig. 3 Flow chart for Y1H with direct transformation and eY1H semiautomated robotic approach

7. Kanamycin [Kan 50]. Stock solution concentration: 50 mg/1 mL H_2O. Working concentration: 50 μg/mL.
8. D-TOPO kit (Invitrogen) (for transcription factors).
9. 5'TOPO kit (Invitrogen) (for promoters).
10. LR Clonase II (Invitrogen).
11. pMW2, pMW3, and pDESTAD-2μ plasmids (Addgene).
12. Phusion® Taq polymerase.
13. 10 mM dATP.
14. 10 mM dNTP.
15. Taq polymerase.

16. Taq polymerase buffer.
17. Primers.

Vector	Primers	Primer sequence
5'TOPO/D-TOPO	M13 Forward M13 Reverse	GTAAAACGACGGCCAG CAGGAAACAGCTATGAC
pMW2	M13 Forward His232 Reverse	GTAAAACGACGGCCAG GGGACCACCCTTTAAAGAGA
pMW3	p1HI Forward LacZ592 Reverse	GTTCGGAGATTACCGAATCAA ATGCGCTCAGGTCAAATTCAGA
pDESTAD-2μ	AD Forward Term Reverse	CGCGTTTGGAATCACTACAGGG GGAGACTTGACCAAACCTCTGGCG

2.2 Yeast Transformation, Mating, and Activation Testing

1. Yeast extract peptone dextrose with adenine (YPDA) agar: 35 g YPDA agar medium (Clontech) in 500 mL dH_2O (*see* **Note 1**).
2. YPDA liquid: 50 g YPDA medium (Clontech) in 1 L of dH_2O.
3. Amino acid selective (agar) media (-HIS/URA): In 350 mL dH_2O mix 10 g dextrose, 6.7 g nitrogen base without amino acids, 0.16 g adenine hemisulfate. pH to 5.8 using 1 M NaOH. Fill to 500 mL with dH_2O. Add to flask 0.375 g -HIS/-URA dropout supplement (Clontech) and 10 g Bacto agar. Pour media into large round plates if conducting activation test. Pour media rectangular plates if media is used for yeast mating.
4. Amino acid selective (agar) media (-TRP): same preparation as the -HIS/-URA selective media as in #3, except instead of the -HIS/-URA dropout supplement, add 0.37 g -TRP dropout supplement (Clontech).
5. Amino acid selective (agar) media (-HIS/-URA/-TRP): same preparation as in #3 except with 0.35 g -HIS/-TRP/-URA dropout supplement (Clontech).
6. 3-Amino-1,2,4-triazole (3-AT) 1 M stock: Measure 42.04 g of 3-Amino-1,2,4-triazole (stored in −20 °C freezer) in fume hood. Add 450 mL dH_2O in a beaker and stir to dissolve. Bring volume to 500 mL with dH_2O. Filter sterilize in Nalgene tissue culture filter unit, 0.45 μm filter diameter. Store at 4 °C and wrap bottle in foil to keep out light (*see* **Note 2**).
7. 3-AT selective agar media: Concentrations of 3-AT used: 10 mM, 20 mM, 30 mM, 40 mM, 60 mM, 80 mM, and 100 mM. Pour contents of beaker into a 500 mL graduated cylinder and fill specified volume. 3-AT is added after autoclaving: 10 mM: 495 mL dH_2O + 5 mL 1 M 3-AT, 20 mM: 490 mL dH_2O + 10 mL 1 M 3-AT, 30 mM: 485 mL dH_2O + 15 mL 1 M 3-AT, 40 mM: 480 mL dH_2O + 20 mL 1 M 3-AT, 60 mM: 470 mL dH_2O + 30 mL 1 M 3-AT, 80 mM: 460 mL dH_2O + 40 mL 1 M 3-AT, 100 mM: 450 mL dH_2O + 50 mL 1 M 3-AT.

8. 50% (W/V) PEG 3350 in H_2O, filter sterilized, 0.45 μm filter diameter. Store at 4 °C.
9. TE/LiAC/H_2O (10 mL): 1 mL 10× TE (pH 8.0), 1 mL 1 M LiAc, 8 mL sterile H_2O.
10. TE/LiAc/PEG (10 mL): 1 mL 10× TE (pH 8.0), 1 mL 1 M LiAc, 8 mL 50% PEG 3350.
11. Sheared salmon sperm DNA (Applied Biosystems).
12. Beta-mercaptoethanol.
13. Z-buffer (1 L): 8.52 g Na_2HPO_4 (anhydrous), 8.28 g $NaH_2PO_4 \bullet H_2O$, 0.746 g KCl, 0.246 g $MgSO_4 \bullet 7H_2O$. pH to 7.0 with 10 M NaOH.
14. 4% X-gal: 0.4 g X-gal, 10 mL Dimethylformamide. Vortex to dissolve. Store at −20° wrapped in foil.
15. Yα1867 and YM4271 yeast strains.
16. 96-deep-well culture plates (USA Scientific).
17. 96-well V-bottom plates, sterile (Greiner Bio-One)
18. 384-well microtiter plates with lids, sterile (Greiner Bio-One).
19. 9 × 9 inch pieces of velveteen sterilized by autoclaving for replica plating (*see* **Note 3**).
20. 6-inch replica block with collar (Cora Styles).
21. Nitrocellulose membrane (0.45 μm, 137 mm) (GVS North America).
22. Whatman filter paper 46 × 57 cm (Whatman).
23. Rectangular PlusPates® for robot: 128 mm × 85 mm (Singer Instruments).
24. RePads® 384 Short Box of 1000 short pin, 384-density pads for solid microbial array manipulation (Singer Instruments).
25. RePads® 384 Long Box of 200 long pin, 384-density pads for liquid and solid microbial array manipulation (Singer Instruments).
26. Acid washed glass beads 425–600 μm (Sigma-Aldrich).
27. Glass beads 4 mm (Fisher Scientific).
28. Plate roller (Sigma-Aldrich).

3 Methods

For the given protocols, the following guidelines can be considered as a reasonable amount of work for one person:

Yeast transformation for promoters: 20 per day.
Yeast transformations for TFs: 4 × 96-well plates per day.

Direct transformations for screening: 4 × 96-well plates per day.
Yeast screens using mating: # promoter × # TF plates = 70 plates per screen.
X-gal assay for screens: A set of 35 plates per day.

3.1 Cloning

3.1.1 Promoter (Bait) Cloning (See Note 4)

Utilize the 5'TOPO kit from Life Technologies in order to generate a pENTR plasmid containing the promoter of your gene of interest. This promoter is then recombined with the pMW2 (HIS3) and the pMW3 (LacZ) reporter plasmids to generate promoter/reporter fusions that will be transformed into yeast.

3.1.2 Primer Design for Promoter Cloning

1. Type the AGI ID of your gene of interest into the www.arabidopsis.org search bar.
2. Click on <AGI ID> name under locus column.
3. Scroll down to <Select Sequence viewer>.
4. Your gene of interest will be highlighted yellow. Note if the arrow is pointing left or right.
5. Scroll over the highlighted sequence and click <nucleotide seq view> in the pop-up window.
6. If the arrow was pointing left on the previous screen, then change the strand view to complement strand. (Forward strand is for the arrow pointing right).
7. Select the 2000 base pairs immediately upstream of the gene start site (*see* **Note 5**). Design primers 18–28 base pairs long at the beginning and end of the segment in order to generate the amplicon for further use (*see* **Notes 6** and 7).

3.1.3 PCR and A-Tailing Reaction

1. Conduct the PCR using your primers
   PCR reaction (*see* **Note 8**):

Water	16 μL
Forward primer 10 μM	2.5 μL
Reverse primer 10 μM	2.5 μL
Genomic DNA [40 ng/μL]	4 μL
2× Phusion Master Mix	25 μL
Total	50 μL

PCR Cycle Conditions (*see* **Note 9**):

98°	30 s	×1
[[98°	10 s	
55°	30 s	×35
72°	2 min]]	
72°	7 min	×1
4°	hold	

2. A-tail the PCR product as the Phusion polymerase generates an amplicon with a blunt end. This is necessary to successfully recombine with the 5'TOPO plasmid.

   A-tailing reaction mix:

20 μL gel extracted or purified PCR product
4.5 μL Taq buffer
0.4 μL dATP (*see* **Note 10**)
0.1 μL Taq

3. Place A-tailing reaction mix in a thermocycler and incubate at 72 °C for 15 min.

3.1.4 Recombination with 5'TOPO (Invitrogen)

1. Incubate reaction mix (4 μL A-tailing product with 1 μL salt solution and 1 μL 5′-TOPO vector from the kit) at room temperature for 20 min.
2. Transform into competent *E. coli* cells.
3. Select for transformants on LB kanamycin agar plates.
4. Sequence clones from both directions with the M13 universal primers contained in the 5′-TOPO kit to check that promoter is amplified correctly and recombined into the TOPO vector in the correct orientation. Use a custom-designed internal sequencing primer to check the sequencing of the entire promoter if over 1 KB.

3.1.5 Recombination of the pENTR Vector with pMW2 and pMW3

1. Grow pMW2 and pMW3 vector stocks in LB + ampicillin + chloramphenicol and extract fresh pMW2 and pMW3 plasmid using a miniprep (*see* **Note 11**).
2. Combine 1 μL fresh destination plasmid [150 ng/μL] (either pMW2 or pMW3), 3 μL of pENTR with promoter [~150 ng/μL], and 1 μL LR Clonase II.
3. Incubate reaction overnight at 25 °C in thermocycler or heat block.
4. Transform entire LR reaction mix into competent *E. coli* cells.
5. Select for transformants on LB ampicillin agar plates.
6. Use a combination of a gene-specific and a vector-specific primer for colony PCR verification.
7. Sequence clones from forward or reverse direction with the appropriate primers: For pMW2: M13F, His232R. For pMW3: p1H1F, LacZ592R (*see* **Note 12**).

3.2 Yeast Transformation

3.2.1 Vector Linearization (See Note 13)

1. Select the appropriate enzyme to linearize the pMW2 and pMW3 plasmids for yeast transformation and chromosomal integration. For pMW2: AflII and XhoI. For pMW3: NcoI and ApaI. It is important to make sure that the selected enzyme does not have a cut site in your promoter of interest or in the reporter gene.

2. Prepare for each vector a reaction with 16 μL of plasmid DNA, 2 μL of enzyme, and 2 μL of buffer.
3. Incubate at temperatures according to the restriction enzyme manufacturer for 3 h to overnight.
4. Heat inactivate the enzyme prior to transformation according to the restriction enzyme manufacturer's instructions.
5. Combine the pMW2 and pMW3 restriction enzyme digests in one Eppendorf tube for yeast transformation.

3.2.2 Transformation of Yeast Strain YM4271 with pMW2 and pMW3 Promoter:Reporter Constructs

This protocol is to transform 1–3 promoters. Adjust the YPDA culture amount and buffers for more than three transformations by scaling accordingly. All steps should be performed in a sterile environment.

1. Spread YM4271 yeast from a glycerol stock onto a YPDA agar plate to make a lawn.
2. Incubate plate for 1–1.5 days at 30 °C.
3. Inoculate the 50 mL liquid YPDA yeast culture: Take 2 mL of YPDA liquid and aliquot 1 mL each into two Eppendorf tubes: 1 mL is used for a control (blanking) in the spectrophotometer and 1 mL is for inoculating the culture to generate sufficient yeast for transformation. Add a pea-sized glob of yeast cells to a tube containing 1 mL of YPDA. Vortex for 30 s to ensure the yeast is thoroughly resuspended. Add ~100 μL of this resuspended yeast to a 50 mL culture in an Erlenmeyer flask. Add more yeast if necessary to get an OD_{600} ≈0.15–0.20.
4. Shake the flask for 2 h at 30 °C, 210 rpm. Grow until the OD_{600} reaches ≈0.4–0.6. If not, incubate longer (*see* **Note 14**).
5. Turn a heating block on and set temperature to 95 °C and a water bath to 42 °C.
6. Transfer the culture from the Erlenmeyer flask to a 50 mL falcon tube. Centrifuge at 1850 × *g* for 10 min at room temperature.
7. Boil a tube of sheared salmon sperm DNA for 5–10 min in the heating block and then immediately put on ice.
8. Discard the YPDA supernatant to maintain the yeast pellet. Gently shake tube to resuspend the cells with 5 mL of sterile dH_2O. Centrifuge at 1850 × *g* for 5 min at room temperature.
9. Discard the supernatant as in **step 8**. Gently shake the tube to resuspend the cells with 1 mL Te/LiAc/H_2O solution. Centrifuge at 1850 × *g* for 5 min.
10. Discard the supernatant. Gently shake the tube to resuspend cells in 250 μL TE/LiAc/H_2O.

11. Add 25 μL of cold, sheared salmon sperm DNA to the cells. This serves as a blocking agent. Gently shake.
12. Per promoter that will be transformed, add 100 μL of the yeast solution to a 1.5 mL Eppendorf tube containing the digested pMW2 and pMW3 constructs.
13. Add 600 μL TE/LiAc/PEG and mix by pipetting up and down ten times carefully.
14. Incubate the yeast transformations for at least 30 min at 30 °C.
15. Heat shock the cells for exactly 20 min in a 42 °C water bath.
16. Centrifuge the tube for 5 s in a microfuge at full speed. Remove supernatant.
17. Resuspend the cells in 150 μL sterile water by pipetting.
18. Spread the yeast transformation with glass beads onto dry -HIS/-URA agar plate until dry (*see* **Note 15**).
19. Incubate at 30 °C; the transformants will grow after 3–5 days.

3.3 Assays to Confirm and Test Activity of Yeast Promoter Transformation

When yeast bait (promoter) colonies have grown to 2–3 mm in diameter, patch up to eight clones per promoter onto large (150 mm × 15 mm) -HIS/-URA plate. Make sure to allow enough space (approximately 1 cm) between patches to prevent cross-contamination in the replica-plating step. Incubate at 30 °C for 2 days.

3.3.1 Autoactivation Test of Promoter Clone(s)

1. Label plates to ensure the correct orientation. Set up a sterile work area.
2. In sterile conditions, lay a velvet square on replica block and secure velvet square with collar, ensuring the velvet square is stretched over the replica block smoothly and tautly.
3. Carefully transfer the yeast colonies from the patched plate to the top of the velvet which is now on top of the collared steel plate. To do this, apply even pressure across the plate with the plate roller to transfer the yeast patches onto velvet.
4. Remove the yeast plate by lifting the plate straight up from the replica block to prevent cross-contamination.
5. The order of the plates to be replica-plated should be the following: -HIS/-URA, YPDA with nitrocellulose membrane, 80 mM 3-AT -HIS/-URA , 60 mM 3-AT -HIS/-URA, 40 mM 3-AT -HIS/-URA, and 20 mM 3-AT -HIS/-URA (*see* **Note 16**).
6. After replica plating the 3-AT series, use a fresh velvet per plate to replica plate the excess yeast so all plates have an even amount of yeast.
7. Incubate the original -HIS/-URA yeast plate and the replicated -HIS/-URA in 30 °C for 2 days. Store at 4 °C (*see* **Note 17**). The remainder of the -HIS/-URA plates to be used for reporter readout need to be incubated at room temperature in the dark for 10 days.

8. Incubate the YPDA plate containing the nitrocellulose membrane with transferred yeast for 2 days at room temperature (*see* **Note 18**).
9. Cut two Whatman filters to fit in the large round petri plates (135 mm diameter) per yeast plate (*see* **Notes 19** and **20**).
10. Mix 6 mL Z-buffer, 11 μL beta-mercaptoethanol, 100 μL 4% X-gal. Add solution into 135 mm diameter petri plate.
11. Add two Whatman filters to soak up solution and smooth out bubbles.
12. Freeze nitrocellulose filter with yeast on it for 10 s in liquid N_2.
13. Slowly lay nitrocellulose filter onto Whatman filters soaked in the solution described in **step 10** to ensure that there are no air bubbles (*see* **Note 21**).
14. Place the petri dish containing the solution described in **step 10**, Whatman filters and nitrocellulose filter in a 37° incubator in the dark, check for reporter activation at 30 min, 1 h, 2 h, 3 h, 4 h, 6–8 h, and 24 h. Thereby, take pictures as soon as you can see colonies turning blue and always take the 6–8 h and 24 h picture.
15. Note the yeast colonies that turn blue by the 8 h time point but not before the 4 h time point (*see* **Note 22**). If all the clones turn blue before 4 h, make a note of the promoter as autoactivating and the time point at which you begin to see blue.
16. After 10 days, take pictures of the 3-AT plates. Choose clones that show some but limited growth in the 3-AT series, and take note of the concentration in which the growth of the yeast clone becomes limited (*see* **Note 23**). Avoid choosing clones that grow well on the 80 mM 3-AT -HIS/-URA plate. The 3-AT concentration where the yeast shows limiting growth should be used for Y1H screening.
17. Select one or two clones that meet both the criteria for **steps 15** and **16** and proceed to yeast genotyping. If no clones have low autoactivation for both reporters, prioritize colonies with lower LacZ autoactivation.

3.3.2 Yeast Genotyping

This is an additional validation step to ensure that the entire promoter fused to the HIS3 reporter and fused to the LacZ reporter plasmid is fully integrated in the yeast genome.

1. Culture the promoter yeast colony from the replica-plated -HIS/-URA plate.
2. Culture yeast clone in 5 mL of YPDA and 3 mL of -HIS/-URA minimal liquid media in 30 °C, shaking at 210 RPM for 2 days.
3. Prepare glycerol stock from the -HIS/-URA culture: Pellet the culture, remove 2 mL of supernatant, resuspend the yeast pellet, and add 500 μL of yeast culture into an Eppendorf tube with equal volume of 60% sterile glycerol.

4. Pellet all the yeast from the YPDA culture in an Eppendorf tube.
5. Add 100 μL of acid-washed glass microbeads (425–600 μm) and the resuspension buffer from plasmid extraction kit.
6. Vortex until the pellet has been fully resuspended.
7. Proceed with the rest of the plasmid extraction protocol.
8. Elute the miniprep with 50 μL water.
9. Perform colony PCR with the yeast miniprep as a template to genotype the colony which has passed your autoactivation tests. Use a gene-specific reverse and a pMW2 or pMW3 vector-specific forward primers. Alternatively you can use a gene-specific forward and a pMW2 or pMW3 reverse primer if the former does not work well. PCR products with the correct size will confirm that the yeast colony contains the promoter and both reporter genes.

   PCR Mix:

Forward primer 10 μM	1 μL
Reverse primer 10 μM	1 μL
dNTP	0.2 μL
Taq buffer	2.5 μL
Taq polymerase	0.1 μL
Yeast miniprep DNA	1 μL
H_2O	19.2 μL
Total	25 μL

PCR Program (*see* **Notes 9** and **24**):

94 °C	5 min	
[[94 °C	15 s	
60 °C	30 s	×15
72 °C	extension time (depends on promoter length, add ~ + 250 bp for pMW2 and ~ + 800 bp for pMW3)]]	
[[94 °C	15 s	
50 °C,	30 s	×15
72 °C	extension time (depends on promoter length, add ~+250 bp for pMW2 and ~+800 bp for pMW3)]]	
72 °C	5 min	
4 °C	hold	

3.4 Prey Transcription Factor Cloning

3.4.1 PCR Amplification of the Prey Gene

1. Design primers to clone entire coding sequence that does not include the stop codon. The pDEST-AD-2μ vector has a stop codon in it.
2. Isolate mRNA from a tissue where you know the gene is expressed to make cDNA (*see* **Note 25**).

PCR mix

1 μL	cDNA (diluted at least 1:50 or 1:100)
1.25 μL	10 μM F primer
1.25 μL	10 μM R primer
8.5 μL	H_2O
12.5 μL	Phusion Master Mix

PCR protocol (*see* **Note 9**)

94 °C	5 min	×1
[[94 °C	15 s	
55 °C	30 s	×30
68 °C	2:30 min]]	
68 °C	10 min	×1
4 °C	hold	

3. Perform gel electrophoresis of the PCR product and extract your band of interest for recombination into the destination vector.

3.4.2 Recombination into D-TOPO (Invitrogen)

1. Set up reaction mix:

4 μL	gel extraction product
1 μL	salt solution
1 μL	vector

2. Incubate reaction overnight at 4 °C.
3. Transform into competent *E. coli* cells.
4. Select for transformants on LB kanamycin agar plates. Patch colonies in 96-well format for colony PCR.
5. Forward and reverse sequence clones with the M13 universal primers to check that TF is amplified correctly and recombined into the D-TOPO vector in the correct orientation. Use an internal sequencing primer to ensure entire TF is present.

3.4.3 Recombination of Prey from pENTR into pDESTAD-2μ

1. Grow fresh pDESTAD-2μ for plasmid miniprep for recombination in LB + ampicillin + chloramphenicol.
2. Combine 1 μL fresh pDESTAD-2μ plasmid [150 ng/μL], 3 μL of pENTR with TF [~150 ng/μL], and 1 μL LR Clonase II (Invitrogen).

3. Incubate overnight at 25 °C in PCR machine or heat block.
4. Transform into competent *E. coli* cells.
5. Select for transformants on LB ampicillin agar plates.
6. Use a combination of a gene-specific and a vector-specific primer for colony PCR verification.
7. After genotyping by colony PCR, grow individual cultures and sequence the clones from both the forward and reverse direction with the appropriate primers: AD and term.

3.4.4 Prey TF Transformation into Yeast Strain Yα1867

This protocol describes transformation of yeast in 96-well plate format. Each well will contain a different TF (*see* **Note 26**). This protocol is similar to the promoter transformation protocol but uses 250 mL liquid YPDA media for culture.

1. Spread Yα1867 yeast from a glycerol stock onto a YPDA agar plate in a lawn.
2. Incubate plate for 1–1.5 days at 30 °C.
3. Inoculate the 250 mL liquid YPDA yeast culture: Take 2 mL of YPDA liquid and separate 1 mL into two Eppendorf tubes: 1 mL is used for blanking in the spectrophotometer and 1 mL is to inoculate culture. Add a pea-sized glob of cells to tube. Vortex for 30 s to ensure yeast is thoroughly resuspended. Add ~300 μL of this resuspended yeast to 250 mL culture. Add more yeast if necessary to get an OD_{600} ≈0.15–0.20.
4. Shake flask for 2 h at 30 °C at 210 rpm. Grow until OD_{600} ≈0.4–0.6. If not, incubate longer.
5. Turn on a heating block to 95 °C and the water bath to 42 °C.
6. Transfer the culture to a 50 mL falcon tube. Centrifuge at 1850 × *g* for 10 min at room temperature.
7. Boil a tube of sheared salmon sperm DNA for 5–10 min and then put on ice.
8. Discard the YPDA supernatant. Resuspend the cells with 25 mL of sterile dH_2O and shake gently to resuspend. Centrifuge at 1850 × *g* for 5 min at room temperature.
9. Discard the supernatant as in **step 8**. Gently shake the tube to resuspend the cells with 5 mL Te/LiAc/H_2O solution. Centrifuge at 1850 × *g* for 5 min.
10. Discard the supernatant. Resuspend cells in 1 mL TE/LiAc/H_2O and gently shake.
11. Add 250 μL salmon sperm to the cells. This serves as a blocking agent. Gently shake.
12. Add up to 4–8 μL of the TFs in the pDESTAD-2μ vector to each well in the 96-well V-bottom plate.
13. Add 25 μL of the yeast solution to each well.

14. Add 100 μL TE/LiAc/PEG to each well and mix by pipetting up and down ten times carefully.
15. Incubate the yeast transformations for at least 30 min at 30 °C.
16. Heat shock the cells for exactly 20 min in a 42 °C water bath.
17. Centrifuge the plate at 1850 × *g* for 5 min. Remove supernatant.
18. Double check that the large round -TRP plate is completely dry.
19. Resuspend the cells in 8 μL sterile water and plate 5–6 μL in same format as the TF transformation plate onto large round -TRP plate. This is easiest with a 12-well multichannel pipette. Let colony spots dry before moving plate to the incubator. Make sure to keep track of plate orientation.
20. Incubate plate at 30 °C; colonies should take 2–4 days to grow.

3.4.5 Direct Transformation of Preys into Yeast Containing a Bait Promoter (See Note 26)

Direct transformation is a lower throughput substitute for the mating assay described below. This can be used if a mating robot is not available or if testing for a small number of interactions. This protocol is to transform one promoter with a 96-well plate of TFs. Each well will contain a different TF. Two control wells are needed in this assay: one well with no vector and one well with a pDESTAD-2μ plasmid with no TF clone.

1. Spread yeast with promoter of interest on a small YPDA agar plate in a lawn.
2. Incubate plate for 1–1.5 days at 30 °C.
3. Inoculate the 250 mL liquid YPDA yeast culture: Take 2 mL of YPDA liquid and separate 1 mL into two Eppendorf tubes: 1 mL is used for blanking in the spectrophotometer and 1 mL is to inoculate culture. Add a pea-sized glob of cells to the tube. Vortex for 30 s to ensure yeast is thoroughly resuspended. Add ~300 μL of this resuspended yeast to a 250 mL culture. Add more yeast if necessary to get an OD_{600} ≈0.15–0.20.
4. Shake flask for 2 h at 30 °C, 210 rpm. Grow until OD_{600} ≈0.4–0.6. If not, incubate longer.
5. Turn on a heating block to 95 °C and the water bath to 42 °C.
6. Transfer the culture to 50 mL falcon tube. Centrifuge at 1850 × *g* for 10 min at room temperature.
7. Boil a tube of sheared salmon sperm DNA for 5–10 min and then put on ice.
8. Discard YPDA supernatant to maintain yeast pellet. Gently shake tube to resuspend the cells with 25 mL of sterile dH_2O. Centrifuge at 1850 × *g* for 5 min at room temperature.

9. Discard supernatant as in **step 8**. Gently shake tube to resuspend the cells with 5 mL Te/LiAc/H_2O solution. Centrifuge at 1850 × *g* for 5 min.
10. Discard supernatant. Gently shake tube to resuspend cells in 1 mL TE/LiAc/H_2O.
11. Add 250 μL salmon sperm to the cells. Gently shake.
12. Add up to 4–8 μL of the TFs in the pDESTAD-2μ vector to each well in the plate. Be sure to include an empty well control and an empty pDESTAD-2μ vector control.
13. Add 25 μL of the yeast solution to each well.
14. Add 100 μL TE/LiAc/PEG to each well and mix by pipetting up and down ten times carefully.
15. Incubate yeast transformations for at least 30 min at 30 °C.
16. Heat shock the cells for exactly 20 min in a 42 °C water bath.
17. Centrifuge plate at 1850 × *g* for 5 min. Remove supernatant.
18. Double check that the -HIS/-URA/-TRP agar plate is completely dry.
19. Resuspend the cells in 8 μL sterile water and plate 5–6 μL in the same format as the TF transformation plate onto the -HIS/-URA/-TRP agar plate. This is easiest with a 12-well multichannel pipette. Let colony spots dry before moving plate to the incubator.
20. Incubate plate at 30 °C; colonies should take 2–4 days to grow.

3.4.6 Direct Transformation Reporter Assays to Determine Positive Interactions Between Bait and Prey

1. Replica plate the transformation plate from above onto the following (*see* **Note 16**):
   (a) -HIS/-URA/-TRP
   (b) YPDA large agar plate with nitrocellulose membrane
   (c) 80 mM 3-AT -HIS/-URA/-TRP
   (d) 60 mM 3-AT -HIS/-URA/-TRP
   (e) 40 mM 3-AT -HIS/-URA/-TRP
   (f) 20 mM 3-AT -HIS/-URA/-TRP
2. After replica plating the 3-AT series, use a fresh velvet per plate to replica plate to remove the excess yeast so all plates have an even amount of yeast.
3. Incubate the 3-AT -HIS/-URA/-TRP plate series and YPDA plate at room temperature.
4. Incubate the YPDA agar with nitrocellulose membrane plate for 2 days at room temperature. Then perform the LacZ assay as described in LacZ/X-gal assay section.
5. Incubate the 3-AT plates in the dark at room temperature for 10 days. Take pictures of the plates.

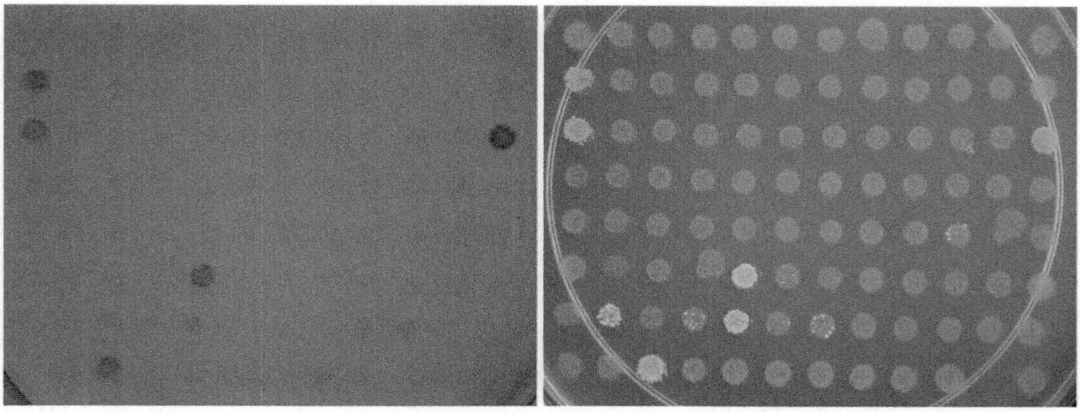

Fig. 4 LacZ and HIS3 reporter readout from Y1H direct transformation assay. *Left: Blue* colonies from LacZ assay show a positive TF–DNA interaction. *Right:* Colony growth on 3-AT-containing media shows a positive TF–DNA interaction

6. To determine the TFs that bind to and activate the reporters (positive binding interactions), two people are needed to review the pictures and come to a consensus. Colony color (for LacZ reporter) and colony growth (for His232 reporter) are compared to the pDESTAD-2μ plasmid control with no TF (Fig. 4).

3.5 Y1H Screening Using Singer ROTOR HDA Robot to Determine Bait × Prey Interactions Using Mating

This section describes the semi-automated mating approach to screen for interactions using the yeast one hybrid approach. This protocol is designed to test thousands of TF-promoter interactions per screen with the use of the ROTOR HDA robot and a large library of TF clones in an arrayed format.

Schedule outline for Y1H screen (*see* **Note 27**).

Day 1: Culture TF plates from glycerol stocks.

Day 2: Prepare YPDA, -TRP, and -HIS/-URA rectangular agar plates (*see* **Note 28**).

Day 3: Transfer TFs into 384-well format plate and spot onto YPDA. Culture yeast with promoter.

Day 4: Finish media prep.

Day 5: Transfer TFs onto -TRP rectangular agar plates. Plate promoter lawn.

Day 6: Prepare YPDA rectangular agar plates.

Day 7: Yeast mating.

Day 8: Prepare -HIS/-URA/-TRP rectangular agar plates (*see* **Note 28**).

Day 9: Diploid selection.

Day 10: Prepare 3-AT -HIS/-URA/-TRP and YPDA rectangular agar plates (*see* **Note 28**). Prepare nitrocellulose filters and Whatman paper.

Day 11: Reporter plating set 1 (first half of plates).

Day 12: Reporter plating set 2 (second half of plates).

Day 13: X-gal assay set 1.

Day 14: X-gal assay set 2.

Day 15–20: Check 3-AT plates.

Day 21: Take pictures of 3-AT plates for set 1.

Day 22: Take pictures of 3-AT plates for set 2.

3.5.1 Prepping for the Screen

1. Make -HIS/-URA rectangle agar plates equal to the number of promoters being screened (*see* **Note 29**).
2. Make 3–5 mL culture of each promoter in liquid -HIS/-URA. Grow at 30 °C, shaking at 210 rpm for 48 h. Culture should be dense (cultures started from glycerol stocks take 1–2 nights, cultures started from streaks on plates take 1 night—regularly check cultures for sufficient density until a pellet is formed).
3. Make -TRP and YPDA agar rectangular plates as needed. The number of TF plates to mate may vary between screens. Depending on the number of promoters to be screened, make multiple -TRP plates per TF plate so that you can transfer your yeast to keep them fresh. This is to avoid depleting the yeast when mating (make enough -TRP plates per TF plate so that one -TRP plate can spot onto 5–8 YPDA plates during mating—do not do more than eight or yeast might mate poorly).
4. Culture TF collection in 96 deep well culture plates from glycerol stocks with multichannel pipette, 315 μL of -TRP per well (*see* **Note 30**). Let grow two nights; you need to see yeast pelleting at bottom before you proceed.

3.5.2 Promoter and TF Plating

1. Pellet -HIS/-URA cultures for 5 min at 1850 × *g* and remove all but 1–1.5 mL of liquid to concentrate cells.
2. In a sterile hood, spread 510 μL of promoter culture onto -HIS/-URA plate with at least 15 glass beads (*see* **Note 15**).
3. Inspect plate for missed sections (*see* **Note 31**). Culture should be spread as evenly as possible. Run beads along the side of plate to ensure even spreading across sides and corners. Remove beads before plate begins to dry.
4. Incubate promoter lawn plates for 2 nights at 30 °C. After 2 nights fix any gaps in yeast lawn (*see* **Note 32**). With a 200 μL beveled pipette tip, pick up yeast from the corner and sides of the plate and carefully spread yeast on regions without yeast.
5. Spin 96-well TF plates down for 5 min at 1850 × *g*. Remove 150 μL of supernatant with a multichannel pipette, carefully not disturbing the pellet. Resuspend the yeast in the remaining supernatant by pipetting up and down. Transfer to 384-well culture plate and array in duplicate (combine 2 × 96-well plates

per one 384-well plate, plate one being in the upper half and plate two as the lower half of the 384-well plate). Each of the wells in the 384 well plate should have 50 μL of culture. Use robot to immediately pin out TFs on dry -TRP plates.

3.5.3 Initializing the ROTOR HDA

1. Open the rollover cover and turn on the power switch.
2. Turn on the compressor by flipping the switch to *Auto* on the air compressor.
3. Select *OK* on touch screen or use *Fast Buttons* on the front face of the robot—(I) is *OK* and (O) is *Cancel* (*see* **Note 33**).
4. Clear all objects from the ROTOR stage, so nothing will obstruct the initialization routine.
5. Select *OK On-screen* or use the *Fast Buttons.*
6. The robot will perform a start-up routine.
7. To sterilize, close rollover cover and select *UV* on right side panel on the touch screen—run UV light for at least 10 min prior to running robot.

3.5.4 Plating TFs from 384 Liquid Culture onto YPDA Agar with the Robot

1. Initialize the robot as specified in Subheading 3.5.3.
2. Choose *Select and run stored programs.*
3. Uncheck the box for *Using V2 plates* and select *Source: Multi-Well → 384.*
4. Check *Using V2 plates* box if using Singer plates and select *Target: PlusPlates® → 384 (Agar).*
5. Select *Pads: Long pin → 384.*
6. Under Singer programs, select *Spot Many.*
7. Under *General*: *Quantity*, select the number of YPDA agar plates to replicate from 384 liquid culture.
8. Select *General: Recycle → Until Repeat.* Check box under *Revisit Source Mode* → select *OK*. RePads® pins are recycled for the duration of one cycle of the program. Repeat for as many plates as needed for each different TF plate. When changing source plate, restart program to dump pads.
9. Under *General: Plate Protection*, uncheck box for *Use Source Protection.*
10. Under *General: Pairs*, press *+All* three times to increase pairs to ×4 and leave *Use pair repetition?* box checked. This setting will repeat transfer of yeast from the source to a target plate four times.
11. Under *Source: Pinning*, select default for *Speed* (*see* **Note 34**); 0 mm for *Backoff* and default for *Repeat pin.*
12. Under *Source: Wet mix*, select default for *Diameter*; default for *Speed*, four rotations for *Cycles*; and default for *Travel.*

13. Touch the side panel *Selected Options* to expand the panel and touch the *Source Mix* (second to last icon) to turn on source mixing.
14. Under *Target*, select default for all settings.
15. Select *OK* on touch screen or (I) on fast buttons.
16. Load plates as directed (*see* **Note 35**). Place the Source/384-well TF culture plate in the black bay and the Target/YPDA agar plate in the red bay of the turntable. Always double-check plate orientation.
17. Load 384 long pins into hopper and the robot will begin the program.
18. Select *OK* to repeat the plating for all TF plates and change the target quantity if needed under *Press here to select target quantity*. Swap out the Source/384-well TF culture plate for the next TF plate and swap out new YPDA agar plate as instructed on the screen.
19. After the last TF plate, select *Cancel* or (O) to finish the program. Remove the plates as instructed. Press *OK* when done. A program summary page will appear once the program has completed.
20. Turn off the robot by selecting *Off* from the right side panel and follow instructions on-screen.
21. Let TFs grow at least one night (preferably two) on YPDA then transfer to -TRP rectangle plates with robot.

3.5.5 Transfer TFs from YPDA to -TRP (or -TRP to -TRP)

1. Initialize the robot as specified in Subheading 3.5.4.
2. Choose *Select and run stored programs.*
3. Check *Using V2 plates* box and select *Source: PlusPlates® → 384 (Agar).*
4. Check *Using V2 plates* and select *Target: PlusPlates® → 384 (Agar).*
5. Select *Pads: Short pin → 384.*
6. Under Singer Programs, select: *Replicate Many.*
7. Under *General: Quantity*, select the number of -TRP agar plates to replicate from a YPDA/-TRP agar plate containing TFs (*see* **Note 36**).
8. Select *General: Recycle → Until Repeat.* Check box under *Revisit Source Mode* → select *OK.*
9. Uncheck box under *General: Plate Protection.*
10. Under *General: Pairs*, select *Reset.*
11. Under *Source: Offset*, select *No Offset.*
12. Select default for all settings for *Source: Pinning* and *Source: Dry Mix.*
13. Select default for all Target settings.

14. Unselect mixing options (second to last and last icons) from *Selected Options* side panels. Make sure those icons are gray and state "*Off*".
15. Select *OK* when ready.
16. Load plates as directed on-screen. Place the YPDA/-TRP agar plate with TFs in the black bay and -TRP agar plates onto the colored bays of the turntable.
17. Load 384 RePads® short pins into hopper.
18. Repeat for all TF plates. Make sure the pins are being disposed at the correct time, when changing source plate.
19. Turn off the robot by selecting *Off* from the right side panel when done and follow the instructions on-screen.
20. Incubate the replicated plates at 30 °C for 2 nights.

3.5.6 Mating Part 1: Plating TFs onto YPDA Agar (-TRP to YPDA) (See Note 37)

1. Initialize the robot as specified in Subheading 3.5.4.
2. Choose *Select and run stored programs.*
3. Check *Using V2 plates* box and select *Source: PlusPlates® → 384 (Agar).*
4. Check *Using V2 plates* box and select *Target: PlusPlates® → 384 (Agar).*
5. Select *Pads: Short pin → 384.*
6. Under Singer Programs, select: *Replicate Many.*
7. Under *General: Quantity*—select the number of plates based on how many promoters there are to be mated against a given TF plate. If more than four promoters, split up the number of replicates. For example, if there will be ten promoters to be screened, the ten plates will be split into 4 + 4 + 2. The quantity to replicate can be changed when the program asks to repeat *Replicate Many* in **step 14**. After 5–7 replicates, change out TF source plate to maintain sufficient amount of yeast for plating.
8. If there are fewer than four YPDA plates to be replicated, select *General: Recycle → Until Repeat.* If there are more than four YPDA plates to be replicated, select *General: Recycle → Full.* This program requires restarting the program when changing to a new TF source plate. Check box for revisit source mode → select *OK.*
9. Uncheck box under *General: Plate Protection.*
10. Under *General: Pairs*, select *Reset.*
11. Under *Source: Offset*, select *Automatic* and 0.16 mm for *Radius.*
12. Select default for all other Source settings.
13. Select default for all Target settings.
14. Double check settings in *Selected Options* side panel → select *OK* to run. Make sure Source Mixing and Target Mixing are turned off.

15. Load plates as directed on the screen.
16. Load 384 RePads® short pins into hopper.
17. Repeat as needed (continue directly to Subheading 3.5.7 mating part 2).

3.5.7 Mating Part 2: Mating Promoters with TFs

1. Return to the main menu and choose *Select and run stored programs*.
2. Check *Using V2 plates* box and select *Source: Other → Lawn*.
3. Check *Using V2 plates* box and select *Target: PlusPlates® → 384 (Agar)*.
4. Select *Pads: Short pin → 384*.
5. Under Singer Programs, Select: *Replicate Many*. If there is only one TF plate to mate, select *Replicate*.
6. Under *General: Quantity*, select the number based on how many TF plates are to be mated against the promoter. If there are more than four TF plates, split the number of replicates.
7. Select *General: Recycle → None*.
8. Check the box for *General: Plate Protection* (*see* **Note 38**).
9. Under *General: Pairs*, select *Reset*.
10. Under *Source: Offset*, select *Automatic* (an automatic offset is used for each source pinning) (*see* **Note 39**). Select 1.96–2.11 mm for *Radius*. This radius allows mating with approximately eight TF plates (*see* **Note 29**). If there are fewer TF plates, the radius can be smaller.
11. Select default for all settings for *Source: Pinning* and *Source: Dry Mix*.
12. Select default for all settings under *Target: Pinning*.
13. Under *Target: Dry mix*, select default for *Clearance*, 0.5 mm for *Diameter* and four rotations for *Cycles*.
14. Select *Target Mix* on the left side panel and *Selected Options* to turn on target dry mix.
15. Double check settings in *Selected Options* → select *OK*.
16. Load plates as directed. Load the -HIS/-URA agar plate containing the promoter lawn onto the black bay and load the YPDA agar plates replicated with the TFs onto the colored bays on the stage.
17. Load 384 RePads® short pins into hopper.
18. Run program and make sure that RePads® pins are discarded after each plating.
19. Repeat for each promoter.
20. Turn off the robot by selecting *Off* from the right side panel when done and follow the instructions on-screen.

21. Grow mated yeast on YPDA plates for 2 nights at 30 °C. Place the -TRP plates with the TFs back at 30 °C.
22. Spot out an extra -TRP plate (s) for each TF plate to maintain the TF collection without having to re-culture from the glycerol stocks (*see* **Note 40**).

3.5.8 Diploid Selection: Replica Plating from YPDA to -HIS/-URA/-TRP Selection

1. Initialize the robot as specified in Subheading 3.5.4.
2. Choose *Select and run stored programs.*
3. Check *Using V2 plates* box and select *Source: PlusPlates® → 384 (Agar).*
4. Check *Using V2 plates* box and select *Target: PlusPlates® → 384 (Agar).*
5. Select *Pads: Short pin → 384.*
6. Under Singer Programs, select *Replicate.*
7. Select *General: Recycle → None.*
8. Uncheck the box for *General: Plate Protection.*
9. Under *General: Pairs*, select *Reset.*
10. Under *Source: Offset*, select *No offset.*
11. Set all other Source settings to default.
12. Set all Target settings to default.
13. Unselect source and target mixing options from left side panel *Selected Options* → Select *OK.*
14. Load plates as directed. Load the YPDA agar plate containing the mated yeast on the red or yellow bays and the -HIS/-URA/-TRP agar plate on the blue or green bays.
15. Load 384 RePads® short pins into hopper.
16. Repeat for all plates. Make sure RePads® pins are discarded after each plating.
17. Turn off the robot by selecting *Off* from the right side panel when done and follow the instructions on-screen.
18. Let yeast plates grow 2 nights on -HIS/-URA/-TRP at 30 °C.

3.5.9 Reporter Plating: Plating from -HIS/-URA/-TRP to YPDA with Nitrocellulose Filter and 3-AT Plates (See Notes 41 and 42).

1. Initialize the robot as specified in Subheading 3.5.4.
2. Choose *Select and run stored programs.*
3. Check *Using V2 plates* box and select *Source: PlusPlates® → 384 (Agar).*
4. Check *Using V2 plates* box and select *Target: PlusPlates® → 384 (Agar).*
5. Select *Pads: Short pin → 384.*
6. Under Singer Programs, select *Replicate Many.*
7. Under *General: Quantity*, select 2.

8. Under *General: Recycle*, select *Until Repeat* and check *Revisit Source Mode* box.
9. Uncheck box for *General: Plate protection.*
10. Under *General: Pairs*, select *Reset.*
11. Under *Source: Offset*, select *No Offset.*
12. Set all other Source settings to default.
13. Select default for all Target settings.
14. Unselect mixing options from *Selected Options* in the left side panel and select *OK.*
15. Load plates as directed on screen. Load the -HIS/-URA/-TRP agar plate containing diploid yeast in black bay, the 3-AT -HIS/-URA/-TRP agar plate in red bay, and the YPDA plate in the blue bay.
16. Load 384 RePads® short pins.
17. Repeat for the remaining -HIS/-URA/-TRP diploid yeast plates.
18. Turn off the robot by selecting *Off* from the right side panel when done and follow the instructions on-screen.
19. Let yeast plate grow for 2 nights at room temperature and proceed with LacZ/X-gal assay (Subheading 3.5.10). Take pictures of the 3-AT -HIS/-URA/-TRP agar plates 10 days after reporter plating.

3.5.10 LacZ/X-Gal Assay Protocol for Mating OR Direct Transformation Assays

1. Calculate total solution amount for 5.5 mL per plate (plus an extra 10 mL). Make solution by mixing 6 mL Z-buffer, 11 μL BME, 100 μL 4% X-gal.
2. Add two Whatman filters to each 135 mm diameter petri plate.
3. Add 5.5 mL solution into petri plates and smooth out bubbles.
4. Freeze nitrocellulose filter with yeast on it for 10 s in liquid N_2.
5. Slowly lay nitrocellulose filter onto filters soaked in the solution to ensure that there are no air bubbles.
6. After a few minutes, get rid of air bubbles under the nitrocellulose filter.
7. Put in 37° incubator, check for reporter activation at 30 min, 1 h, 2 h, 3 h, 4 h, 6–8 h, and 24 h Thereby, take pictures as soon as you can see colonies turning blue and always take the 6–8 h and 24 h picture.

3.5.11 Calling Positive Interactions for Reporter Assays

To determine the TFs that activate the reporters and are called positive binding interactions, two people are needed to review the pictures and come to a consensus. Colony color (for LacZ reporter) and colony growth (for His232 reporter) are compared to the pDESTAD-2μ plasmid control with no TF (Fig. 5).

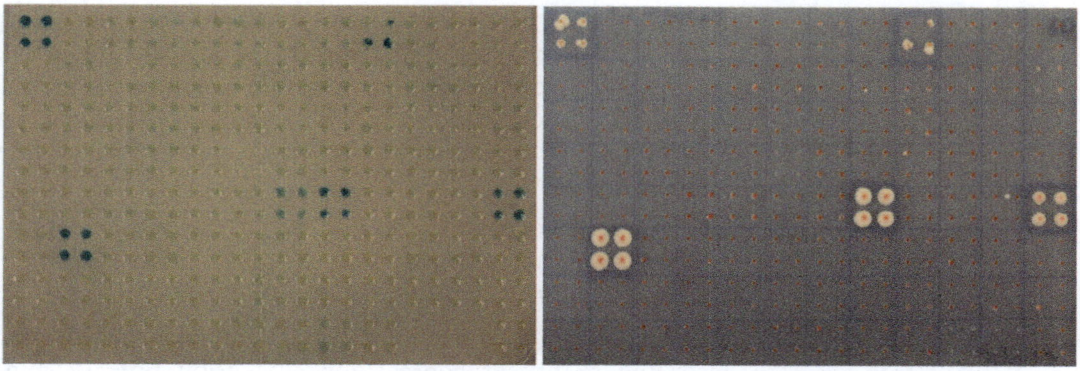

Fig. 5 LacZ and HIS3 reporter readout from eY1H mating assay. TFs are represented in quadruple replicates in the shape of a square. *Left: Blue* colonies from LacZ assay show a positive TF–DNA interaction. *Right:* Colony growth on 3-AT-containing media shows a positive TF–DNA interaction

4 Notes

1. Depending on the purpose, this amount of media will make the equivalent of 10–12 rectangular plates (128 mm × 85 mm from Singer Instruments), 20–24 small round plates (100 mm × 15 mm), and 7–8 large round plates (150 mm × 15 mm).
2. 3-AT is a reproductive toxin. Please follow your institution's Environmental Health and Safety Guidelines for proper disposal of 3-AT agar plates.
3. Velvet squares for replica plating may be purchased or can be made by cutting velvet fabric to at least 9 × 9 inches in dimensions. Hemming the edges will make the velvet squares more durable and less prone to unraveling when washing. Velvet squares are reusable. Autoclave used velvet squares and then machine wash in warm water, without any detergent or fabric softener. Tumble dry low without any fabric softener. Autoclave velvets wrapped in foil once more before replica plating.
4. This section has been optimized for cloning promoters from *Arabidopsis thaliana*. In other species we recommend using species-specific databases to determine the correct promoter sequence and using up to 2000 bp of promoter sequence.
5. Length can be varied, also if the nearest gene is closer than 2000 base pairs; select the entire intergenic region.
6. If it is known, it is advisable to include a promoter fragment that has been demonstrated as sufficient to complement a mutant with its respective coding region. If the primers are a poor match, nested PCR can be performed, adding 200 base pairs on either side of the desired promoter region.
7. Lee et al. [6] demonstrated that utilizing 3000 bp of upstream regulatory sequence was sufficient to drive expression of the

associated gene nearly 80% of the time. However, including this length of promoter sequence can increase the amount of time it takes to clone the promoter of interest into the respective pENTR and pMW2 or pMW3 vectors. No change has been observed in *in planta* validation with promoter length of 2000 bp.

8. Use Phusion polymerase because it is a high-fidelity polymerase. If a mutation in a promoter sequence arises in cloning, then it may influence which transcription factors it binds.
9. Primer annealing temperature (55 °C) should be adjusted according to primer composition. Extension time (2 min) should be adjusted according to amplicon length.
10. Make sure dATPs are fresh (do not use after aliquot has been freeze/thawed more than 20 times).
11. Do not use a maxiprep plasmid isolation protocol as this interferes with Gateway recombination reactions.
12. After colony PCR verification, a single sequencing reaction is sufficient to confirm promoter or TF presence in plasmid.
13. It is possible to use undigested/unlinearized plasmid if all enzymes cut the promoter sequence and are thus insufficient for digestion and linear integration. Use 20 μL of undigested plasmid in this case.
14. If the culture is very dense at 2 h, it is likely contaminated.
15. Glass beads are washed with soap, rinsed with dH_20, and soaked in 95% ethanol (kept overnight in a sterile flow hood until ethanol is evaporated). Beads are then autoclaved in a glass jar prior to use.
16. The yeast is plated in this order to ensure that there is enough yeast transferred to the YPDA plate for the LacZ assay and the minimal media dropout plate as a control for transformation success.
17. Transfer -HIS/-URA patch plate to 4 °C for storage until autoactivation assay is completed. Select colonies for glycerol stocks from this plate. 3-AT plates are grown at room temperature because room temperature is similar to Arabidopsis growth conditions; therefore, protein folding should be similar at these conditions.
18. Yeast on LacZ reporter plates cannot be stored at 4 °C and one must proceed with X-gal assays 2 days after diploid selection. Temperature changes will affect protein folding and can affect the results of the X-gal assay. Room temperature is similar to Arabidopsis growth conditions; therefore, we hypothesize that protein folding should be similar at these conditions.

19. Whatman paper will expand when wet. If cut too large, the filters won't sit flat on the plate and would introduce air bubbles underneath, reducing contact of nitrocellulose membrane with the X-gal assay solution.
20. Large petri dishes for X-gal assays may be reused to reduce cost. Dispose of the used nitrocellulose and Whatman filters according to your institution's Environmental Health and Safety Guidelines. The petri dishes can then be washed with dish detergent and warm water, scrubbing with a soft cloth or paper towels. Rinse the petri dishes with deionized water before drying at 37 °C.
21. After a few minutes, get rid of air bubbles under the nitrocellulose filter by gently rolling a glass pipette on its side from the center of the filter and rolling outward.
22. If possible, choose colonies that have slight autoactivation but are not blue before 4 h. These colonies usually perform the best in the Y1H screen.
23. The yeast will sometimes turn red due to depleting adenine in the media agar after 10 days on the plate. This phenotype should not be taken into account for selecting clones. Selection of clones should be based only on the degree of growth on different 3-AT concentrations.
24. Step-down PCR has the best success when performing genotyping PCR.
25. If the gene is over 2000 base pair in length or the PCR reaction will not amplify the target sequence, use targeted cDNA synthesis with specific primers for your gene.
26. Include the pDESTAD-2μ vector with no TF in the TF collection transformation as a control for the Y1H mating assay or in the direct transformation assay. In our experience we prefer this control to be in the H12 position.
27. Yeast on selective dropout media may be stored at 4 °C before proceeding to the next stage. Yeast on YPDA agar cannot be stored any longer than 2 days at room temperature. Any longer and the yeast may lose the episomal plasmid containing the TFs.
28. The media in the Singer plates needs to be poured as evenly as possible and without bubbles. Stack five plates on top of each other and start with the bottom plate, working up the stack. Use about 45–50 mL of media per plate, about halfway up the plate. Make sure that the bottom of the next plate stacks neatly and aligned with the lid of the previous plate to ensure the agar media solidify evenly. Do not make plates too thin as they tend to dry up faster and be uneven and the contact with the robot pad head may not be sufficient for transferring the amount needed of the yeast.

29. Each promoter lawn can be used to mate with approximately eight TF plates. If the screen contains more than eight TF plates, make multiple -HIS/-URA agar lawns for each promoter.
30. The amount of -TRP liquid culture can vary between 300 μL and 1 mL, depending on the amount of TF plates needed. Have a bigger starting volume if there is a need to prepare new glycerol stock plates. Keep the glycerol stock plates on dry ice at all times and do not allow them to melt. To make a glycerol stock, use 50 μL of concentrated yeast culture and 50 μL of 60% glycerol. Include an empty well in position H11 and the control (empty pDESTAD-2μ vector) on position H12 on a 96-well plate. Seal with foil tape.
31. If the plate dries before yeast culture is spread evenly across the entire plate, add more liquid yeast culture to the plate to ensure complete coverage.
32. Agar with yeast growth on it would appear matte. Gaps in the yeast lawn will be shinier and more reflective due to the exposed agar media.
33. Every few weeks you will be prompted to check the oil level and drain fluid in the tank. The oil level can be checked by looking at the back of the compressor. To drain fluid in the tank, attach tubing to spigot on compressor front, place tubing into container to collect fluid, turn drain cock, and let run until fluid stops flowing out. Failure to do this draining of fluids can void warranty if damage is caused.
34. If the speed is too fast and causes liquid foaming, reduce the speed to 10.
35. Target plate orientation: A1 is on right lower corner. Source plate orientation while loading the plates: A1 is on left top corner.
36. A single -TRP agar plate containing TFs can be the source plate for up to eight replicates. If the yeast screen contains more than six promoters, replicate each TF plate twice in Subheading 3.5.5 to ensure yeast will not be depleted during part 1 of yeast mating (Subheading 3.5.6).
37. If screen has a large number of promoters or TFs to mate, screen can be split into 2 days and screen half of the promoters on the first day and half of the promoters on the second day.
38. Plate protection: By protecting the source plates, you can ensure that lids are only removed when it is vital to do so. This will increase the time it takes to run each program, but each source plate will be exposed for less time, and the printhead will never move over a source plate without a lid on, unless it is pinning from it.
39. Alternative options for source during the mating part 2:

(a) Offset—Manual (This will allow you to select location of where the pins touch on the promoter lawn before each source pinning.)

(b) Offset—Random (A random offset is used for each source pinning. Change radius up to around 2 mm.)

40. The same TF plate can be used up to 4 weeks during the screen. Plates must be refreshed once a week on a new -TRP plates. If viability is reduced, start new cultures from the glycerol stocks.

41. Nitrocellulose filters for the YPDA reporter plates are cut and autoclaved prior to reporter plating. Cut the filters using a template (115 mm × 77 mm) that makes the filter fit perfectly on the plates. Cut one corner at an angle to recognize the plate orientation. Due to the static nature of the nitrocellulose filters, the nitrocellulose filters come sandwiched between blue filter papers to reduce the static charge. It is best to cut at most three nitrocellulose filters with the blue filter papers at a time to prevent warping and miscutting of the filters. While cutting the filters, hold the filters as flat as possible to maintain precise cutting with the template. Wrap nitrocellulose filters along with the blue filter paper in foil, autoclave them lying flat with dry cycle, and make sure the filters are dry before using them. Place the filters on the YPDA plates in a sterile flow hood by hand and using a sterile tweezers if needed. Make sure no bubbles are left between the filter and the media surface. Bubbles under the filter reduce yeast growth as the colony cannot get contact with the media and causes it to dry.

42. -HIS/-URA/-TRP plates or diploid selection plates should be kept in 4 °C until reporter assays are completed in case any reporter plating and assays need to be repeated.

References

1. Deplancke B, Dupuy D, Vidal M, Walhout AJM (2004) A gateway-compatible yeast one-hybrid system. Genome Res 14:2093–2102
2. Deplancke B et al (2006) A gene-centered *C. elegans* protein-DNA interaction network. Cell 125:1193–1205. doi:S0092-8674(06)00624-6 [pii] 10.1016/j.cell.2006.04.038
3. Deplancke B, Vermeirssen V, Arda HE, Martinez NJ, Walhout AJM (2006) Gateway-compatible yeast one-hybrid screens. Cold Spring Harb Protoc. doi:10.1101/pdb.prot4590
4. Reece-Hoyes JS et al (2011) Yeast one-hybrid assays for gene-centered human gene regulatory network mapping. Nat Methods 8:1050–1052. doi: nmeth.1764 [pii] 10.1038/nmeth.1764
5. Gaudinier A et al (2011) Enhanced Y1H assays for arabidopsis. Nat Methods 8:1053–1055. doi:10.1038/nmeth.1750
6. Lee JY, Colinas J, Wang JY, Mace D, Ohler U, Benfey PN (2006) Transcriptional and posttranscriptional regulation of transcription factor expression in Arabidopsis roots. Proc Natl Acad Sci 15:6055–6060. doi:10.1073/pnas.0510607103

Chapter 14

Mapping Protein-Protein Interaction Using High-Throughput Yeast 2-Hybrid

Jessica Lopez and M. Shahid Mukhtar

Abstract

A tremendous asset to the analysis of protein-protein interactions is the yeast-2-hybrid (Y2H) method. The Y2H assay is a heterologous system that is expanding network biology knowledge via in vivo investigations of binary protein-protein interactions. Traditionally, the Y2H protocol entails the mating or co-transformation of yeast in solid agar media followed by visual analysis for diploid selection. Having played a key role in identifying protein-protein interactions for nearly three decades in a wide range of biological systems, the Y2H system assays the interaction between two proteins of interest which results in a reconstituted and/or activation of transcription factor allowing a reporter gene to be transcribed. Overall, the Y2H method takes advantage of two factors: (1) the auxotrophic yeast requires expression of the reporter gene to grow in media purposefully designed to lack one or more essential amino acids, and (2) the DNA-binding (DB) domain of transcription factor *GAL4* is unable to initiate transcription unless it is physically associated with an activating domain (AD), which, together, DBs and ADs are fused to proteins of interest that must interact with each other to reconstitute the transcription factor and activate the reporter gene. The applications of Y2H are broad, entailing fields such as drug discovery, clinical trials for human disease including cancer and neurodegenerative disease, and extend even into synthetic biology applications and cellular engineering. This chapter begins with an introduction to the fundamental mechanics of Y2H utilizing a genetically engineered strain of yeast and proceeds with an in-depth look at the different types of Y2H and turn our focus particularly to the *GAL4*-based Y2H system to map protein-protein interactions. We will then provide a step-by-step protocol for the Y2H experimentation preceded by a materials listing while simultaneously including key notes throughout the entire experimental process of biological-mechanistic and historical understandings of the steps.

Key words Yeast-2-hybrid, High-throughput, Protein-protein interactions, Network biology, Systems biology, Reporter gene

1 Introduction

The complexity of life from vertebrates to plants reiterates the need to study gene and protein interactions in a more simplistic model system. In attempts to unravel the many mysteries, researchers have turned to one of the most widely used genetic models to analyze protein interactions—the budding yeast *Saccharomyces*

Wolfgang Busch (ed.), *Plant Genomics: Methods and Protocols*, Methods in Molecular Biology, vol. 1610, DOI 10.1007/978-1-4939-7003-2_14, *cerevisiae* (*S. cerevisiae*), a single-cell organism, that has easily measurable growth rates which provide bountiful insight into fitness while serving a robust, quantitative focus. As messengers, signalers, regulators, defenders, and many more, protein interactions are involved in all cellular processes and are crucial for survival [1, 2]. While numerous studies have committed to elucidating protein-protein interactions, nearly all organisms' genomes still harbor a significant number of proteins with unidentified functions. This poses a deficit for the scientific community as multi-protein complexes, not individual proteins are increasingly being recognized as the molecular basis of cellular processes emphasizing a greater need for systems level understanding [3]. More broadly, the alteration of protein-protein interactions is a leading cause of many diseases; therefore, manipulation of such interactions may be possible to deter and cure diseases [4]. Understanding the entire set of proteins that the genome expresses, referred to as proteomics, is leading to innovations in systems biology by mapping protein interactions within a single species or across multiple species, a popular method for studying interactions.

The Y2H technique utilizing the yeast *Saccharomyces cerevisiae* was originally described by Stanley Fields and Ok-Kyu Song in 1989 [5]. This system allows for the identification of protein interactions and enables researchers to assess the strength of the interaction. Having played a key role in revealing protein interactions since 1989, Y2H will continue to prove a valuable mechanism for elucidating protein-protein interaction networks [5]. Further, it can be automated for high-throughput studies of protein interactions on a genome-wide level, confirming its value to understand systems-level interactions [6].

At the individual protein level, interaction maps have played pivotal roles in revealing and characterizing the structure and function of gene products [7]. At the broad systems level, such interactome webs are facilitating investigations of how biological functions of proteins are carried about by regulatory circuits and global network properties. Currently two high-throughput methods dominate the landscape of identifying protein-protein networks: binary protein-protein interaction assays, which are used to detect pairwise interactions, and the coupling of affinity purification with mass spectrometry to detect stable, co-purifying protein complexes of direct and indirect protein association. Here, we describe binary interaction assays, which have, for nearly 30 years, been centered on the Y2H method [5]. In 2009, a considerable number of studies were published demonstrating the reliability of interactome datasets at the whole proteome level. Provided that all experimental steps are thoroughly executed in accordance with protocols and all necessary controls and quality standards are included, this helps to debunk the myths that high-throughput models produce low-quality information [8, 9].

1.1 Fundamental Principle of Y2H Systems

A fundamental principle for Y2H assays is the interaction between two proteins of interest that result in a reconstituted transcription factor allowing a reporter gene to be transcribed. Eukaryotic transcription factors contain at least two different domains, a DNA domain and a regulatory domain [10, 11] (Fig. 1). A DNA-binding domain is an independently folded protein domain that recognizes DNA via a sequence motif and interacts with nucleotides of the DNA [12, 13]. The regulatory domain serves to activate or repress the transcription from a proteome through coactivators and corepressors, which directly services the transcriptional machinery of the cell [14] (Fig. 1). The two-hybrid technique exploits the fact that the DNA-binding domain of GAL4 is incapable of activating transcription unless physically, but not necessary covalently associated with an activating domain [15].

The Y2H method is often performed with a genetically engineered strain of yeast that has been designed to lack the ability to synthesize specified amino acids or nucleic acids. When yeast is placed on a medium lacking that particular nutrient, it cannot grow. However, Y2H typically takes advantage of plasmid insertions, engineering a unique plasmid that can produce a protein product in which the DB fragment is fused onto a protein, while another

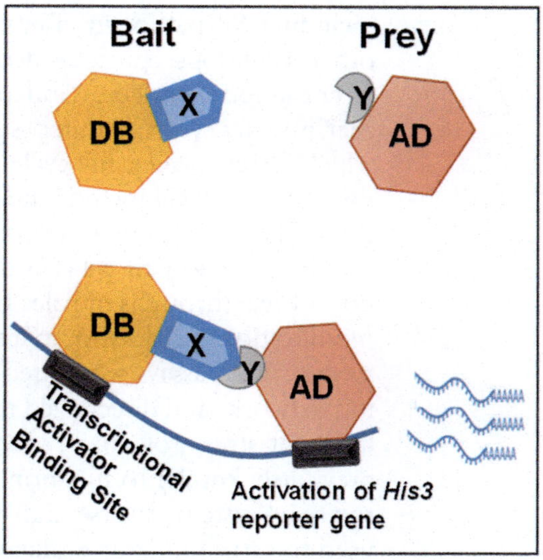

Fig. 1 *GAL4*-based Y2H system. *GAL4* DNA-binding domain (DB) and activating domain (AD) fused with protein X and Y are designated as bait and prey, respectively. The DB binds to the specific transcription factor binding site upstream of the reporter gene. The DB-target protein fusion is called the bait. The AD recruits a transcription initiator complex to the promoter sequence, therefore causing expression of the target reporter gene. The AD-binding partner fusion is called the prey. Overall, the Y2H method takes advantage of two factors: (1) that auxotrophic yeast require expression of the reporter gene to grow in media intentionally created to lack required amino acids, and (2) the DB of transcription factor *GAL4* is unable to initiate transcription unless it is physically associated with an AD

plasmid is engineered to produce a protein product in which the activating domain AD fragment is fused onto another protein. The bait fuses to the DB part of the transcription factor and the prey fuses to the AD, and this movement of the AD and DB coming into proximity of each other reconstitutes a transcription factor and leads to expression of a reporter genes such as histidine (His), LacZ, and GFP (Fig. 1) [10, 16–18]. The GAL4 system-Y2H has played a key role in elucidating protein-protein interactions. One striking example of the value of Y2H is the generation of the network map of *Arabidopsis thaliana* that revealed over 6200 interactions between 2700 plan proteins [3]. With community analyses, researchers were able to more carefully evaluate proteins and map their interactions, linkages they have with other proteins, and how linkages play valuable roles. This study helped pave the way for another landmark Y2H project that systematically mapped binary protein-protein interactions between 552 plant-immune and pathogen proteins and the ~8000 full-length Arabidopsis proteins [3]. This was then used to generate the Arabidopsis Interactome version 1 (AI-1) using the same yeast two-hybrid-based pipeline [3, 19]. From this project, a plant-pathogen immune network was developed utilizing Y2H and bioinformatics procedures for mapping that was found to contain 21,358 interactions among 926 proteins, including 83 pathogen effectors, 170 immune proteins, and 673 other Arabidopsis proteins demonstrating that pathogens from different kingdoms deploy independently evolved virulence proteins that interact with a limited set of highly connected cellular hubs which fosters diverse life cycle strategies [6]. Other examples of the utilization of Y2H include the identification of regulators and effectors of heterotrimeric G-protein subunits in *Arabidopsis* which led to the detection of 544 interactions between 434 proteins and novel breakthroughs in roles of G-proteins, particularly in cell wall modifications [20]. Also, in a study involving the rice plant (*Oryza sativa*), Y2H assays were used to elucidated stress response signaling networks and directly led to the identification of ten novel regulators of stress tolerance, including two from protein classes not previously known to function in stress responses [21]. In addition to GAL4 system, diverse high-throughput Y2H methods have been used to carry out large-scale interactomes (*see* **Note 1**).

The protocol that we outline here is a well-established method for carrying out Y2H and was the basis for Dreze et al. [19]. We also note that there are numerous other reviews and method papers that report in great detail the different types of available strains, plasmids, and experimental procedures for achieving results [22, 23]. Additionally, many companies offer kits for Y2H experiments including Dualsystems Biotech (DUALhybrid kit), Invitrogen (ProQuest), Clontech (Matchmaker Gold), Agilent Technologies (HybriZAP), and Promega (CheckMate). Finally, two-hybrid services are available from companies like Hybrigenics and Dualsystems Biotech [24–27].

2 Materials

Prepare all solution using deioninzed water, and store reagents at room temperature. Follow waste disposal regulations when disposing of yeast and other reagents.

2.1 Transformation of Yeast

1. Costar® 96-well round-bottom culture plates (Corning).
2. Yeast strains: Y8800-MATa and Y8930-MATα strains.
3. Tris-EDTA-lithium acetate (TE/LiAc).
4. TE/LiAc polyethyleneglycol (PEG).
5. Multichannel pipette.
6. Sterile pipette tips.
7. SD-Leu dropout mix (Clontech).
8. SD-Trp dropout mix (Clontech).
9. Sterile 45 ml tube.
10. Sterile 30% glycerol.
11. YEPD media: 20 g of yeast etract, 40 g of bacto-peptone, water, 65 mM adenine solution per liter of medium.
12. 96-well Thermowell Sealing Tape (Corning).
13. Incubator.
14. Spectrophotometer with capacity to read $OD_{600.}$
15. Shaker with capacity to perform 120 rpm shaking.
16. Laboratory micro-centrifuge.
17. Carrier DNA (salmon sperm DNA, Sigma-D9156).
18. Sterile toothpicks.

2.2 Identification of Autoactivators and Y2H Primary Screening

1. YEPD media: 20 g of yeast extract, 40 g of bacto-peptone, water, 65 mM adenine solution per liter of medium.
2. SD-Leu dropout mix (Clontech).
3. SD-Trp dropout mix (Clontech).
4. SD-Leu-Trp dropout mix (Clontech).
5. SD-Leu-Trp-His dropout mix (Clontech).
6. 1 mM of 3-amino-1,2,4-triazole (3AT) (*see* **Note 2**).
7. SD-Leu-His.
8. 1 mg/l cycloheximide (CHX) solution (*see* **Note 2**).

3 Methods

Carry out all procedures at room temperature unless otherwise noted.

3.1 Transformation of Yeast

Following the cloning of the DB-X and AD-Y expression plasmids, the yeast should be transformed. This step is where the DB-X and AD-Y expression plasmid constructs are individually transformed into competent Y8930 (MATα) and Y8800 (MATa) strains, respectively. To maximize the efficiency of the transfomations, two freshly prepared solutions of Tris-EDTA-lithium acetate (TE/LiAc) and TE/LiAc polyethyleneglycol (PEG) solutions should be prepared. The following protocol is optimized for 96 individual transformations in 96-well plate. Use extra reagents as necessary.

1. Streak Y8800 and Y8930 on separate YEPD plates, and incubate at 30 °C for 24–48 h to isolate colonies.
2. Inoculate 20 ml of YEPD with a single colony for each strain, and incubate at 30 °C on a shaker for 14–18 h.
3. Measure the OD_{600}, which should be around 5.0. Dilute these cells into 100 ml YEPD media to obtain a final OD_{600} of 0.1. Incubate for 4–6 h at 30 °C on a shaker at 120 rpm to obtain the optimum OD_{600} of 0.6.
4. Centrifuge at 800 × *g* for 5 min. Decant the supernatant and resuspend cells gently in 10 ml of sterile water.
5. Centrifuge cells again as described above in **step 4**, resuspend cells in 10 ml of TE/LiAc solution, centrifuge and discard the supernatant.
6. Resuspend the cells with 2 ml of TE/LiAc, 10 ml of TE/LiAc/PEG, and 200 μl of parboiled carrier DNA. Aliquote 120 μl of these competent cells into a 96-well round-bottom culture plate.
7. Add 10 μl of plasmid DNA (1 μg/μl) to the competent yeast, and mix by pipetting up and down several times. Seal the plates with adhesive foil and incubate for 30 min at 30 °C.
8. After 30 min, centrifuge the 96-well round-bottom culture plate for 5 min at 800 × *g*, and remove the supernatant carefully with a multichannel pipette.
9. Add 100 μl of sterile water and resuspend the cell pellets by pipetting up and down.
10. Centrifuge the 96-well round-bottom culture plate for 5 min at 800 × *g*, and then carefully remove 90 ml of water from each well using a multichannel pipette.
11. Resuspend the cell pellets by vortexing the 96-well plate on a shaker at 120 rpm.
12. Spot 5 μl/well of cell suspension onto an appropriate selective plate (SD-Trp for AD-Y, SD-Leu for DB-X). Incubate at 30 °C for 72 h.
13. Using a sterile toothpick, pick the transformed yeast colonies and (optional) create archived stocks by combining 80 μl of yeast culture with 80 μl of 40% autoclaved glycerol in 96-well round-bottom culture plate and store at −80 °C (*see* **Note 3**).

3.2 Identification and Removal of Autoactivators

1. Prepare one YEPD plate, one SD-Leu-Trp plate, and one SD-Leu-Trp-His +1 mM 3AT plate for each of the 96-well plates. It's helpful to prepare the YEPD plates considerably in advance so they have time to dry.
2. Add 160 μl of SD-Leu media to all 96 wells in a 96-well round-bottom culture plate, and then add 5 μl of glycerol stocks of each DB-X yeast strain to each well.
3. For each plate of DB-X yeast strains that will be tested, inoculate a tube containing 0.55 ml of SD-Trp media with Y8800 transformed with empty pDEST-AD.
4. Incubate at 30 °C for 14–18 h on a shaker.
5. Very precisely, spot 5 μl of DB-X liquid cultures onto a YEPD plate, being careful to not allow droplets to run together, and allow the spots to dry for 30–60 min.
6. Aliquot the pDEST-AD transformed Y8800 yeast culture into a 96-well round-bottom culture plate.
7. Very precisely, spot 5 μl of pDEST-AD transformed Y8800 on top of the already-spotted DB-X spots. Again, use caution to ensure droplets do not run together.
8. Spotting Y2H controls at the bottom of the plate is optional, but highly recommended for visual comparisons after the incubation time.
9. Incubate the mating plates at 30 °C for 14–18 h.
10. Replica-plate onto SD-Leu-Trp media to select for diploid cells.
11. Incubate at 30 °C for 14–18 h.
12. Replica-plate from the SD-Leu-Trp media onto SD-Leu-Trp-His +1 mM 3AT media. Yeast cells that are not autoactive will be prevented from activating the GAL1::HIS3 reporter gene, so no growth should be spotted on this particular media.
13. Incubate at 30 °C for 14–18 h, and proceed directly to steps outlined in Subheading 3.3.

3.3 Y2H Pairwise Screening

These processes can be performed by hand using a multichannel pipette with considerable attention to detail and accuracy or by using a liquid-handling robot (*see* **Note 4**).

3.3.1 Inoculation

1. Thaw the glycerol stocks of DB-X and AD-Y strains to inoculate 96-well round-bottom culture plate that contain a selective media.
2. Add the 5 μl each of DB and AD strains to 160 μl of SD-Leu and SD-Trp media, respectively. Seal the plates, and incubate at 30 °C on a shaker (approximately 120 rpm) for 24 h.

3.3.2 Mating

1. Combine 5–7 μl of both AD and DB yeast strains, as described in Subheading 3.3.1 on top of each other onto a YEPD agar plate. This can be achieved by aliquoting 5–7 μl of the AD strain and carefully pipetting the same amount of the DB strain precisely on top, creating a larger droplet.
2. Incubate at 30 °C, for 24 h.

3.4 Diploid Selection and Positive Interactions

1. Prepare a plate of media SD-Leu-Trp also termed diploid selection plates.
2. Aliquot 5–7 μl of culture from the mating plate that incubated for 24 h as described in Subheading 3.3.2.
3. Incubate at 30 °C for 14–18 h.
4. Replica-plate yeast from these mating plates onto SD-Leu-Trp-His +1 mM 3AT (positive selection plate) and SD-Leu-His +1 mM 3AT + 1 mg/l CHX (de novo autoactivation detection plate) (*see* **Note 2**).
5. Incubate at 30 °C for 14–18 h, and proceed directly to 3.4

3.5 Phenotyping and Selection for Positive Interactions (Scoring)

1. After 48–72 h of growth on their respective plates, visual scoring should be conducted independently by at least two observers who visually inspect the plates by removing the lid under a laminar flow hood and carefully examining the yeast growth (*see* **Note 2**). Growth will be evident to the naked eye without any need for magnification.
2. Visually score each SD-Leu-Trp-His +1 mM 3AT (positive selection plate), and compare the yeast growth with the corresponding SD-Leu-His +1 mM 3AT + 1 mg/l CHX (de novo autoactivation detection plate) (Fig. 2).
3. Mark the interaction as "positive," if yeast growth is observed on positive selection plate but not on de novo autoactivation detection plate (Fig. 2).
4. Following the scoring of yeast, the method of Y2H is complete; however, numerous retests are necessary for verification. Multiple effective ways for retesting Y2H interactions are outlined in **Note 5**.

4 Notes

1. The GAL4 system is just one example of a type of Y2H system. Another very commonly used system is known as the LexA system, in which the DB used is the bacteria protein LexA that expresses an acidic peptide, which, upon fusing with the DNA-binding domain, acts as a transcriptional activator in the yeast [15, 28]. The LexA system utilizes the *S. cerevisiae* L40 reporter strain, which contains two integrated reporters, the

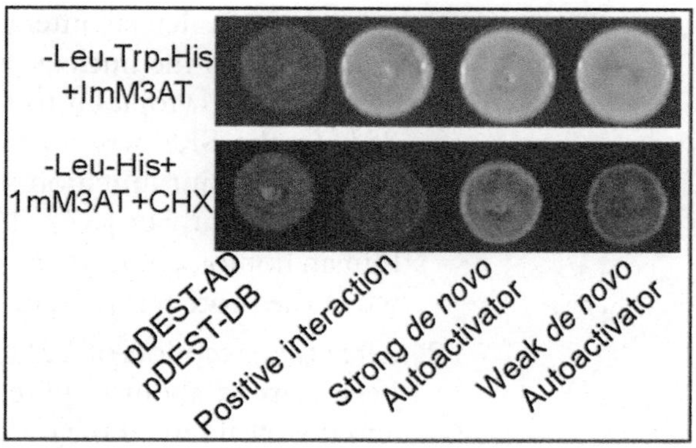

Fig. 2 Detection of interaction patterns using Y2H system. Four yeast strains expressing baits and preys are spotted onto SD-Leu-Trp-His +1 mM 3AT (positive selection) and SD-Leu-His +1 mM 3AT + 1 mg/l CHX (de novo autoactivation detection) media. pDEST-AD and pDEST-DB expressing yeast strains were used as negative control. Positive interaction can be identified based on the growth of yeast on positive selction media but not onto de novo autoactivation detection. Two de novo autoactivation pairs with different quantities of yeast growth are shown. Increasing dose of 3AT might eliminate the growth of yeast on de novo autoactivation plates

bacterial *lacZ* gene and the yeast *HIS3* gene, which are both controlled by the LexA operators [29, 30]. Another novel reporter system that has been created for Y2H assays utilizes an enhanced green fluorescent protein (EGFP)-based reporter in order to characterize interactions. This system represents the first application of the split EGFP system as a readout in a yeast two-hybrid assay [31]. Upon reconstitution, the EGFP has been observed to evidence a restored fluorescence upon protein-protein interactions in yeast, allowing for the characterization of its interactor.

With standard Y2H assays, detection of an interaction requires that two proteins localize to the nucleus [32, 33]. In 2007, a yeast-based genetic technology for the in vivo detection of membrane protein interactors was developed and is referred to as the split-ubiquitin membrane yeast two-hybrid system. Here, one integral membrane protein (the bait) is fused at its N- or C-terminus to the carboxy-terminal half of ubiquitin (Cub) and a LexA-VP16 transcription factor (TF). A second protein, either integral membrane or cytosolic (the prey), is fused at either its N- or C-terminus to the amino-terminal half of ubiquitin bearing an Ile13Gly mutation (NubG-Prey or Prey-NubG) [34, 35]. If the bait and prey interact, quasi-native ubiquitin is reconstituted, which is a target of ubiquitin-specific proteases that cleave at the C-terminus of the Cub. This cleavage releases the TF, allowing it to enter the nucleus and activate transcription of the HIS3 and lacZ reporter genes [34, 36]. Furthermore, the Sis recruitment system (SRS) detects interactions occurring in the cytoplasm as opposed to the nucleus, allowing proteins from virus replication in the cytoplasm to be tested in a natural context [37, 38]. In the SRS system, a strain of *S.cerevisiae* is used that contains a point mutation in the CDC25 gene, causing temperature sensitivity [37]. This defect is complemented by a human homolog known as son of sevenless homolog 1 (Sos1), when the protein is recruited to the cell membrane [39].

2. After the inception of Y2H, over the last 20 years, much has been learned about requirements of the appropriate controls and the reliability of the outcomes. Today, Y2H can be considered a very efficient, reliable binary interaction assay for interactome mapping from small to large scales. As the quality of any dataset can be adversely affected by false negatives and/or false positives, the same holds true for Y2H. False negatives are defined as true interactions that are missing from the screen. The stochastic false-negative rate for an entire screen may be estimated as $\sum_i[\hat{k}_i - (w_i - \hat{f}_i)]/\sum_i\hat{k}_i$, where the numerator represents the total number of true interactions minus the observed interactions and the denominator represents the total number of true-positive interactions [40, 41]. In contrast, false positives are Y2H strains that are transformed with plasmids that should not cause reconstitution of the transcription factor, but somehow still express the reporter protein. False positives can occur in yeast strains transformed with plasmids with only a bait or a prey protein fused to the AD and DB or in yeast strains transformed with plasmids expressing noninteracting proteins fused to the AD and DB [8, 42]. Currently, there is no way to eliminate the false positives entirely, but it is controlled for by adding 3AT. Both false negatives and false positives can impede the generation of a network map [8, 42]. Despite the issues of false negatives and false positives, Y2H can be considered a very efficient, reliable method of achieving binary interaction assays available for interaction method at small or large levels. In an effort to combat the false positives and false negatives by minimizing risks, researchers tend to promote a rigorous implementation of procedures that can help minimize, detect, and eliminate false negatives and false positives. Below, we define steps that can be taken to help implement a high-quality Y2H to ensure high-quality data.

   Using low copy number yeast expression vectors, in conjunction with weak promoters, expressing DB-X and AD-Y hybrid proteins can help minimize artifacts such as false positive and false negatives by mass action [19]. While high copy number vectors can increase DB-X and AD-Y protein expression, thereby strengthening the sensitivity of the assay, the negative side is that it increases the detection of unspecific interactions [8]. As it pertains to yeast strains, the following two strains are highly published, highly cited strains as effective for Y2H assays: Y8800-MATα and Y8930- MATα strains [43]. These two strains are derived from the PJ69–4 which have the genotype *leu2–3112 trp1–901 his3–200 ura3–52 gal4D gal80D GAL2-ADE2 LYS2::GAL1-HIS3 MET2::GAL7-lacZ cyh2R*. These two haploid strains of opposite mating types enable the use of mating to efficiently combine large collections of DB-X and AD-Y constructs. The Y8800-MATα and Y8930-MATα strains are transformed with AD-Y and DB-X constructs, respectively. Utilizing a yeast strain that is highly reported as a valuable role-player in Y2H experimentation in conjunction with low DB-X and AD-Y hybrid protein expression aids in minimizing distractors like false negatives and false positives. Additionally, to help determine which candidate clones likely represent genuine biophysical interactors, it is recommended to add controls to the master plates of Y2H experiments. A collection of diploid control strains, controls mirroring those of Dreze et al., that are established negative vectors or positive controls can aid in proper classification of an interaction type [19]. In Y2H mating experiments, autoactivation of a Y2H-inducible reporter gene is common. Autoactivation occurs when transcription in a Y2H gene is initiated exclusively by DB, without assistance from AD [15]. Autoactivation can cause serious experimental problems in cases where an entire DNA library is fused to the DNA-binding domain, in turn stimulating a considerable amount of false-positive results, as libraries, or fragments, are more prone to latent activating activity [15, 44]. The issue of false-positive autoactivation by the bait protein can be resolved by removing the domain that induces the autoactivation or by increasing 3AT levels when using HIS3, a protein-encoding gene in *Saccharomyces cerevisiae*, as a marker [22, 45]. 3AT lowers the threshold for a visible growth result. Because some bait constructs produce extensive background growth independent of the prey used, an interacting pair may be difficult to visually detect. 3AT can help to reduce such background growth. However, excessive amounts, over 50 mM, of 3AT can lead to the abolishment of growth. We use 1 mM concentration of 3AT.

As discussed previously, autoactivators often present themselves, so in an effort to maintain rigorous preventative measures, the identification of autoactivators can be achieved in the diploid yeast strains obtained by mating DB-X yeast strains with the Y8800 yeast strain transformed with the AD encoding plasmid containing no insert (empty pDEST-AD). All diploid yeast strains showing a growth phenotype (stronger than a "noninteracting" control) are considered autoactivators. This reiterates the importance of establishing controls that can serve as a visual reference in helping to detect growth from potential autoactivators. After following select protocols for removing autoactivators, the Y2H pipeline extends into critical mating stages that fulfill the ultimate purpose of this method.

Selective media are used for maintaining the AD-Y and DB-X plasmids and detection of reporter activity. Prototrophic markers are used for selection on plates that lack the appropriate amino acid. In the traditional system aligning with Dreze et al. that is defined here, the DB-expressing plasmid contains the selectable marker LEU2 which enables growth of the Y8800/Y8930 yeast strains on plates lacking leucine, while the AD-expressing plasmid contains the TRP1 marker which enables growth on plates lacking tryptophan. The other two prototrophic markers (HIS3 and ADE2) are used as reporter genes. These markers will be expressed on plates that lack histidine that also will have added 3AT, a competitive inhibitor of the product of the *HIS3* gene. In contrast, the plates used to detect for autoactivators, are often termed negative selection plates. Autoactivators are detected by yeast growth on SD-Leu-His containing 1 mg/L of cyclohexamide (CHX) containing media. CHX is an inhibitor of protein synthesis and acts on the product of the gene *CYH2*, which is located on the AD plasmid. When CHX is present, the AD vector cannot yield the viable protein complex necessary to bind with the DB. Therefore, if reporter gene expression occurs, it can be attributed to autoactivation since the AD vector is not transcribed. On the solid agar plates, autoactivators can be detected from visually inspecting the agar plates of SD-Leu-His media type.

3. The selection of plasmids in yeast results from the auxotrophic mutant strains that cannot grow without a particular component to the media. The result of the transformation with a mutated gene plasmid allows the transformant to grow on the medium lacking the required component. Hence, the dropout medium supplements select for transformed yeast cultures.

4. In parallel with the evolution of technical aspects of the Y2H system, several additional strategies have been developed and optimized for screening experiments. Based on the research objectives and availability of the materials, genomic DNA, cDNA, normalized cDNA, full-length cDNA, or open reading frame (ORF) libraries can be utilized for generating Y2H libraries. While originally Y2H screening involved the simultaneous transformation of a two-hybrid strain with the bait plasmid and a prey library, an alternative strategy consists of the construction of mating type a and α two-hybrid strains with the bait and prey plasmids, respectively [36, 40]. This method of using mating type a and α is still very commonplace in Y2H experimentation today. In this case, the screening step is performed by mating both strains on a medium that is selective for an interaction.

5. Additionally, complementary assays can help to assess precision. One validation assay is the yellow fluorescent protein (YFP)-based protein complementation assay in which two non-fluorescent fragments of YFP are attached to ORFs that code for the two proteins under investigation. If the two proteins interact, functional YFP can be reconstituted and detected by fluorescence-activated cell sorting. Another validation assay is the nucleic acid programmable protein array (NAPPA) [46]. In this assay, the two proteins under investigation are fused to a glutathione S-transferase (GST) tag and to an HA epitope tag, respectively, and expressed in a coupled transcription/translation reticulocyte lysate. The GST-tagged protein is captured using an anti-GST antibody that is immobilized at the bottom of a 96-well microtiter plate. If the two proteins do interact, the interaction can be detected with an anti-HA antibody.

Acknowledgments

This work was supported by the National Science Foundation (IOS-1557796) to MSM.

References

1. Mao L et al (2009) Arabidopsis gene co-expression network and its functional modules. BMC Bioinformatics 10:346
2. Mochida K, Shinozaki K (2011) Advances in omics and bioinformatics tools for systems analyses of plant functions. Plant Cell Physiol 52(12):2017–2038
3. Arabidopsis Interactome Mapping, C (2011) Evidence for network evolution in an Arabidopsis interactome map. Science 333(6042):601–607
4. Gonzalez MW, Kann MG (2012) Chapter 4: protein interactions and disease. PLoS Comput Biol 8(12):e1002819
5. Fields S, Song O (1989) A novel genetic system to detect protein-protein interactions. Nature 340(6230):245–246
6. Mukhtar MS et al (2011) Independently evolved virulence effectors converge onto hubs in a plant immune system network. Science 333(6042):596–601
7. Oliver S (2000) Guilt-by-association goes global. Nature 403(6770):601–603
8. Braun P et al (2009) An experimentally derived confidence score for binary protein-protein interactions. Nat Methods 6(1):91–97
9. Simonis N et al (2009) Empirically controlled mapping of the *Caenorhabditis elegans* protein-protein interactome network. Nat Methods 6(1):47–54
10. Bruckner A et al (2009) Yeast two-hybrid, a powerful tool for systems biology. Int J Mol Sci 10(6):2763–2788
11. Verschure PJ, Visser AE, Rots MG (2006) Step out of the groove: epigenetic gene control systems and engineered transcription factors. Adv Genet 56:163–204
12. Lilley DM (1992) DNA--protein interactions. HMG has DNA wrapped up. Nature 357(6376):282–283
13. Ptashne M, Gann A (1997) Transcriptional activation by recruitment. Nature 386(6625):569–577
14. Warnmark A et al (2003) Activation functions 1 and 2 of nuclear receptors: molecular strategies for transcriptional activation. Mol Endocrinol 17(10):1901–1909
15. Van Criekinge W, Beyaert R (1999) Yeast two-hybrid: state of the art. Biol Proced Online 2:1–38
16. Dai W et al (2013) Establishment of recombinant lac Z reporter gene-transformed yeast cells for bioassay of androgen-like compounds in environment. Xi Bao Yu Fen Zi Mian Yi Xue Za Zhi 29(3):303–309
17. Liang ST, Dennis PP, Bremer H (1998) Expression of lacZ from the promoter of the *Escherichia coli* spc operon cloned into vectors carrying the W205 trp-lac fusion. J Bacteriol 180(23):6090–6100

18. Whitaker RD, Walt DR (2007) Fiber-based single cell analysis of reporter gene expression in yeast two-hybrid systems. Anal Biochem 360(1):63–74
19. Dreze M et al (2010) High-quality binary interactome mapping. Methods Enzymol 470:281–315
20. Klopffleisch K et al (2011) Arabidopsis G-protein interactome reveals connections to cell wall carbohydrates and morphogenesis. Mol Syst Biol 7:532
21. Seo YS et al (2011) Towards establishment of a rice stress response interactome. PLoS Genet 7(4):e1002020
22. Stynen B et al (2012) Diversity in genetic in vivo methods for protein-protein interaction studies: from the yeast two-hybrid system to the mammalian split-luciferase system. Microbiol Mol Biol Rev 76(2):331–382
23. Yu H et al (2011) Next-generation sequencing to generate interactome datasets. Nat Methods 8(6):478–480
24. Clontech, Matchmaker Clontech, Matchmaker® Gold Yeast Two-Hybrid System
25. Gangloff S et al (1994) The yeast type I topoisomerase Top3 interacts with Sgs1, a DNA helicase homolog: a potential eukaryotic reverse gyrase. Mol Cell Biol 14(12):8391–8398
26. Inc, B., Biocompare, HybriZAP®-2.1 XR cDNA Synthesis Kit from Agilent Technologies
27. Inc, P., promega, CheckMate™ Mammalian Two-Hybrid System Protocol
28. Brent R, Ptashne M (1985) A eukaryotic transcriptional activator bearing the DNA specificity of a prokaryotic repressor. Cell 43(3 Pt 2):729–736
29. Estojak J, Brent R, Golemis EA (1995) Correlation of two-hybrid affinity data with in vitro measurements. Mol Cell Biol 15(10):5820–5829
30. Fashena SJ, Serebriiskii IG, Golemis EA (2000) LexA-based two-hybrid systems. Methods Enzymol 328:14–26
31. Park K et al (2007) A split enhanced green fluorescent protein-based reporter in yeast two-hybrid system. Protein J 26(2):107–116
32. Ito T et al (2001) A comprehensive two-hybrid analysis to explore the yeast protein interactome. Proc Natl Acad Sci U S A 98(8):4569–4574
33. Ito T et al (2000) Toward a protein-protein interaction map of the budding yeast: a comprehensive system to examine two-hybrid interactions in all possible combinations between the yeast proteins. Proc Natl Acad Sci U S A 97(3):1143–1147
34. Paumi CM et al (2007) Mapping protein-protein interactions for the yeast ABC transporter Ycf1p by integrated split-ubiquitin membrane yeast two-hybrid analysis. Mol Cell 26(1):15–25
35. Stagljar I et al (1998) A genetic system based on split-ubiquitin for the analysis of interactions between membrane proteins in vivo. Proc Natl Acad Sci U S A 95(9):5187–5192
36. Suter B, Auerbach D, Stagljar I (2006) Yeast-based functional genomics and proteomics technologies: the first 15 years and beyond. Biotechniques 40(5):625–644
37. Aronheim A et al (1997) Isolation of an AP-1 repressor by a novel method for detecting protein-protein interactions. Mol Cell Biol 17(6):3094–3102
38. Moerdyk-Schauwecker M et al (2011) Detecting protein-protein interactions in vesicular stomatitis virus using a cytoplasmic yeast two hybrid system. J Virol Methods 173(2):203–212
39. Simon MA et al (1991) Ras1 and a putative guanine nucleotide exchange factor perform crucial steps in signaling by the sevenless protein tyrosine kinase. Cell 67(4):701–716
40. Huang H, Jedynak BM, Bader JS (2007) Where have all the interactions gone? Estimating the coverage of two-hybrid protein interaction maps. PLoS Comput Biol 3(11):e214
41. Stumpf MP, Wiuf C, May RM (2005) Subnets of scale-free networks are not scale-free: sampling properties of networks. Proc Natl Acad Sci U S A 102(12):4221–4224
42. Hengen PN (1997) False positives from the yeast two-hybrid system. Trends Biochem Sci 22(1):33–34
43. Vidal M et al (1996) Reverse two-hybrid and one-hybrid systems to detect dissociation of protein-protein and DNA-protein interactions. Proc Natl Acad Sci U S A 93(19): 10315–10320
44. Beranger F et al (1997) Getting more from the two-hybrid system: N-terminal fusions to LexA are efficient and sensitive baits for two-hybrid studies. Nucleic Acids Res 25(10):2035–2036
45. Struhl K, Davis RW (1977) Production of a functional eukaryotic enzyme in *Escherichia coli*: cloning and expression of the yeast structural gene for imidazole-glycerolphosphate dehydratase (his3). Proc Natl Acad Sci U S A 74(12):5255–5259
46. Miersch S, LaBaer J (2011) Nucleic Acid programmable protein arrays: versatile tools for array-based functional protein studies. Curr Protoc Protein Sci Chapter 27:Unit27 2

Chapter 15

Mapping Protein–Protein Interactions Using Affinity Purification and Mass Spectrometry

Chin-Mei Lee, Christopher Adamchek, Ann Feke, Dmitri A. Nusinow, and Joshua M. Gendron

Abstract

The mapping of protein–protein interaction (PPI) networks and their dynamics are crucial steps to deciphering the function of a protein and its role in cellular pathways, making it critical to have comprehensive knowledge of a protein's interactome. Advances in affinity purification and mass spectrometry technology (AP-MS) have provided a powerful and unbiased method to capture higher-order protein complexes and decipher dynamic PPIs. However, the unbiased calling of nonspecific interactions and the ability to detect transient interactions remains challenging when using AP-MS, thereby hampering the detection of biologically meaningful complexes. Additionally, there are plant-specific challenges with AP-MS, such as a lack of protein-specific antibodies, which must be overcome to successfully identify PPIs. Here we discuss and describe a protocol designed to bypass the traditional challenges of AP-MS and provide a roadmap to identify bona fide PPIs in plants.

Key words Affinity purification, Mass spectrometry, Binding affinity, Engineered bait protein, Protein–protein interaction, Interactomes, Protein complex

1 Introduction

Proteins interface with each other to form complexes that dramatically increase their potential to perform biological functions (Fig. 1). Deciphering protein–protein interaction (PPI) networks and their dynamics is not only pivotal to understanding protein functions but also the basic framework of biological pathways in cells [1–3]. PPIs occur on protein interfaces and display a wide range of affinities [4]. High affinity interactions are driven by lower dissociation rates and result in longer dwell times, while low affinity interactions have higher dissociation rates and shorter dwell times [5]. Both are biologically relevant, but low affinity interactions can be difficult to detect due to rapid dissociation, which makes systematic mapping of comprehensive PPI networks difficult [6].

Wolfgang Busch (ed.), *Plant Genomics: Methods and Protocols*, Methods in Molecular Biology, vol. 1610, DOI 10.1007/978-1-4939-7003-2_15, © Springer Science+Business Media LLC 2017

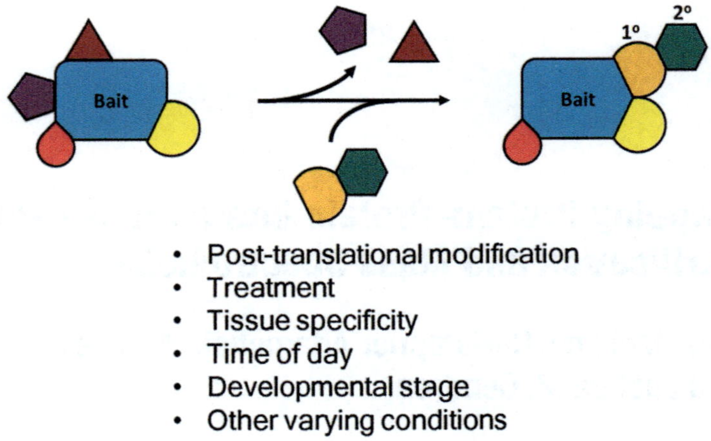

Fig. 1 Dynamics of protein–protein interactions are biologically relevant. The composition of protein complexes of the protein of interest or bait is dynamic depending on the conditions (post-translational modification, time of day, tissue specificity, etc.) which represent different biological events. Therefore, it is critical to determine the biological questions, experimental designs, and sample preparation for successful AP-MS experiments

Many methodologies have been developed to detect PPIs [1, 7–9]. Some of the most widely used techniques to test PPIs between plant proteins include yeast two-hybrid (Y2H) and its variants [10, 11], in vitro binding assays [12], bimolecular fluorescence complementation (BiFC) [13, 14], fluorescence resonance energy transfer (FRET) [15], and co-immunoprecipitation followed by western blot analyses [7, 12]. Each of these techniques has significant and unique strengths for detecting PPIs, but none can currently provide an unbiased and comprehensive analysis of a dynamic PPI network [1, 7]. Therefore, AP-MS retains the distinct advantages in that it can identify higher-order protein complexes, capture spatially and temporally dynamic complexes, and permit the experimenter to search for possible posttranslational modifications that may be critical for biological functions [16–18]. Another benefit of AP-MS is that it can identify indirect interactors and will provide an unbiased analysis of the entire protein complex (Fig. 2). This, however, can also be a significant hurdle when determining which of the identified proteins directly interact with your proteins of interest (POIs) or baits. Thus, any putative interactors must be further investigated using aforementioned techniques such as in vitro binding, Y2H, or BiFC to determine that an interaction is direct between two proteins. The rapid evolution of mass spectrometers for detecting amino acid peptides [19–21], the emergence of free or inexpensive mass spectrometry and AP-MS analysis software [22–25], and the ability to modify POIs with a variety of affinity tags that work well in plants make AP-MS a viable and inexpensive experimental option for virtually any laboratory interested in deciphering the PPI networks of their favorite protein [26].

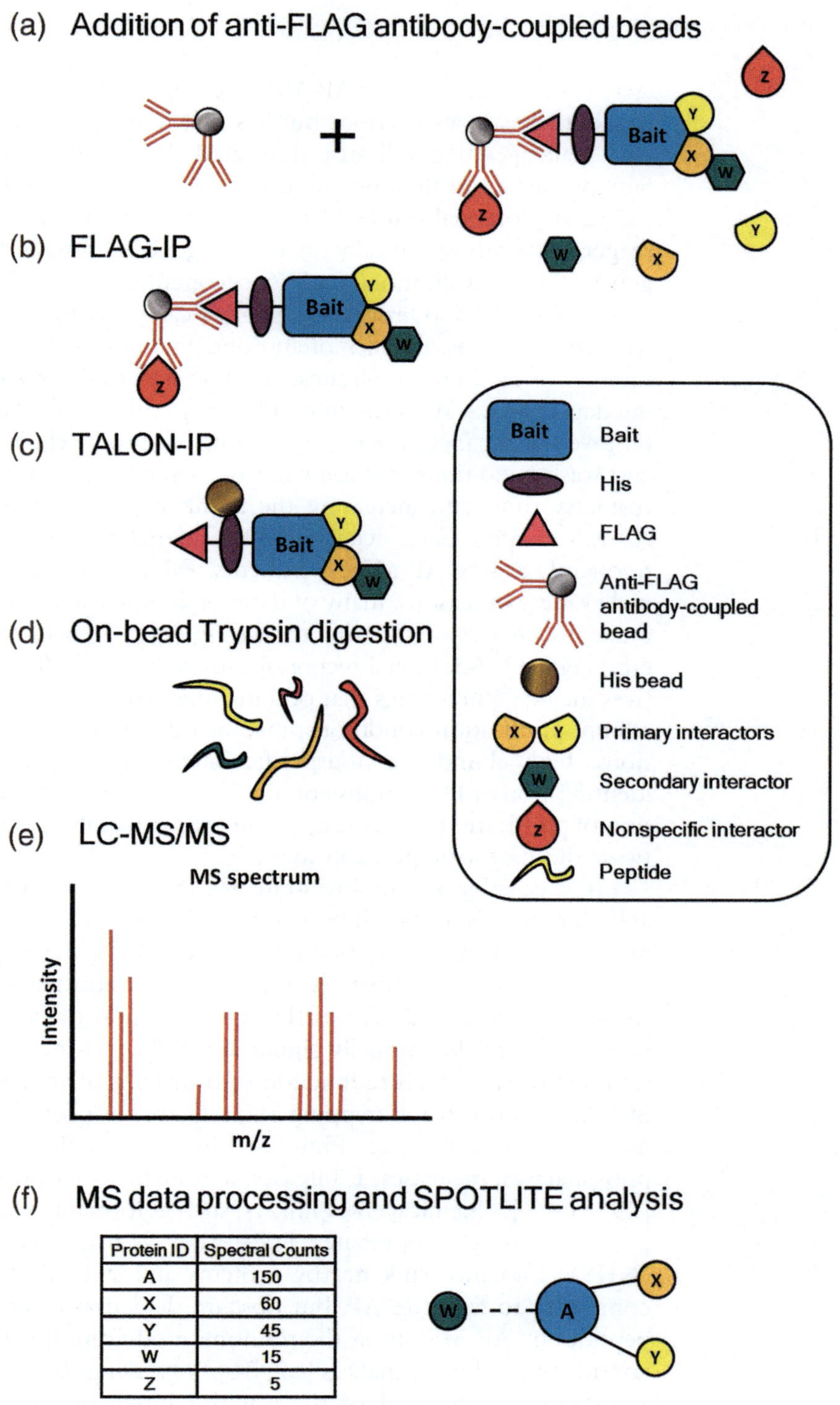

Fig. 2 Overview of the AP-MS procedure. Protein extracts from wildtype or bait-expressing plants are first immunoprecipitated with anti-FLAG antibody coupled to protein G magnetic beads (**a**, **b**). After elution from the beads, eluates are incubated with TALON magnetic beads to immunoprecipitate His-tagged protein complexes (**c**). The protein complexes are subjected to on-bead trypsin digestion (**d**), and LC-MS/MS is performed on the resultant peptides to give spectra of mass-to-charge ratios (*m*/*z*) (**e**). Raw MS spectrum data are processed with MASCOT or other interpretation software to identify proteins and spectral counts. Protein lists and spectral counts are consolidated to generate an input file (Table 1) for SPOTLITE analysis to find high confidence primary and secondary interactors (**f**)

While the strengths of AP-MS encourage wider use of the technique, there are also various hurdles that must be overcome. The two challenges that will arise during AP-MS are the detection of false positives and the discarding of true positives (false negatives) [27, 28]. Potential causes of false positives are the use of affinity reagents that nonspecifically retain endogenous proteins, low stringency affinity purification conditions, contamination arising during tissue collection and processing, peptide carryover from mass spectrometry columns, the lack of appropriate controls, an insufficient number of biological replicates, and low statistical thresholds during data analyses. With this in mind, false positives can be effectively removed by the inclusion of proper controls, proper choice of resins and wash conditions, randomizing the samples during mass spectrometry runs, and increasing the number of biological replicate AP-MS samples. False negatives, while harder to track, also occur frequently during AP-MS experiments. False negatives occur for a wide variety of reasons, many of them biological (i.e., inappropriate tissue, tissue harvested during the wrong time of day or conditions, etc.) (Fig. 1). Additional technical causes that underlie false negatives include affinity tags that occlude interactions, overly stringent affinity purification conditions, incompatibility with buffer conditions, high abundance nonspecific interactors that obscure MS identification of POIs, transient interactions compared to the duration of purification procedure, and improper handling of biological tissue during the preparation steps.

It is possible to employ strategies that increase the strength and duration of PPIs while maintaining a reduced false-positive rate. Choosing the appropriate time, developmental stage, treatment condition, and tissue is of paramount importance (Fig. 1). Modifying protein structures through mutation or domain deletion can reveal functionally significant PPIs and increase understanding of the structure function relationships of your POIs [29, 30]. Mechanism-based trapping of interactors by protein engineering can increase binding affinity and has been applied to kinases, phosphatases, proteases, E3 ligases, and many other protein types [29, 31–33]. Chemical cross-linkers, such as formaldehyde, dithiobis succinimidyl propionate (DSP), and *N*-hydroxysuccinimide (NHS), covalently link nearby proteins and stabilize interaction complexes to facilitate AP, but obscure detection of cross-linked peptides by AP-MS unless the resulting modification is taken into consideration during analysis [27, 34]. Increasing the cellular concentrations of the POI or the putative interacting partners can result in more protein captured, but will likely increase the number of false positives identified and potentially obscure cellular dynamics [35]. Constitutive overexpression is a common method to attempt to increase cellular protein concentrations by increasing mRNA expression levels. Additionally, an increase in protein level can be achieved through chemical treatments that block protein degradation by the proteasome (MG132, epoxomicin, etc.), cellular proteases (aprotinin, serpin, etc.), or other lytic pathways (brefeldin A) [21, 24]. As a note, when cellular protein concentrations are modified, special consideration must be taken during analyses and during follow-up experiments to avoid the pursuit of artifacts due to the chosen methodology.

While there is much to consider when planning AP-MS experiments for your favorite protein, in the following sections, we provide help with the design, execution, and analytical stages of an AP-MS experiment (Fig. 2). We have also included a detailed protocol that has been successful for the detection of PPIs [36, 37] that any plant molecular biology laboratory can execute.

1.1 Experimental Design

As with all experiments, design is the most important step. Mistakes made here will be amplified across the experimental regime and in the worst case cause experimental failure. Below is the appropriate sequence to design an AP-MS experiment from the ground up to increase the detection of "true" PPIs and decrease the detection of "false" PPIs:

1. Choosing a POI or POIs for AP-MS: We encourage AP-MS experimental design to include as many diverse baits as possible. Optimally, choosing multiple bait proteins with diverse function increases the statistical analysis power and can provide increased filtering of false positives to reveal biologically significant interactors [38, 39]. These baits should obviously include controls, but can also include other proteins that the laboratory is interested in investigating. If a single bait protein is to be used during AP-MS, we encourage that multiple control types (**step 4** below) and as many biological replicates (**step 7** below) as possible are employed [7, 22, 39, 40].
2. Choosing the affinity tags: The optimal resource for AP-MS is an antibody that is highly specific to your POI. These are frequently not available, but if this resource exists, we encourage its use. If an antibody to your POI does not exist, we frequently use a combination 3XFLAG and 6XHis affinity tag system [36]. The FLAG (DYKDDDDK) tag was chosen because it is a small synthetic peptide epitope with low cross-reactivity to endogenous plant proteins [7, 27]. The His tag can be pulled down using a nonprotein-based affinity column, allowing on-column digestion for high-throughput applications [36, 41]. The dual-tag system allows the researcher to choose a highly selective, highly stringent two-step AP protocol or lower stringency single AP protocol when needed.
3. Choosing the expression system: Native promoter, constitutive expression, or tissue-specific promoters are widely used [36, 42–45]. The benefits and drawbacks of each type of expression system must be carefully considered, and when possible the use of native

promoters is encouraged to reduce mis-expression artifacts [35]. That being said, there are many justifiable reasons to use alternate expression strategies (as mentioned in the introduction). The emergence of CRISPR has allowed the insertion of affinity tags directly into genomic loci, which will likely become a commonly used method to translationally fuse affinity tags to proteins while maintaining the genomic context of the gene [46].

4. Choosing control lines: It is critical to use multiple controls to assist in the removal of false positives during the analytical stages [7, 22, 39, 40]. We encourage using a non-transgenic control line (the parent wildtype or mutant line) to detect any proteins that interact with the affinity resin or antibody. We also strongly recommend using a tagged version of a plant-inert protein, such as GFP, to allow for the removal of proteins that associate with the affinity tag [40]. It is critical that the same affinity tag be used for the control protein, and it can be beneficial to use the same promoter when possible.
5. Choosing an appropriate tissue: PPIs vary across tissues (Fig. 1) [47, 48]. It is critical to determine which tissue is pertinent to your biological question, and tissue amount oftentimes is limiting. We use approximately 2 mL of ground plant powder as starting tissue for each AP, and we compare the amount of our POI to a control sample that has given good AP-MS results in the past (this control material can be provided upon request) to adjust our starting amount.
6. Choosing appropriate biological conditions: Adding biological complexity through timepoints, treatments, or other biological conditions will increase your ability to identify biologically relevant interacting proteins (Fig. 1). For instance, an interacting protein that decreases upon a specific chemical treatment may provide more functional insight into your POI than the one that stays associated constitutively throughout the treatment time course.
7. Including biological replicates: As with many experiments, the balance of cost with statistical significance must be weighed at the outset. At least three biological replicates are suggested, although two biological replicates are often used when additional follow-up PPI experiments are conducted. Technical replicates of MS analyses can also increase the probability of identifying low abundance interactors, but are not a substitute for biological replicates [36, 38, 49, 50].

1.2 Affinity Purification–Mass Spectrometry

The next major consideration is whether to perform a one-step or two-step AP procedure. As mentioned above, a one-step procedure will result in retaining more of your bait POI and interacting proteins but will also give a higher false-positive rate. A two-step

AP will reduce false positives but decrease detection of low-affinity interactors. An additional consideration is that the two-step AP will increase the total cost of the experiment.

Following is a brief overview of our AP procedure using the His-FLAG system (Fig. 2a–e).

1. FLAG-tagged bait proteins are affinity purified using anti-FLAG antibodies coupled to protein G-coated magnetic beads. The magnetic beads are advantageous because they have low nonspecific binding and allow for fast isolation of protein complexes (Fig. 2a, b). The immunoprecipitated protein complexes can be eluted easily from the magnetic beads with FLAG peptides dissolved in a wash buffer that needs to be appropriate for binding to the metal affinity column in the next step.
2. The second AP step entails immunoprecipitation of the His-tag using a coordinated metal affinity column (we use TALON magnetic beads because of their higher specificity for the His tag) [51]. This second step further eliminates nonspecific interactors with the FLAG column and also efficiently removes the FLAG peptides used for elution (Fig. 2c). We recommend removing the His AP step if a one-step AP is desired for higher bait yield. We have tested direct trypsin digestion from the FLAG column. We observe a 300–400% increase in detection of true interactors (based on a known high affinity interacting protein) in one-step FLAG AP compared to two-step AP. These data are not shown and may not be applicable to all POIs, but this estimates the potential difference between a one-step FLAG AP and a two-step FLAG to His AP.
3. For quality control, it is recommended that all AP experiments are checked by way of western blot (WB) analyses for efficient depletion of the bait POI (Fig. 3) and the control proteins. With this, the relative amount of bait protein that is immunoprecipitated can be compared, and it can be checked whether the molecular weight of immunoprecipitated protein is correct. An inexpensive alternative is to perform SDS-PAGE analyses followed by a high sensitivity stain of the gel (i.e., silver stain). This will allow rapid detection of the major band where the POI should be and allow for comparison of any abundant interacting proteins between purifications with the POI and control proteins.
4. After AP, the bait POI complexes bound to TALON beads (of the FLAG column for one-step AP) can be digested directly with trypsin (or other enzymes) to prevent further loss of low abundance interactors caused by traditional elution or desalting before MS analyses (Fig. 2d). The peptides released from the affinity-purified protein complexes are then subjected to liquid chromatography–tandem mass spectrometry (LC-MS/

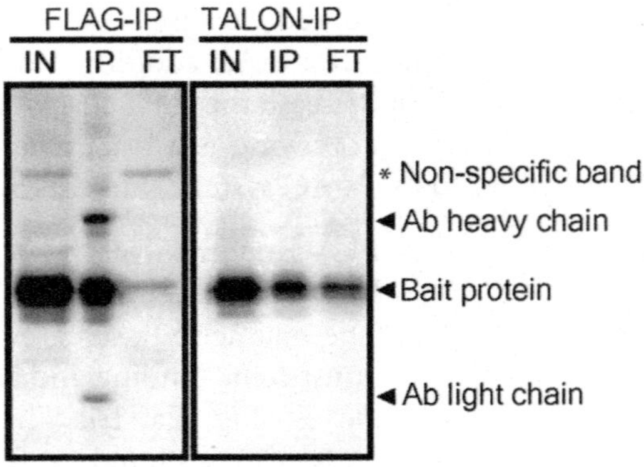

Fig. 3 WB analysis with anti-FLAG antibody to evaluate AP efficiency. The FLAG- and His-tagged bait protein from total protein input (FLAG-IP: IN) binds to anti-FLAG antibody-conjugated protein G beads (FLAG-IP: IP) and is depleted from the flow-through (FLAG-IP: FT). Similar results are seen for His-tag purification with TALON beads (TALON-IP). The anti-FLAG M2 antibodies (heavy and light chain) are eluted from beads in FLAG-IP after adding SDS-PAGE sample loading buffer

MS) analyses. Briefly, peptides are separated by chromatography and then ionized and fragmented for separation in an electromagnetic field resulting in a mass-to-charge ratios (m/z) unique to the individual peptide (Fig. 2e). For further details, see [52].

5. Care should be taken when choosing a mass spectrometry facility or collaborator for your samples. Many facilities have long wait times and can be inflexible with data handling while collaborators may not have sufficient resources to process your data multiple times if changes are requested. We strongly suggest familiarizing yourself with the basics of mass spectrometry and data processing to ensure the final product meets your needs.

1.3 Mass Spectrometry Data Analysis

After the data is collected, the mass spectra can be examined directly, converted to peptide sequences, and mapped to the proteome [24, 25, 53, 54]. Data analysis will also provide ways to quantify peptides (i.e., spectral counts). The next section describes freely available software to statistically analyze AP-MS data (Fig. 2f). We recommend requesting all raw data files to ensure that the data can be reprocessed in the future if necessary or as search methodology improves. Many mass spectrometry data interpretation software packages exist. A fairly comprehensive list can be found via Wikipedia (https://en.wikipedia.org/wiki/List_of_mass_spectrometry_software) or online searches, and many are freely available. One such free package that was developed by the Mann group is called Maxquant (http://www.biochem.mpg.de/5111795/maxquant). Maxquant is widely used and includes a wide array of online tutorials and manuals that can help you get started with analysis of your spectra [25, 53]. In addition, a course is offered yearly that helps new and advanced users familiarize themselves with basic and sophisticated analyses of mass spectra. Our home facility (Keck Proteomics) uses MASCOT, which is a widely used package available for purchase and will be described in more detail below [54].

Following the mapping of your peptides to the proteome, it is now quite straightforward to perform statistical analyses of your AP-MS data using the SPOTLITE analysis package that is freely available as a standalone website (http://cancer.unc.edu/majorlab/software.php) [55]. This site contains tools that can be applied to filter out contaminants and to evaluate the reliability of interactors. SPOTLITE is a web-based openware that implements three popular algorithms, HGSCore [56], CompPASS [38], and SAINT [57, 58], to score the probability of interactions by comparing the spectral counts from your bait protein AP-MS data to the controls or additional variable baits. The benefits of each of these algorithms are discussed in detail in the protocol below. After completing the experiment, it is highly recommended that further experimentation be performed to cross validate interacting proteins and determine whether interactions are direct or indirect.

2 Materials

2.1 Equipment and Supplies

1. Mixer Mill MM400 system and screw-top grinding jars (Retsch). Mortar and pestle can be an alternative.
2. Sonifier cell disruptor (Branson, SFX-250) or other sonicator for cell disruption.
3. Nutator.
4. 30 °C and 37 °C incubator.
5. Centrifuge capable of subjecting 1.7 mL microcentrifuge tubes to 12,000 × *g* at 4 °C.
6. PureProteome™ magnetic stand (EMD Millipore, LSKMAGS08) or equivalent.
7. SpeedVac.
8. NanoDrop 2000c spectrophotometer (Thermo Scientific).
9. Thermo Scientific Orbitrap Elite mass spectrometer or equivalent.
10. 15 mL conical centrifuge tube.
11. 1.7 mL low-adhesion microcentrifuge tubes.
12. 0.22 μm filter.
13. Liquid nitrogen.

2.2 Stock Solution and Reagents

1. SII buffer: 100 mM sodium phosphate (pH 8.0), 150 mM NaCl, 5 mM EDTA, 0.1% Triton X-100 (*see* **Note 1**).
2. FLAG-to-His (FTH) buffer: 100 mM sodium phosphate (pH 8.0), 150 mM NaCl, 0.05% Triton X-100 (*see* **Note 1**).
3. Antibody conjugating buffer: phosphate-buffered saline (PBS, pH 7.4) with 0.1% Tween 20.
4. PureProteome™ protein G magnetic bead system (EMD millipore, LSKMAGG10) (*see* **Note 2**).
5. TALON magnetic beads (Clontech).
6. Monoclonal ANTI-FLAG® M2 antibody produced in mouse, 2 μg/μL (Sigma).
7. 10 mM ammonium bicarbonate.
8. Sequencing grade modified trypsin, 0.5 mg/mL (Promega).
9. Liquid nitrogen.

2.3 Freshly Prepared Solution and Reagents

1. SII-inhibitor buffer: SII buffer with 1 mM phenylmethylsμlfonyl fluoride (PMSF), cOmplete™ EDTA-free protease inhibitor cocktail tablet (Roche), and PhosSTOP tablet (Roche) (*see* **Note 3**).
2. FLAG elution buffer: 100 μg/mL 3× FLAG peptide (Sigma) in FTH buffer (*see* **Note 4**). Prepare 1.6 mL of elution buffer for each AP sample.
3. 25 mM ammonium bicarbonate, filtered through a 0.22 μm filter.
4. Peptide reconstitution buffer: mix 2 μL of 70% formic acid (FA) and 23 μL of 0.1% trifluoroacetic acid (TFA).

2.4 Mass Spectrometry Analysis Software

1. Mass spectrometry interpretation software. See software listed on Wikipedia for an up-to-date list (https://en.wikipedia.org/wiki/List_of_mass_spectrometry_software).
2. SPOTLITE (http://cancer.unc.edu/majorlab/software.php) [38].

3 Methods

3.1 Cryogenic Grinding of Plant Tissue for Affinity Purification

1. Harvest plant tissue (around ten leaves from 6- to 8-week-old Arabidopsis grown on soil) and freeze them immediately in liquid nitrogen.
2. Prechill grinding jars, screw cap, ball mills (or mortar and pestle if chosen), spatula, and 15 mL conical centrifuge tubes in liquid nitrogen.
3. Add the frozen plant materials to the grinding jars and tightly screw on the cap. The screw-top grinding jars are then set into

grinding station of the Mixer Mill MM400 system. Tissue is ground at 30 Hz for 1 min.

4. Open the grinding jars and use a prechilled spatula to collect ground tissue powder in prechilled 15 mL conical centrifuge tubes for AP directly or storage at −80 °C.

3.2 Affinity Purification of FLAG- and His-Tagged Bait Protein

The following protocol is optimized for using 2 mL of plant tissue powder as starting material for each AP-MS sample. The amount of plant tissue and magnetic beads can be adjusted based on the experimental designs and materials (*see* **Note 5**).

3.2.1 Conjugation of Anti-FLAG Antibody to Protein G Magnetic Beads

1. Resuspend the protein G magnetic beads thoroughly by gentle shaking. For each 2 mL of plant tissue powder, aliquot 100 μL bead slurry to a 1.7 mL tube, pull down the beads with a magnetic stand, and remove the solution.
2. Wash the beads by adding 1 mL of antibody conjugating buffer and vortexing for 10 s, place the tube in the magnetic stand for 1 min, and then remove the buffer while the tube is still in the stand. Ensure that all buffer is removed from the cap and the bottom of the tube (use this practice for all the washes in the following steps).
3. Resuspend the beads in 200 μL of antibody conjugating buffer and add 5 μL of monoclonal mouse anti-FLAG antibody (stock concentration 2 μg/μL) (*see* **Note 6**). Incubate the bead antibody mix at room temperature on the nutator for 30 min.
4. After antibody conjugation, wash the beads three times with 1 mL of antibody conjugating buffer using the bead wash protocol from **step 2**.
5. Resuspend the beads in 20 μL of antibody conjugating buffer and keep on ice.

3.2.2 Preparation of TALON Magnetic Beads

1. Resuspend the TALON magnetic beads and aliquot 250 μL of beads to a 1.7 mL tube. Place the tube in the magnetic stand for 1 min and remove the solution.
2. Wash the beads three times with 1 mL of FTH buffer.
3. Resuspend the beads in 100 μL of FTH buffer, and keep on ice.

3.2.3 Preparation of Protein Extract

1. Transfer 2 mL of frozen plant tissue powder to a prechilled 15 mL conical tube. Resuspend the powder in 2 mL of SII-inhibitor buffer and rotate at 4 °C for 10 min. To prevent protein degradation during AP procedure, all subsequent steps should be conducted in the cold room or on ice unless other temperatures are noted.
2. Disrupt the plant tissue with a sonicator twice for 10 s with 10% output power on ice. The output power and the duration

can be varied depending on plant samples and the model of sonicator used.

3. Transfer the protein extract to new 1.7 mL tubes, and clarify the protein extract by centrifugation at 12,000 × *g* for 10 min at 4 °C. Transfer the supernatant to another set of 1.7 mL tubes and repeat the centrifugation a second time.

3.2.4 Two-Step Affinity Purification

1. Transfer the supernatant to a 15 mL conical tube, and collect 2% of the supernatant as "FLAG-input" for WB analysis to evaluate affinity purification (IP) efficiency.
2. Add 20 μL of anti-FLAG antibody-conjugated protein G beads into the supernatant in the conical tube, and incubate on the nutator at 4 °C for 1 h.
3. After incubation, transfer the solution to a 1.7 mL low adhesion centrifuge tube, and place the tube on the magnetic stand. After 1 min, transfer the supernatant to a new tube. Collect 2% of the supernatant (should be the same volume the 2% input) as "FLAG flow-through" for WB analysis.
4. Wash the magnetic beads three times with 1 mL of pre-chilled SII buffer. After adding 1 mL of SII buffer during the third wash, resuspend the beads and collect 2% (20 μL) of the SII bead mix as "FLAG-IP" for WB analysis. Place the tube with remaining beads on the magnetic stand for 1 min and remove the supernatant.
5. Wash the magnetic beads twice with 1 mL of prechilled FTH buffer.
6. Elute the bait proteins with 400 μL of FLAG elution buffer twice by adding the elution buffer and rotating tubes on the nutator at 4 °C for 15 min. Collect the eluates using a magnetic stand. Elute bait proteins an additional two times with 400 μL FLAG elution buffer at 30 °C for 15 min. Combine eluates in a new 1.7 mL low-adhesion tube, and save 2% of the combined eluate as "TALON-input" for WB analysis.
7. Add 100 μL of TALON magnetic beads to the eluate, and incubate at 4 °C on the nutator for 30 min.
8. After incubation, place the tube in the magnetic stand for 1 min, collect 2% of the supernatant as "TALON flow-through" for WB analysis, and remove the supernatant.
9. Wash the TALON beads with 1 mL of prechilled FTH buffer twice.
10. Wash the TALON beads three times with 1 mL of prechilled 25 mM ammonium bicarbonate. During the third wash, resuspend the beads and collect 2% of solution containing beads as "TALON-IP" for WB analysis. Place the tube with the remaining solution on the magnetic stand and remove the supernatant.

On-bead trypsin digestion can be performed directly, or the protocol can be paused here indefinitely by freezing the beads in liquid nitrogen and storing them at −80 °C.

11. Perform WB analysis to evaluate IP efficiency (Fig. 3). The WB protocol can be found in [59]. The FLAG-IP and TALON-IP should completely deplete the bait protein from the FLAG-input and TALON-input (no band in the flow-through lane). If this is not the case, the amount of FLAG or TALON magnetic beads can be increased to get better efficiency of affinity purification.

3.3 On-Bead Trypsin Digestion, Sample Preparation, and Mass Spectrometry

1. Wash the beads with 100 μL of 10 mM ammonium bicarbonate by incubating beads on the nutator for 5 min, capturing the beads using the magnetic stand, and removing the supernatant. Repeat this step once.
2. Digest the protein on the beads with 1/10 estimated protein weight of trypsin enzyme (*see* **Note 7**). For example, use 1 μg of trypsin for 10 μg of protein on the beads. Prepare 0.01 μg/μL trypsin in 10 mM ammonium bicarbonate as trypsin enzyme mix. Add 100 μL trypsin enzyme mix to 10 μg of protein on the beads, and incubate the sample at 37 °C overnight.
3. Pull down the beads using the magnetic stand and transfer the supernatant to a new low-adhesion 1.7 mL microcentrifuge tube.
4. Dry down the supernatant in the SpeedVac at room temperature until all of the solution in the sample is evaporated. Be careful not to dry the pellet for too long because this can make the pellet difficult to resuspend.
5. Reconstitute the digested sample with 25 μL of peptide reconstitution buffer.
6. Measure the peptide concentration using absorbance at 280 nm in a NanoDrop spectrophotometer.
7. Dilute the samples with peptide reconstitution buffer to a final concentration of 0.1 μg/μL. Inject 5 μL of sample into an Orbitrap Elite mass spectrometer or other instrument for LC-MS/MS analyses.

3.4 Analysis of Protein Interacting Networks with Mass Spectrometry Data

1. The m/z spectra from mass spectrometry raw data are initially processed by Thermo Scientific Xcalibur™ Software (https://www.thermofisher.com/order/catalog/product/OPTON-20487) to a .mgf format and then searched against an *Arabidopsis thaliana* protein database with MASCOT [54] or other software (see https://en.wikipedia.org/wiki/List_of_mass_spectrometry_software).
2. The database for the MASCOT search used in this protocol is the Protein SwissProt_2016_05.fasta *Arabidopsis thaliana*

database. The parameters were as follows: oxidation (M) modification, cleavage by trypsin with maximum of two missed cleavages and mass tolerance of fragment ions within 10 ppm for precursor ions, and 0.5 Da for fragments. The parameters can be altered to fit experimental needs.

3. A typical MASCOT result file contains peptide sequences, modifications, peptide spectral counts, and protein IDs associated with the spectra (check the detailed guide at http://www.matrixscience.com/help/msms_summaries_help.html). The "protein ID" represents "UniProtKB_ID" (http://www.uniprot.org/), which can be matched to a specific protein, for example, CB1A_ARATH for LHCB1.1 (AT1G29920). Multiple peptide sequences can match to one protein; therefore, the peptide spectral counts for each protein (protein ID) can be retrieved and combined.
4. Each dataset from the AP-MS samples can be combined into a tab-delimited file (Table 2) with headers: Experiment ID, Experiment Type, Bait, Prey, and Spectral Counts, for SPOTLITE analysis (http://cancer.unc.edu/majorlab/software.php) [55] (*see* **Note 8**). Each experiment or AP-MS sample should have a unique Experiment ID, for example, Exp_1, Exp_2, etc. Experiment type specifies experiment (T) or control (C) sets. Baits are the baits of your POIs in the AP-MS experiments, and preys are the protein ID identified from mass spectrometry. Spectral counts are the peptide counts for each protein ID.
5. After uploading the input file (Table 1) to SPOTLITE, three algorithms, HGSCore [56], CompPASS [38], or SAINT [57, 58], can be used to determine the confidence levels of interacting candidates by comparing between AP-MS test and control datasets. HGSCore is suitable for large datasets, in which both

Table 1
An example of SPOTLITE input file

Experiment ID	Experiment type	Bait	Prey	Spectral count
IP_sample_1	T	POI[a]	TIR1_ARATH	158
IP_sample_1	T	POI[a]	TPL_ARATH	40
IP_sample_2	T	POI[a]	CB1A_ARATH	63
IP_sample_2	T	POI[a]	TIR1_ARATH	82
GFP	C	GFP	IF5A1_ARATH	25
WT	C	None	CB1A_ARATH	49

[a]*POI* protein of interest

Table 2
An example of SAINT analysis output from SPOTLITE

Bait	Prey	SAINT[b]	SAINT *p*-value
POI[a]	TIR1_ARATH	1	3.15E-6
POI[a]	TPL_ARATH	1	3.15E-6
POI[a]	CB1A_ARATH	1	3.15E-6
POI[a]	PORA_ARATH	1	3.15E-6
POI[a]	COPA!_ARATH	1	1.26E-5
POI[a]	ISPG_ARATH	1	1.26E-5

[a] *POI* protein of interest
[b] *SAINT* SAINT score

baits and identified preys are used as baits in the AP-MS experiments. CompPASS also gives good performance with lots of variable baits in the datasets, but the performance is limited when dense connection between baits and preys is expected. Alternatively, SAINT is appropriate for analyses of smaller datasets with fewer baits and multiple control sets. The detailed algorithm can be found on the SPOTLITE website [55]. We used SAINT for the test AP because wildtype and His-FLAG-GFP controls were included and only one bait protein was used. The output is shown in Table 2. A higher SAINT score (falling between 0.5 and 1) with a lower *p*-value indicates interacting protein with higher statistical confidence.

4 Notes

1. 100 mM sodium phosphate (pH 8.0) can be made by mixing 93.2 mL of 1 M Na_2HPO_4 and 6.8 mL of 1 M NaH_2PO_4 and brought up to a volume of 1 L using distilled H_2O. The optimal pH of the SII and FTH buffers for AP is between 7 and 8.
2. In this protocol, protein G magnetic beads were chosen because of their higher affinity to the mouse IgG1-type anti-FLAG M2 antibody. To modify this protocol for use with other antibodies, protein A or protein G magnetic beads can be chosen depending on the isotypes and the origins of the antibodies. Further information can be found in https://www.neb.com/tools-and-resources/selection-charts/affinity-of-protein-ag-for-igg-types-from-different-species.
3. SII-inhibitor buffer should be freshly made prior to experimentation by adding 100 mM PMSF stock, cOmplete™ EDTA-free protease inhibitor cocktail tablets, and PhosSTOP tablets (according to the instructions from the manufacturer)

and kept on ice. As EDTA interferes with the immunoprecipitation of His-tagged proteins, protease inhibitors with EDTA should be avoided.

4. The 5 mg/mL 3X FLAG peptide stock can be prepared by dissolving lyophilized peptide in FTH buffer and stored at −20 °C. The FLAG elution buffer should be made just prior to the elution step and kept on ice.
5. The amount of starting plant tissue can be varied according to the expression level of the bait protein, the affinities and expression levels of the interacting proteins, and the sensitivity of the mass spectrometry instrument. The bait proteins described in this protocol were overexpressed under a 35S promoter, and 2 mL of tissue was sufficient starting material to identify interacting proteins. For lower abundance baits or interacting proteins, more plant tissue is desired. For high abundance bait proteins, the amounts of FLAG beads and TALON beads used for AP may be increased.
6. The ratio of protein G magnetic beads to the conjugated antibody needs to be optimized if different magnetic beads or antibodies are used. Follow the antibody conjugation protocol in the Subheading 3.2.1, but add different quantities of antibody. After conjugation, separate the beads and solution (or flow-through) with a magnetic stand, and collect the solution. Perform a WB on both the beads and the solution to check for the efficiency of conjugation (Fig. 4).

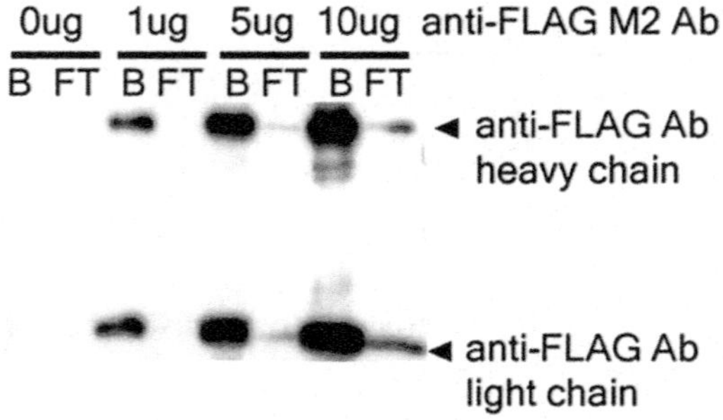

Fig. 4 Optimization of the ratio of antibody to protein G beads for antibody coupling. For 50 μL of protein G beads, 0, 1, 5, or 10 μg of anti-FLAG antibody (Ab) were used for antibody coupling followed the protocol in the Subheading 3.2.1. After antibody coupling, pull down the beads using magnetic stand to separate the conjugated beads (B) from the solution or flow-through (FT). SDS-PAGE sample loading buffer was added to the conjugated beads and the flow-through and then subsequently boiled. SDS-PAGE and WB analyses were performed. The anti-FLAG antibody recognizes the heavy and light chains of the anti-FLAG antibody. If FT contains heavy and light chains, then the beads are oversaturated with antibody. The ideal amount of antibody is the lowest value for which the beads are oversaturated

7. To estimate the amount of protein on the beads, perform SDS-PAGE on 2% of TALON-IP sample along with different amounts of BSA (e.g., 2 μg, 500 ng, 100 ng, 20 ng, and 5 ng) and detect proteins with silver stain. Related protocols can be found in [59, 60]. The protein amount can be roughly estimated based on the intensities of the protein bands compared to BSA.
8. The list of protein IDs identified from raw MS data should pass a stringent cutoff, such as <1% false discovery rate. In addition, control and test sets should use the same AP-MS protocol, reagents, and instruments to give better results.

Acknowledgments

D. A. N. provided the pB7-HFN binary vector used for generating transgenic lines expressing 3XFLAG-6XHis-tagged bait proteins. The trypsin digestion, sample preparation for mass spectrometry, and MASCOT analysis were technically supported by Dr. Jean Kanyo at the W. M. Keck Biotechnology Resource Laboratory at Yale University. We also want to thank Dr. Shirin Bahmanyar, Dr. Man-Wah Li, Dr. Wei Liu, and Olivia Compagna for their useful discussion and comments. This research is supported by NSF (IOS-1456796 to D. A. N. and IOS-1548538 to J. M. G. and DGE-1122492 to AMF), NIH (T32 GM007499 to A.F.), the Gruber Foundataion (A.F.), Yale University Forest B. H. and Elizabeth D. W. Brown Fund Endowed Postdoctoral Fellowship (C.L.) and the Rudolph J. Anderson Postdoctoral Fellowship (C.L.).

References

1. Zhang Y, Gao P, Yuan JS (2010) Plant protein–protein interaction network and interactome. Curr Genomics 11:40–46
2. Dietz KJ, Jacquot JP, Harris G (2010) Hubs and bottlenecks in plant molecular signalling networks. New Phytol 188:919–938
3. Bensimon A, Heck AJR, Aebersold R (2012) Mass spectrometry–based proteomics and network biology. Annu Rev Biochem 81:379–405
4. Chen J, Sawyer N, Regan L (2013) Protein–protein interactions: general trends in the relationship between binding affinity and interfacial buried surface area. Protein Sci 22:510–515
5. LaCava J et al (2015) Affinity proteomics to study endogenous protein complexes: pointers, pitfalls, preferences and perspectives. Biotechniques 58:103–119
6. Perkins JR et al (2010) Transient protein–protein interactions: structural, functional, and network properties. Structure 18:1233–1243
7. Morris JH et al (2014) Affinity purification-mass spectrometry and network analysis to understand protein–protein interactions. Nat Protoc 9:2539–2554
8. Snider J et al (2015) Fundamentals of protein interaction network mapping. Mol Syst Biol 11:848
9. Uzoma I, Zhu H (2013) Interactome mapping: using protein microarray technology to reconstruct diverse protein networks. Genomics Proteomics Bioinformatics 11:18–28
10. Walhout AJM et al (2000) Protein interaction mapping in *C. elegans* using proteins involved in vulval development. Science 287:116–122
11. Snider J et al (2010) Detecting interactions with membrane proteins using a membrane two-hybrid assay in yeast. Nat Protoc 5:1281–1293
12. Miernyk JA, Thelen JJ (2008) Biochemical approaches for discovering protein–protein interactions. Plant J 53:597–609

13. Kerppola TK (2008) Bimolecular fluorescence complementation (BiFC) analysis as a probe of protein interactions in living cells. Annu Rev Biophys 37:465–487
14. Ohad N, Shichrur K, Yalovsky S (2007) The analysis of protein–protein interactions in plants by bimolecular fluorescence complementation. Plant Physiol 145:1090–1099
15. Truong K, Ikura M (2001) The use of FRET imaging microscopy to detect protein–protein interactions and protein conformational changes in vivo. Curr Opin Struct Biol 11:573–578
16. Dedecker M, Van Leene J, De Jaeger G (2015) Unravelling plant molecular machineries through affinity purification coupled to mass spectrometry. Curr Opin Plant Biol 24:1–9
17. Daneri-Castro SN, Svensson B, Roberts TH (2016) Barley germination: spatio-temporal considerations for designing and interpreting 'omics' experiments. J Cereal Sci 70:29–37
18. Nikolov M et al (2011) Chromatin affinity purification and quantitative mass spectrometry defining the interactome of histone modification patterns. Mol Cell Proteomics 10(M110):005371
19. Qu M et al (2016) Qualitative and quantitative characterization of protein biotherapeutics with liquid chromatography mass spectrometry. Mass Spectrom Rev. doi:10.1002/mas.21500
20. Nielsen ML, Savitski MM, Zubarev RA (2005) Improving protein identification using complementary fragmentation techniques in fourier transform mass spectrometry. Mol Cell Proteomics 4:835–845
21. Kim S et al (2010) The generating function of CID, ETD, and CID/ETD pairs of tandem mass spectra: applications to database search. Mol Cell Proteomics 9:2840–2852
22. Armean IM, Lilley KS, Trotter MWB (2013) Popular computational methods to assess multiprotein complexes derived from label-free affinity purification and mass spectrometry (AP-MS) experiments. Mol Cell Proteomics 12:1–13
23. Piehowski PD et al (2013) STEPS: a grid search methodology for optimized peptide identification filtering of MS/MS database search results. Proteomics 13:766–770
24. Xu H, Freitas MA (2009) MassMatrix: a database search program for rapid characterization of proteins and peptides from tandem mass spectrometry data. Proteomics 9:1548–1555
25. Cox J, Mann M (2008) MaxQuant enables high peptide identification rates, individualized p.p.b.-range mass accuracies and proteome-wide protein quantification. Nat Biotechnol 26:1367–1372
26. Van Leene J et al (2015) An improved toolbox to unravel the plant cellular machinery by tandem affinity purification of Arabidopsis protein complexes. Nat Protoc 10:169–187
27. Gingras AC et al (2007) Analysis of protein complexes using mass spectrometry. Nat Rev Mol Cell Biol 8:645–654
28. Banks CA et al (2015) Proteins interacting with cloning scars: a source of false positive protein–protein interactions. Sci Rep 5:8530
29. Wu J et al (2006) Identification of substrates of human protein-tyrosine phosphatase PTPN22. J Biol Chem 281:11002–11010
30. DeBlasio SL et al (2016) Visualization of host-polerovirus interaction topologies using protein interaction reporter technology. J Virol 90:1973–1987
31. Westphal K et al (2012) A trapping approach reveals novel substrates and physiological functions of the essential protease FtsH in *Escherichia coli*. J Biol Chem 287:42962–42971
32. Yumimoto K et al (2012) Comprehensive identification of substrates for F-box proteins by differential proteomics analysis. J Proteome Res 11:3175–3185
33. Rodriguez J et al (2016) Substrate-trapped interactors of PHD3 and FIH cluster in distinct signaling pathways. Cell Rep 14:2745–2760
34. Leitner A et al (2014) Chemical cross-linking/mass spectrometry targeting acidic residues in proteins and protein complexes. Proc Natl Acad Sci U S A 111:9455–9460
35. Keilhauer EC, Hein MY, Mann M (2015) Accurate protein complex retrieval by affinity enrichment mass spectrometry (AE-MS) rather than affinity purification mass spectrometry (AP-MS). Mol Cell Proteomics 14:120–135
36. Huang H et al (2016) Identification of evening complex associated proteins in Arabidopsis by affinity purification and mass spectrometry. Mol Cell Proteomics 15:201–217
37. Huang H et al (2016) PCH1 integrates circadian and light-signaling pathways to control photoperiod-responsive growth in Arabidopsis. Elife 5:e13292
38. Sowa ME et al (2009) Defining the human deubiquitinating enzyme interaction landscape. Cell 138:389–403
39. Nesvizhskii AI (2012) Computational and informatics strategies for identification of specific protein interaction partners in affinity purification mass spectrometry experiments. Proteomics 12:1639–1655
40. Mellacheruvu D et al (2013) The CRAPome: a contaminant repository for affinity purification-mass spectrometry data. Nat Methods 10:730–736
41. Rees JS, Lilley KS, Jackson AP (2015) SILAC-iPAC: a quantitative method for distinguishing genuine from non-specific components of

protein complexes by parallel affinity capture. J Proteomics 115:143–156
42. Tamura K et al (2010) Identification and characterization of nuclear pore complex components in *Arabidopsis thaliana*. Plant Cell 22:4084–4097
43. Choi H et al (2014) The homeodomain-leucine zipper ATHB23, a phytochrome B-interacting protein, is important for phytochrome B-mediated red light signaling. Physiol Plant 150:308–320
44. Book AJ et al (2010) Affinity purification of the Arabidopsis 26 S proteasome reveals a diverse array of plant proteolytic complexes. J Biol Chem 285:25554–25569
45. Li S et al (2014) Identification of the self-incompatibility locus F-box protein-containing complex in *Petunia inflata*. Plant Reprod 27:31–45
46. Lackner DH et al (2015) A generic strategy for CRISPR-Cas9-mediated gene tagging. Nat Commun 6:10237
47. Kotlyar M et al (2015) Integrated interactions database: tissue-specific view of the human and model organism interactomes. Nucleic Acids Res 44(D1):D536–D541
48. Barshir R et al (2012) The TissueNet database of human tissue protein–protein interactions. Nucleic Acids Res 41(D1):D841–D844
49. Song YH et al (2014) Distinct roles of FKF1, GIGANTEA, and ZEITLUPE proteins in the regulation of CONSTANS stability in Arabidopsis photoperiodic flowering. Proc Natl Acad Sci U S A 111:17672–17677
50. Rohila JS et al (2006) Protein–protein interactions of tandem affinity purification-tagged protein kinases in rice. Plant J 46:1–13
51. Ueda EK et al (2003) Current and prospective applications of metal ion-protein binding. J Chromatogr A 988:1–23
52. Pitt JJ (2009) Principles and applications of liquid chromatography-mass spectrometry in clinical biochemistry. Clin Biochem Rev 30:19–34
53. Tyanova S et al (2015) Visualization of LC-MS/MS proteomics data in MaxQuant. Proteomics 15:1453–1456
54. Perkins DN et al (1999) Probability-based protein identification by searching sequence databases using mass spectrometry data. Electrophoresis 20:3551–3567
55. Goldfarb D et al (2014) Spotlite: web application and augmented algorithms for predicting co-complexed proteins from affinity purification—mass spectrometry data. J Proteome Res 13:5944–5955
56. Guruharsha KG et al (2011) A protein complex network of *Drosophila melanogaster*. Cell 147:690–703
57. Choi H et al (2011) SAINT: probabilistic scoring of affinity purification-mass spectrometry data. Nat Methods 8:70–73
58. Teo G et al (2014) SAINTexpress: improvements and additional features in significance analysis of INTeractome software. J Proteomics 100:37
59. Blancher C, Jones A (2001) Analysis of cells and tissues. In: Brooks SA, Schumacher U (eds) Metastasis research protocols, vol 1. Humana Press, Totowa, NJ, pp 145–162
60. Dunn MJ (2002) Detection of proteins in polyacrylamide gels by silver staining. In: Walker JM (ed) The protein protocols handbook. Humana Press, Totowa, NJ, pp 265–271

Chapter 16

Measuring Protein Movement, Oligomerization State, and Protein–Protein Interaction in *Arabidopsis* Roots Using Scanning Fluorescence Correlation Spectroscopy (Scanning FCS)

Natalie M. Clark and Rosangela Sozzani

Abstract

Scanning fluorescence correlation spectroscopy (scanning FCS) can be used to determine protein movement, oligomerization state, and protein–protein interaction. Here, we describe how to use the scanning FCS techniques of raster image correlation spectroscopy (RICS) and pair correlation function (pCF) to determine the rate and direction of protein movement. In addition, we detail how number and brightness (N&B) and cross-correlation analyses can be used to determine oligomerization state and binding ratios of protein complexes. We specifically describe how to acquire suitable images for scanning FCS analysis using the model plant *Arabidopsis* and how to perform the various analyses using the SimFCS software.

Key words Scanning fluorescence correlation spectroscopy, Diffusion coefficient, Protein movement, Oligomerization state, Protein stoichiometry

1 Introduction

Scanning fluorescence correlation spectroscopy (scanning FCS) is a set of imaging techniques that uses spatial and temporal information to quantify kinetic parameters such as protein movement, oligomerization state, and protein–protein interaction in vivo [1, 2]. The ability to quantify protein movement is key to study developmental networks, as many are controlled by moving proteins that cross cell boundaries [3–5]. Further, some transcription factors must form higher-order oligomers or bind other proteins in specific ratios in order to function [6, 7]. Scanning FCS techniques have been used to quantify protein mobility, oligomeric state, and protein association in cell cultures as well as in the multicellular plant *Arabidopsis* [8–17].

Wolfgang Busch (ed.), *Plant Genomics: Methods and Protocols*, Methods in Molecular Biology, vol. 1610,
DOI 10.1007/978-1-4939-7003-2_16,

Here, we describe how to use the scanning FCS techniques of raster image correlation spectroscopy (RICS), pair correlation function (pCF), and number and brightness (N&B) to measure protein movement, oligomerization state, and protein–protein interaction. We use RICS and pCF to quantify the rate and direction, respectively, of protein movement [1, 8–10, 12–19]. The RICS analysis uses raster images and a 3D Gaussian diffusion model to calculate the average diffusion coefficient of a population of cells [8, 9]. We use pCF analysis on line scans to measure the directionality of movement within or between cells [14]. Further, we use N&B and cross-correlation analyses to characterize oligomeric state and the stoichiometry of protein–protein complexes [11, 17, 20, 21]. N&B uses the mean and variance of the fluorescence intensity distribution to determine the oligomeric state of proteins, while cross N&B determines protein–protein binding by cross-correlating pixels over space and time. Using these techniques together allows us to determine key kinetic parameters, increasing our understanding of various biological processes.

2 Materials

2.1 Plant Growth

1. Plates: We recommend 13 by 13 cm^2 plates, but any plates suitable for growing *Arabidopsis* seedlings may be used.
2. 1 × MS, 1% sucrose plant media: For 500 mL of media, add 5 g of granulated agar to a 1 L flask. In a separate large beaker, add 400 mL water, 5 g sucrose, 0.25 g MES, and 2.165 g MS salts. Mix solution and then pH to 5.7 using 1 M KOH. Add 100 mL of water to the beaker and stir. Aliquot the solution into the 1 L flask. Cover with foil and autoclave on 20 min liquid cycle. After media cools, pour into plates. Plates should be stored at 4 °C.
3. Bleach.
4. 1 M hydrochloric acid (HCl).
5. 1.5 mL tubes.
6. Container with lid.
7. Sterile water.
8. P1000 micropipette and sterile 1 mL tips, or sterile Pasteur pipettes.
9. Parafilm.

2.2 Imaging and Image Analysis

1. Slides (25 × 75 mm).
2. Coverslips (22 × 40 mm).
3. Micropipette and tips.

4. Propidium iodide (PI): Dissolve 10 mg of PI powder in 10 mL of water. Store at 4 °C in the dark. For imaging, aliquot 200 mL of this stock solution into 10 mL of water.
5. Tissue wipers.
6. Forceps.
7. 0.6 μM GFP solution: Prepare by diluting 1 M rGFP protein (Clontech #632373) in water.
8. Heptane glue solution: Prepare according to [22] (*see* **Note 1**).
9. SimFCS software (www.lfd.uci.edu/globals).
10. Commercial confocal microscope, such as the Zeiss LSM 710, LSM 780, or LSM 880.

3 Methods

3.1 Plating Arabidopsis Seedlings and Mounting for Imaging

1. Aliquot seeds to be imaged in a 1.5 mL tubes. Place open tubes into a container with a lid. Place container in chemical hood.
2. Aliquot 50 mL of bleach into a beaker and place in the container. Add approximately 3 mL of 1 M HCl to the bleach until the solution bubbles and releases gas.
3. Cover container and leave seeds in the gas for 2 h.
4. After 2 h, add 500–700 μL of sterile water to tubes in a sterile hood. Keep imbibed seeds at 4 °C in the dark for 2 days.
5. After 2 days, plate seeds on plates containing 1× MS, 1% sucrose media under a sterile hood. Using either a P1000 micropipette or sterile Pasteur pipettes, pipette seeds one at a time onto the plate. Plates should then be wrapped in Parafilm and grown at desired conditions.
6. Once seedlings have reached desired age, mount 1–3 plants on one slide. Place plants on slide using forceps and add water. Then, place coverslip on top of the roots. The shoots should be outside of the coverslip (*see* **Note 1**).

3.2 Point Spread Function (PSF) Calculation

1. Acquire raster images of a 0.6 μM aqueous solution of EGFP with the objective to be used for imaging (typically 40× water objective, NA 1.2). Set gain to 800, digital offset to 0, and digital gain to 1. The laser intensity should be set high enough so that the GFP is visible, but not so high as to photobleach the sample (*see* **Note 2**). Set frame size to 256 × 256 pixels, pixel dwell time to 3.15 μs, and pixel size to 50 nm.
2. Acquire a time series of 100 images (*see* **Note 2**).
3. After acquiring images, start the SimFCS software and click on the RICS button.

4. In the RICS screen, load an image file by selecting File → Open multiple images.
5. Calculate the autocorrelation function (ACF) by selecting Tools → RICS →. Subtract moving average.
6. Click "Fit" to fit a diffusion model to the ACF. Set pixel time to 3.15, line time to 1.89, pixel size to 0.05, and D1 to 78. Click the checkbox so that pixel size and D1 are fixed as previously shown [13] (*see* **Note 3**).
7. Click "Perform fit." The value returned in the Wo box is the PSF that best fits the data (*see* **Note 4**).

3.3 Raster Image Correlation Spectroscopy (RICS)

1. Acquire Raster Images of Plants of Interest. Set Gain and Laser as described previously (Subheading 3.2, **step 1**). Zoom in on a Region of Interest such that the pixel size is between 50 and 100 nm. Set frame size to 256 × 256 pixels and pixel dwell time to 12.61 μs (Fig. 1).
2. Collect a time series of 100 raster scans (*see* **Note 2**).
3. After acquiring images, start the SimFCS software and click on the RICS button.
4. In the RICS screen, load an image file by selecting File → Open multiple images.
5. Select the channel corresponding to your fluorescent protein, i.e., if GFP was imaged in channel 1, select Ch1.
6. Mark the "Use moving average" box and set the moving average to 10 (*see* **Note 5**).
7. Calculate the autocorrelation function (ACF) by selecting Tools → RICS →. Subtract moving average.
8. In the top menu, click "Fit" to fit a diffusion model to the ACF. Set pixel time to 12.61 μs, line time to 7.53 ms, pixel size to the pixel size of your images (usually between 0.05 and 0.1 μm), and Wo to the PSF of the objective. Click the checkbox so that pixel size and Wo are fixed (Fig. 1).
9. Click "Perform fit." A plot will appear that shows the diffusion model fit to the ACF (Fig. 1). The value returned in the D1 box is the diffusion coefficient that best fits the data (*see* **Note 6**).

3.4 Pair Correlation Function (pCF)

1. Set imaging parameters (Subheading 3.3, **step 1**).
2. Select a region of interest of the root that contains 2–3 cells. The area should be cropped such that the edge of the frame is flush with the cell wall. Additionally, the middle of the frame should not be on top of a cell wall, as this is where the line scans (Fig. 2).
3. Take a snapshot of the region you are going to image as a reference.

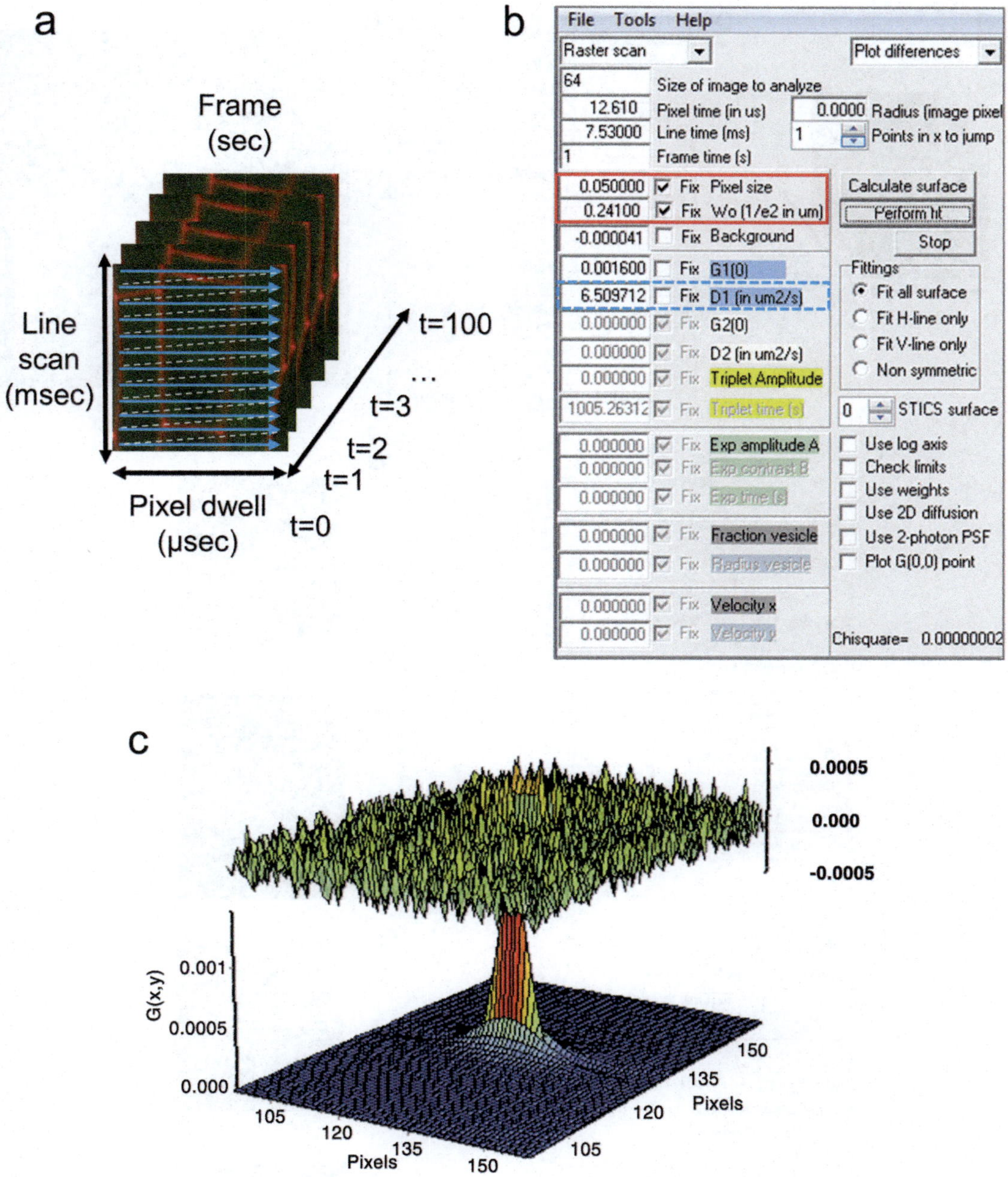

Fig. 1 RICS analysis using the SimFCS software. (**a**) Image acquisition for RICS. (**b**) RICS analysis window. Parameters outlined with a *red*, *solid box* are set before performing the fit. *Blue*, *dashed box* highlights the value of the diffusion coefficient that best fits the data. (**c**) Diffusion model fits to the autocorrelation function. Residuals are shown at the top of the graph

4. Change the image type from a frame to a line. Set the line to 32 pixels long (Fig. 2).
5. Collect a time series of 200,000 frames.

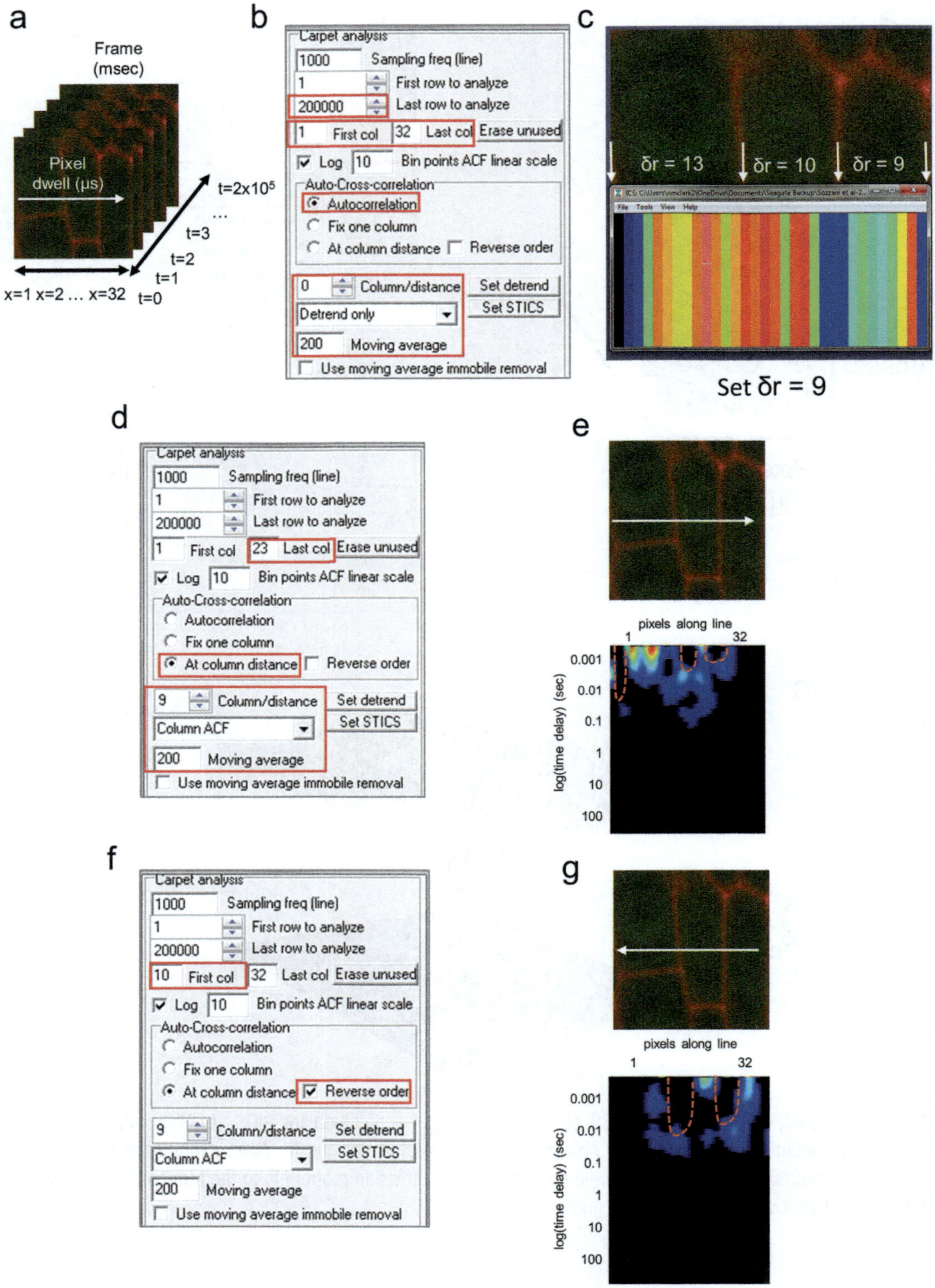

Fig. 2 pCF analysis using the SimFCS software. (**a**) Image acquisition for pCF. (**b**) Calculation of the autocorrelation carpet of the image. *Red* boxes highlight parameters that are set before calculation. (**c**) Setting of the column distance (δr). Cell walls are marked in *red* using propidium iodide (PI). *White* arrows show cell wall boundaries. The pCF carpet is aligned with the middle of the reference image, where the laser scans along a line.

6. After the line scan has finished, reuse the settings from the reference image to take another snapshot of the region. This second image is used to check if the sample moved during imaging. If it appears that the sample moved, repeat **steps 3** and **4** (*see* **Note 5**).
7. After acquiring images, open the SimFCS software and select SimFCS. Then select Analysis → Scanning FCS.
8. Open an image file by selecting File → Open .tif line file. A smaller window will appear with image loading settings. Close the window to apply default settings. The image is loaded as a carpet on the left side of the window.
9. Apply an average to the carpet by setting points per period to 32 and periods to average to 800. Then click "Calculate Image" to obtain the averaged carpet.
10. Calculate the autocorrelation of the carpet by selecting "Carpet analysis." In the pop-up window, set last row to analyze to 200,000 and last col. to 32. Select the autocorrelation button. Set column distance to 0, select "Detrend only," and set moving average to 200. Then click "Analyze." The resulting image is 32 columns, where the color of each column represents the autocorrelation function at that pixel (Fig. 2).
11. Choose the column distance (δr) by aligning the carpet with the reference image. To align the carpet, select "Hold image 1" in the drop-down box on the bottom right to obtain a pop-up window of the carpet. The distance should be close to the width of one of the cells in the image (*see* **Note 7**) (Fig. 2).
12. Perform pCF by selecting "Carpet analysis." In the pop-up window, set the last col. to $32-\delta r$. Select the "at column distance" button. Set column distance to δr, select "Column ACF," and set moving average to 200. Then click "Analyze" to obtain the pCF carpet (Fig. 2).
13. Under the "Smooth" tab, click "Smooth columns" twice and then "Smooth surface" twice.
14. Adjust the color scale by selecting "Scale." Uncheck "Autoscale," and then drag the top slider to the right until the large area of blue near the bottom turns black. Then, drag the bottom slider to the left to highlight areas of yellow or red that have high autocorrelation.

Fig. 2 (continued) Calculation of the width of each cell. δr is the smallest of these distances (in this example, $\delta r = 9$). (**d**) Performing pCF analysis on the carpet. *Red* boxes highlight parameters that are set before calculation. The last column is set to $32-\delta r$. (**e**) Analysis of the pCF carpet for movement. The *orange* arches represent delayed movement across the cell wall in the direction of the *white* arrow. (**f**) Analysis of the line scan in the opposite direction. The first column is set to $1 + \delta r$ and the "reverse order" box is checked. (**g**) Analysis of the pCF carpet for movement in the opposite direction. *Orange* arches represent delayed movement

15. Select "hold image 1" from the drop-down menu to obtain the pCF carpet. The carpet can be saved as an image by selecting File → Copy bitmap to clipboard, pasting the image into Paint or a similar program, and then saving the image as a bitmap.
16. To analyze the image for protein movement, align the pCF carpet with the reference image. If there is an arch directly underneath a barrier, such as a cell wall, then there is delayed movement. If there is no arch but rather a vertical, black line, there is no protein movement (Fig. 2).
17. To analyze the same line scan in the reverse direction, at **step 11**, check the "Reverse order" checkbox. Set the last col. to 32 and the first col. to 1 + δr. Then, repeat **steps 11–15** (Fig. 2).

3.5 Number and Brightness (N&B)

1. First, acquire images of plants that express a monomeric form of the fluorophore of interest. For example, use 35S:GFP as a control for plants that express GFP. The image should contain a region of the plant expressing the fluorophore and a region that is outside of the plant (referred to as the background). Set imaging parameters as previously described (Subheading 3.3, **step 1**) (Fig. 3).
2. Collect a time series of 100 raster scans (*see* **Note 2**).
3. After acquiring control images, start the SimFCS software and click on the RICS button.
4. In the RICS screen, load an image file by selecting File → Load multiple images.
5. Calculate the autocorrelation function (ACF) by selecting Tools → RICS →. Subtract moving average.
6. Perform N&B analysis by selecting Tools → N and B → Detrend.
7. Set Υ min and Υ max such that all of the pixels are visible in the graph (*see* **Note 8**). Uncheck the "autoscale" box and then click "Recalculate plots." Under the Filter tab, a filter can be applied to reduce noise (*see* **Note 9**) (Fig. 3).
8. To calibrate the *S*-factor using the background image, change cursor type to "rectangle" and increase the size such that it covers the background region of the image. The area selected by the cursor will be highlighted in the selection map.
9. Under the Cal tab, adjust *S*_factor 1 such that the background region is centered at $y = 1$. Each time the value is changed, click "Recalculate plots" to update the image (*see* **Notes 10** and **11**).
10. Once the *S*-factor is adjusted, repeat **steps 1** and **2** to acquire images of the monomeric form of the fluorophore in the region of interest, without any background (Fig. 3).

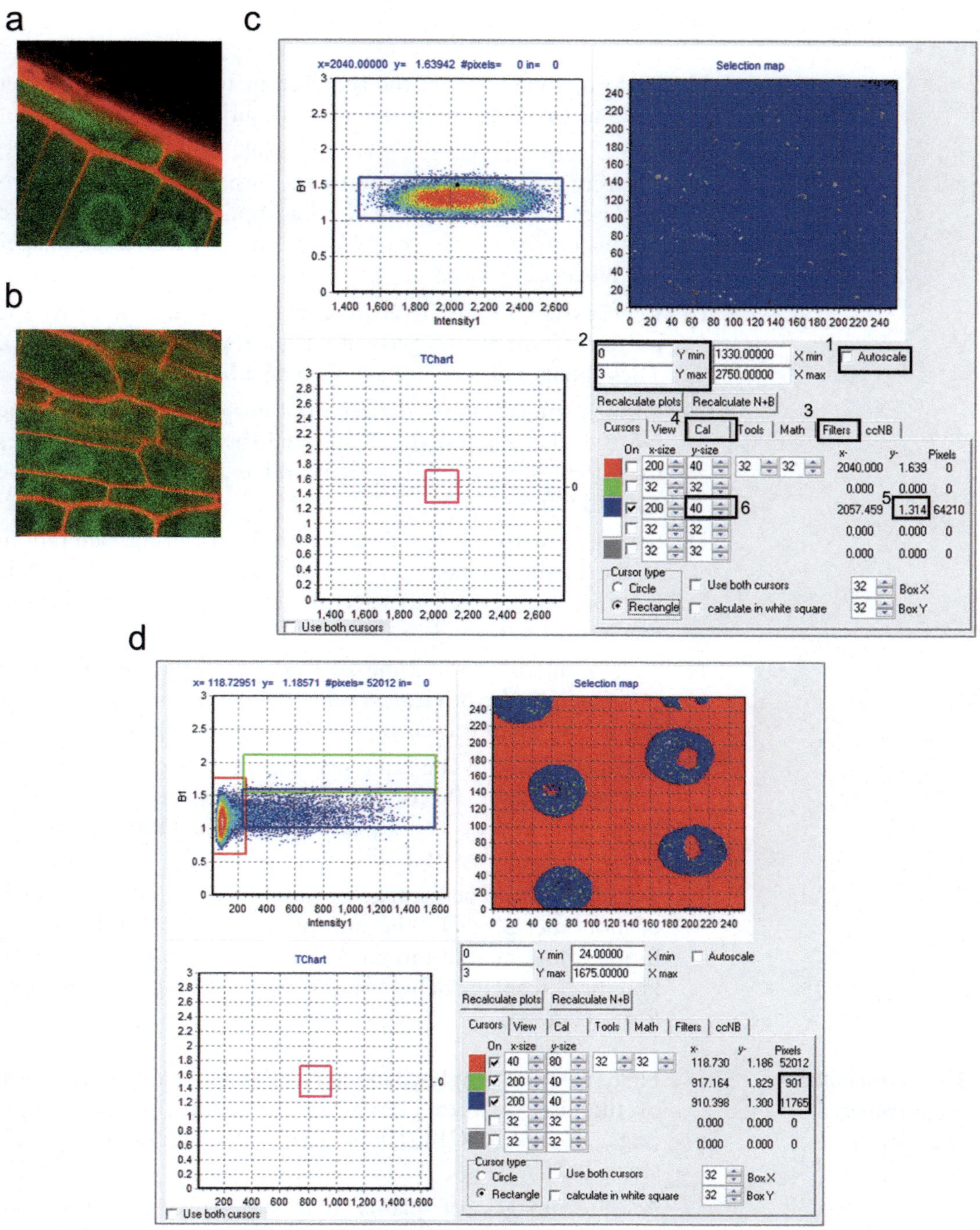

Fig. 3 N&B analysis using the SimFCS software. (**a**) Image used for *S*-factor calibration. Cell walls are marked in *red* using PI. The image contains a region of the frame that does not have any fluorescence (referred to as "background" or "immobile fraction"). (**b**) Image to determine brightness of the monomeric fluorophore. Unlike panel (**a**), this image contains no background. (**c**) Measuring brightness of monomeric fluorophore. The selection map is the image in (**b**) false colored according to the brightness vs. intensity graph. *Blue* represents monomeric protein. (*1*) The autoscale box is unchecked to set *Y* min and *Y* max (*2*). (*3*) Image is smoothed using the filter. (*4*) The *S*-factor is set to the calibrated value under the Cal Table (*5*) The monomer cursor (*blue*) is positioned at the center of the distribution and the size (*6*) is set. The brightness of the monomer is the value in box (*5*) minus 1. (**c**) N&B analysis of an image. In the false-colored image, *red* represents background, *blue* represents monomer, and *green* represents homodimer. Pixel values in the *black* box are used to determine the proportion of monomer and homodimer

11. Perform **steps 3–7** on the acquired images. After **step 7**, set the *S*-factor to the value determined during calibration.
12. Use a cursor to cover all of the pixels in the brightness vs. intensity graph. The *y*-position of the cursor is used to calculate the brightness of the fluorophore. Record the size of the cursor to use when analyzing experimental images (*see* **Note 12**) (Fig. 3).
13. Repeat **steps 1** and **2** to acquire images of plants that have the protein of interest tagged with a fluorophore (*see* **Note 13**). These images do not need to include a background region.
14. Perform **steps 3–6** on the acquired images. After starting the N&B analysis, adjust the *Υ* min and the *Υ* max to the values used for calibrating the system and reduce noise using the median filter (**step 7**).
15. Set the *S*-factor to the value determined during calibration. If the image contains an area of background or immobile fraction, use a cursor to select it and exclude it from analysis (*see* **Note 14**) (Fig. 3).
16. Take a cursor of the same size used during calibration and position it at the brightness value of monomeric protein. Any pixels that fall in this cursor are monomers (Fig. 3).
17. Another cursor of the same size can be placed on top of the monomer cursor. Pixels that fall inside this box are homodimers. This process can repeated to detect homotrimers and higher oligomeric states (Fig. 3).
18. To calculate the percentage of monomer, divide the number of monomer pixels by the sum of the pixels of all oligomeric forms (monomer, homodimer, etc.). This can be repeated to find the percentage of homodimer and other higher oligomeric forms (Fig. 3).

3.6 Cross Number and Brightness (Cross N&B)

1. First, perform single-channel N&B on a monomeric form of the fluorophores present in the double-tagged sample (e.g., 35S:GFP, UBQ10:mCherry, etc.) (Subheading 3.5, **steps 1–10**).
2. Acquire images of plants that have two proteins each tagged with a separate fluorophore as previously described (Subheading 3.5, **steps 1** and **2**) (Fig. 4) (*see* **Note 12**).
3. After acquiring images, start the SimFCS software and click on the RICS button.
4. In the RICS screen, load an image file by selecting File → Load multiple images.
5. Change the active channel to "Ch2-Ch1 (B1–B2 map)." Then select Tools → N and B → Detrend.

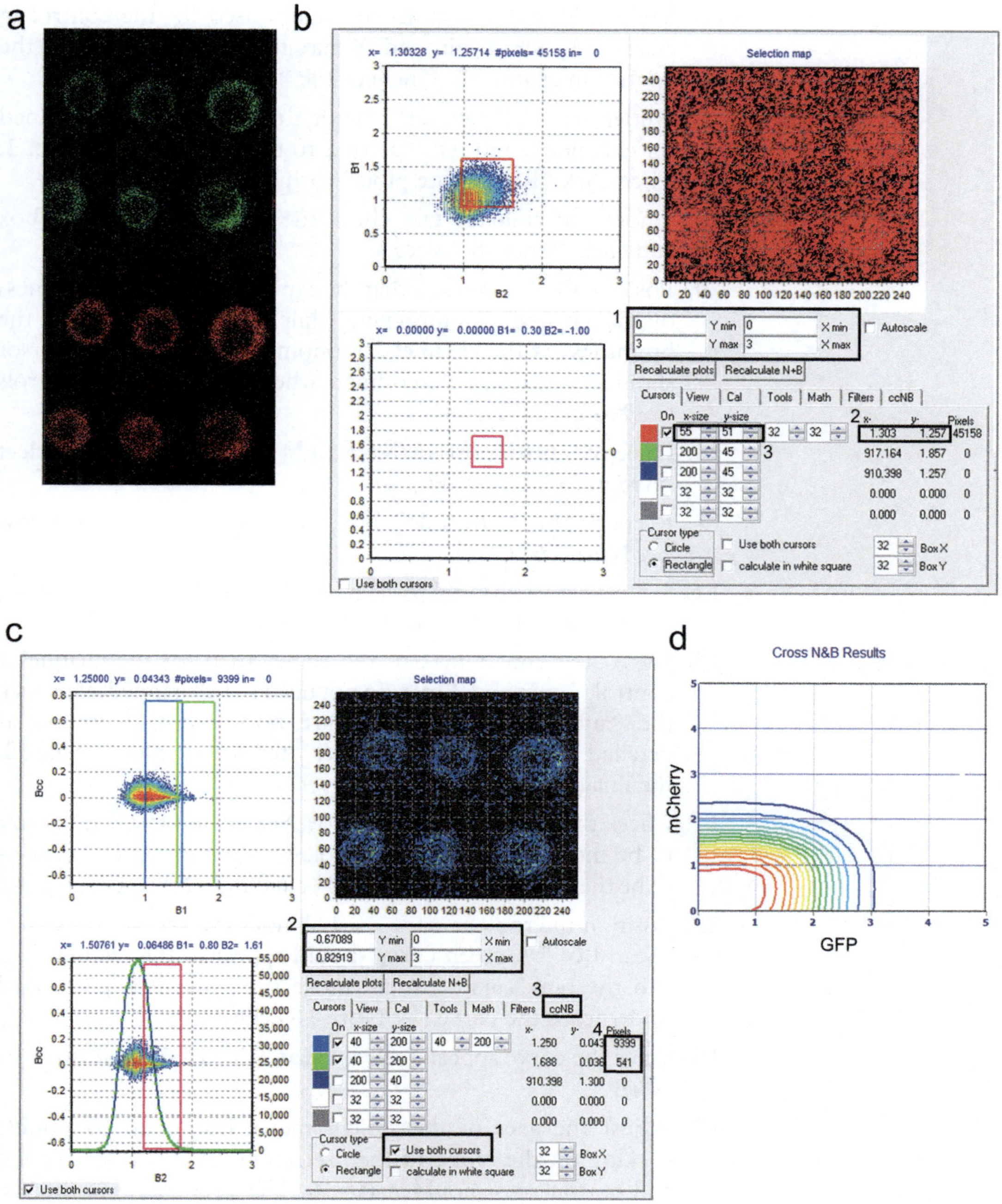

Fig. 4 Cross N&B analysis using the SimFCS software. (**a**) Two proteins are marked with different fluorophores (one with GFP, the other with mCherry). (**b**) First, the B1–B2 area is set. (*1*) *X*min/*Y*min and *X*max/*Y*max are set based on N&B analysis of controls. (*2*) The cursor position is based on the brightness of GFP monomer (*y*-axis) and the brightness of mCherry monomer (*x*-axis). (*3*) The cursor size is expanded to include most of the pixels to the left and right of the monomer position. (**c**) Cross N&B analysis. In the false-colored image, *blue* represents GFP monomer binding mCherry monomer, while *green* represents GFP homodimer binding mCherry monomer. (*1*) Both the top and bottom cursors are used for the analysis. (*2*) *X*min and *X*max are set, but *Y*min and *Y*max are free. (*3*) The cross-correlation curve (*green*) in the bottom graph is calculated using the ccNB Table (*4*) Pixel values are used to determine proportion of complexes (monomer to monomer, homodimer to monomer, etc.). (**d**) Stoichiometry histogram from cross N&B analysis. The colors of the level curves represent the proportions of each complex

6. Set *Y* min and *Y* max to the values used for the control in channel 1. Set *X* min and *X* max to the values used for the control in channel 2. Uncheck the "autoscale" box.
7. Under the "Cal" tab, set *S*_factor 1 to the *S* factor determined for channel 1, and set *S*-factor 2 to the *S* factor for channel 2. Then click "Recalculate plots" to apply changes.
8. Under the "Filters" tab, check the "Use median filter" box and click "Smooth" twice.
9. Position the cursor such that the *x*-position is at the brightness of the channel 1 monomer, while the *y*-position is at the brightness of the channel 2 monomer. The size of the cursor should be the size determined when analyzing the controls (*see* **Note 15**) (Fig. 4).
10. Once the cursor position is set, right click the cursor and select "Assign to B1-B2" (Fig. 4).
11. Recheck the "autoscale" box, and then close the window (*see* **Note 16**).
12. Set the active channel to "Ch1-Ch2 (Bcc)" map in the drop-down menu. Then select Tools → N and B → Detrend.
13. Set *X* min and *X* max to the values used for the channel 1 control. *Y*min and *Y*max do not need to be changed. Uncheck the "autoscale" box. Make sure both *S*-factors are set as detailed in **step 7**, and then click "recalculate plots." Smooth the image as in **step 8**.
14. Check the "use both cursors" box. Set the top (B1) cursor size to be the same as for the control. Set the bottom (B2) cursor to be the same size as the top (B1) cursor (*see* **Note 17**) (Fig. 4).
15. Position the top cursor at $x = 1$. Under the "ccNB" tab, click "Scan Bcc." A green curve should appear in the second plot, and the pink cursor should move such that it is positioned underneath the curve (*see* **Note 18**) (Fig. 4).
16. Once the curve appears, select "Scan B1-B2" to perform cross N&B.
17. Adjust the axes of the stoichiometry histogram by double clicking on the plot and then selecting "Axis." Both the left and bottom axis should start at 0. The level curves in the histogram represent the different stoichiometries of the protein complex (*see* **Note 19**) (Fig. 4).
18. To calculate the percentages of each complex, repeat **step 15**. After the pink cursor has been automatically repositioned, move the top cursor to the position of the monomer on the *x*-axis. The highlighted pixels are the pixels that cross-correlate between the two channels. Turn on a second cursor and move it to the position of the homodimer. This can be repeated for higher oligomeric states. The percentage can then be calculated (*see* Subheading 3.5, **step 15**).

4 Notes

1. When mounting plants for imaging, use as little water as possible so that the roots are covered in water. Use tissue wipers to absorb any excess water before imaging. If the coverslip is able to easily move around, less water should be used. If there are issues with sample movement during imaging, apply 30 μL of the heptane glue solution to the slide. Allow to dry; then place plants on top of the dried glue. Add water and then place the coverslip on top. If plants still move, use more glue.
2. We recommend using a line with a known diffusion coefficient, such as 35S:GFP [17], to set the best imaging parameters such as pixel dwell time, pixel size, and laser intensity for the system. These imaging parameters vary depending on the confocal model and objective, so setting them using a known control can save time when imaging experimental samples. Suggested ranges of imaging parameters for the Zeiss 710 and 780 confocal models have been previously described [17]. For RICS, we recommend a smaller pixel size (typically 50 nm). In addition, while 100 frames are typically used, the number can range from 50 to 300 frames, where a higher number of frames minimize the error in the diffusion model fit to the ACF [18].
3. GFP has a known diffusion coefficient of 78 $\mu m^2/s$ [23].
4. The expected PSF should be between 0.2 and 0.4 μm [13].
5. The moving average is used to correct for sample movement and photobleaching during image acquisition [8]. Typically the moving average is calculated as the floor of the number of frames divided by 10. If the moving average is an odd number, subtract 1 from the value. For example, a time series of 70 frames would use a moving average of 6. While the moving average can correct for slight movement and photobleaching, images where the region of interest shifts outside of the field of view or that have severe photobleaching should not be used for any of analyses described here. If images near the end of the time series are severely photobleached or have high movement, they can be eliminated from the analysis by changing the last frame. For example, if the last 20 images for RICS are severely photobleached, change the last frame to 80 and recalculate the moving average.
6. When the diffusion model is fit to the ACF during the RICS analysis, the residuals of the fit are displayed at the top of the curve (Fig. 1). These residuals can be used to determine the quality of the image. We recommend that the residuals not exceed ½ of the amplitude of the ACF curve (denoted G(0) or G0 in the software). If the residuals are larger than this, there is likely an issue with imaging parameters or sample movement.

7. Due to irregularities in cell wall orientation and cell size, we recommend repeating pCF analysis with a column distance of $\delta r \pm 2$ as previously described [17].
8. For the N&B plots, we recommend a Υ min (X min) of 0 and Υ max (X max) of 3 or 4.
9. A filter may be used on images for N&B to remove noise: however, this is not essential for the analysis and can be skipped. We recommend using the median filter for 8-bit images and the mean (default) filter for 12-bit images.
10. When adjusting the *S*-factor, if the background is above 1, increase the *S*-factor. If the background is below 1, decrease the *S*-factor. The cursor can be fixed in place by unchecking the "Change cursor position on refresh" box under the View tab. Recheck the box when the *S*-factor calibration is finished.
11. The *S*-factor does not need to be calibrated for confocal models higher than the Zeiss LSM 780, i.e., Zeiss LSM 880 and above.
12. Cursor size refers to the height of the cursor. A typical cursor size is between 30 and 60 pixels. The cursor should be wide enough to select all of the pixels. (Fig. 3).
13. The monomeric form of the fluorophore fused to the protein(s) of interest is needed as a control for N&B and cross N&B.
14. Pixels that represent background fluorescence or immobile structures should not be used for the N&B analysis. Any background or immobile fraction of the image will have low intensity and a brightness close to 1. When positioning the cursor, only the region of the image with no fluorescence should be marked as background. If there is no background region of the image, this step may be skipped.
15. When setting the B1–B2 cross correlation area, if there are pixels to the right or top of the cursor, the cursor should be enlarged to include them. These pixels represent higher oligomeric forms and should be included in the cross N&B analysis.
16. When performing N&B analysis, we recommend rechecking the "autoscale" box before closing the plotting window. Sometimes, failure to check this box can require restarting the software.
17. When performing the cross N&B analysis, the Bcc vs. B1 graph has brightness on the *x*-axis instead of the *y*-axis as in single-channel N&B. Thus, the cursor size refers to the width instead of the height of the cursor.
18. During cross N&B analysis, if the curve does not appear above the bottom (B2) cursor, try moving the top (B1) cursor to $x > 1$. Then, click "Scan Bcc" again. Repeat until the green curve appears. Reposition the top (B1) cursor at $x = 1$ before continuing the analysis.

19. The color of the level curves is used to analyze the stoichiometry histogram. Red and orange indicate a high amount of binding. Yellow and green indicate a low amount of binding. Blue indicates that little to no complex of that ratio exists.

Acknowledgments

N. M. C. is supported by a NSF GRF (DGE-1252376). This work was funded by a NSF CAREER grant (MCB-1453130) to R. S.

References

1. Digman MA, Gratton E (2011) Lessons in fluctuation correlation spectroscopy. Annu Rev Phys Chem 62:645–668. doi:10.1146/annurev-physchem-032210-103424.Lessons
2. Petrasek Z, Schwille P (2008) Precise measurement of diffusion coefficients using scanning fluorescence correlation spectroscopy. Biophys J 94:1437–1448. doi:10.1529/biophysj.107.108811
3. Gallagher KL, Paquette AJ, Nakajima K, Benfey PN (2004) Mechanisms regulating SHORT-ROOT intercellular movement. Curr Biol 14: 1847–1851. doi:10.1016/j.cub.2004.09.081
4. Koizumi K, Wu S, MacRae-Crerar A, Gallagher KL (2011) An essential protein that interacts with endosomes and promotes movement of the SHORT-ROOT transcription factor. Curr Biol 21:1559–1564. doi:10.1016/j.cub.2011.08.013
5. Xu XM, Wang J, Xuan Z et al (2011) Chaperonins facilitate KNOTTED1 cell-to-cell trafficking and stem cell function. Science 333:1141–1144
6. Nakashima K, Takasaki H, Mizoi J (2012) NAC transcription factors in plant abiotic stress responses. Biochim Biophys Acta 1819: 97–103. doi:10.1016/j.bbagrm.2011.10.005
7. Sornaraj P, Luang S, Lopato S, Hrmova M (2016) Basic leucine zipper (bZIP) transcription factors involved in abiotic stresses: a molecular model of a wheat bZIP factor and implications of its structure in function. Biochim Biophys Acta 1860:46–56. doi:10.1016/j.bbagen.2015.10.014
8. Digman MA, Brown CM, Sengupta P et al (2005) Measuring fast dynamics in solutions and cells with a laser scanning microscope. Biophys J 89:1317–1327. doi:10.1529/biophysj.105.062836
9. Digman MA, Sengupta P, Wiseman PW et al (2005) Fluctuation correlation spectroscopy with a laser-scanning microscope: exploiting the hidden time structure. Biophys J 88:L33–L36. doi:10.1529/biophysj.105.061788
10. Digman MA, Gratton E (2009) Analysis of diffusion and binding in cells using the RICS approach. Microsc Res Tech 332:323–332. doi:10.1002/jemt.20655
11. Digman MA, Wiseman PW, Choi C et al (2009) Stoichiometry of molecular complexes at adhesions in living cells. Proc Natl Acad Sci U S A 106:2170–2175
12. Jameson DM, Ross JA, Albanesi JP (2009) Fluorescence fluctuation spectroscopy: ushering in a new age of enlightenment for cellular dynamics. Biophys Rev 1:105–118
13. Rossow MJ, Sasaki JM, Digman MA, Gratton E (2010) Raster image correlation spectroscopy in live cells. Nat Protoc 5:1761–1774. doi:10.1038/nprot.2010.122
14. Hinde E, Cardarelli F, Digman MA, Gratton E (2010) In vivo pair correlation analysis of EGFP intranuclear diffusion reveals DNA-dependent molecular flow. PNAS 107:16560–16565. doi:10.1073/pnas.1006731107
15. Hinde E, Cardarelli F (2011) Measuring the flow of molecules in cells. Biophys Rev 3:119–129
16. Vetri V, Ossato G, Militello V et al (2011) Fluctuation methods to study protein aggregation in live cells: Concanavalin A oligomers formation. Biophys J 100:774–783. doi:10.1016/j.bpj.2010.11.089
17. Clark NM, Hinde E, Winter CM et al (2016) Tracking transcription factor mobility and interaction in Arabidopsis roots with fluorescence correlation spectroscopy. Elife. doi:10.7554/eLife.14770

18. Brown CM, Dalal RB, Hebert B et al (2008) Raster image correlation spectroscopy (RICS) for measuring fast protein dynamics and concentrations with a commercial laser scanning confocal microscope. J Microsc 229:78–91
19. Digman MA, Gratton E (2009) Imaging barriers to diffusion by pair correlation functions. Biophys J 97:665–673. doi:10.1016/j.bpj.2009.04.048
20. Digman MA, Dalal R, Horwitz AF, Gratton E (2008) Mapping the number of molecules and brightness in the laser scanning microscope. Biophys J 94:2320–2332. doi:10.1529/biophysj.107.114645
21. Digman MA, Wiseman PW, Horwitz AR, Gratton E (2009) Detecting protein complexes in living cells from laser scanning confocal image sequences by the cross correlation raster image spectroscopy method. Biophys J 96:707–716. doi:10.1016/j.bpj.2008.09.051
22. Supatto W, Mcmahon A, Fraser S, Stathopoulos A (2009) Quantitative imaging of collective cell migration during Drosophila gastrulation: multiphoton microscopy and computational analysis. Nat Protoc 4:1397–1412. doi:10.1038/nprot.2009.130
23. Chen Y, Mu JD, Ruan Q, Gratton E (2002) Molecular brightness characterization of EGFP in vivo by fluorescence fluctuation spectroscopy. Biophys J 82:133–144. doi:10.1016/S0006-3495(02)75380-0

Chapter 17

Studying Protein–Protein Interactions *In Planta* Using Advanced Fluorescence Microscopy

Marc Somssich and Rüdiger Simon

Abstract

The formation of protein complexes through direct protein–protein interaction is essential for virtually every biological process, and accordingly the ability to determine the interaction properties of specific proteins is important to understand these processes. Förster resonance energy transfer (FRET) measurements are state-of-the-art confocal fluorescence microscopy- and imaging-based techniques that allow the analysis of protein interactions in vivo and *in planta*, in specific compartments of single cells or tissues. Here we provide a step-by-step guide to perform FRET measurements by acceptor photobleaching (APB) and fluorescence lifetime imaging microscopy (FLIM) in the plant expression system *Nicotiana benthamiana*.

Key words Confocal fluorescence microscopy, Live cell imaging, Protein–protein interaction, FRET, FLIM, APB, Acceptor photobleaching, Fluorescence lifetime, *Nicotiana benthamiana*

1 Introduction

Studying direct protein–protein interactions is a prerequisite in modern molecular biology to analyze and understand different biological processes. The physical phenomenon Förster resonance energy transfer (FRET) serves as the basis for several techniques that allow the detection and quantification of protein interactions in vivo and in near physiological conditions [1]. FRET describes the emission-free transfer of energy between two fluorophores through intermolecular long-range dipole–dipole coupling [2]. For this, three prerequisites have to be met: first, the emission spectrum of the fluorophore that emits the energy (named the donor fluorophore) must overlap with the excitation spectrum of the fluorophore that will absorb the energy (named the acceptor fluorophore) [2]. Second, the FRET efficiency depends on the distance between the two fluorophores with an inverse sixth power law, i.e., the closer they are, the stronger the FRET. The distance at which 50% of the donor energy is transferred to the acceptor is

Wolfgang Busch (ed.), *Plant Genomics: Methods and Protocols*, Methods in Molecular Biology, vol. 1610,
DOI 10.1007/978-1-4939-7003-2_17, © Springer Science+Business Media LLC 2017

referred to as the Förster distance, which is specific for every fluorophore pair and has been determined for GFP and mCherry to be about 51 Å or 5.1 nm in an in vitro situation [2, 3]. Third, the donor emission dipole moment has to be aligned with the acceptor absorption dipole moment [2, 4]. If these conditions are fulfilled, FRET can occur between the donor and the acceptor. While in this FRET state, the fluorescence of the donor will be reduced ("quenched"), due to the energy being transferred to the acceptor, without emission, through FRET, while the acceptor gets excited by the resonance energy and its fluorescence will increase.

In FRET measurements, these characteristics of the FRET-pair fluorophores are exploited to study interactions between proteins. If the two free fluorophores are expressed in a cell, they will not interact with each other as they do not have a natural affinity to form complexes, and accordingly no FRET between them is detectable. To test if two proteins can interact, one of the proteins is fused to the FRET-donor GFP, the other to the FRET-acceptor mCherry. Then, the two fusion proteins are co-expressed in a cell. If the two proteins interact to form a complex, the fused fluorophores will be brought into close proximity, and FRET can occur between them. This FRET can then be measured and quantified by a variety of different techniques [5], out of which we will focus on FRET-acceptor photobleaching (APB) and FRET-fluorescence lifetime imaging microscopy (FLIM) [6, 7].

FRET-APB is a simple, fast, and effective technique to measure stable interactions between two proteins. It is based on the dequenching (i.e., increase) of the donor fluorescence following photobleaching of the acceptor fluorophore [6]. To this end, the acceptor fluorophore is irreversibly photobleached by strong irradiation with a laser that is specific for the excitation spectrum of the acceptor but will not affect the donor. Once the acceptor is photobleached, the donor fluorescence is no longer quenched, which will be reflected in a quantifiable increase in fluorescence. The benefits of this technique are low hard- and software requirements and fast execution. It is, however, a crude application with limited sensitivity when it comes to interaction dynamics and strength, and due to the invasive photobleaching step, it is not suitable for time-lapse experiments. For these needs, **FRET-FLIM** measurements are the preferred choice. The fluorescence lifetime describes the time that a fluorophore remains in the excited state after it was excited by a laser pulse and before it rids itself of the acquired energy by emitting a photon, i.e., fluorescence [7]. This lifetime is specific for every fluorophore and dependent on its FRET state. To measure it, the elapsed time is determined from the moment the fluorophore is excited by a short laser pulse until the emitted photon is detected. FRET between a donor and an acceptor fluorophore will reduce the donor's average fluorescence lifetime; thus in a FRET sample, the average donor lifetime will be shorter, relative to a donor-only sample or a FRET-negative sample. This technique is almost noninvasive due to very low laser power used, and accordingly it can also be used to measure interaction dynamics over time in the same cell. Furthermore, it is sensitive enough to allow for quantification of FRET efficiency and, i.e., interaction strength [8]. However, these advantages go along with more technically demanding measurements and high hard- and software requirements.

The expression system *Nicotiana benthamiana* allows the analysis of two plant proteins for their ability to interact in a plant context, rather than switching to an in vitro situation or a completely different expression system, such as *Saccharomyces cerevisiae*, which is commonly used for yeast two-hybrid interaction studies [9, 10]. However, the use of strong promoters, such as the cauliflower mosaic virus (CaMV) 35S promoter, can result in artifacts due to overexpression of the proteins tested, affecting not only their intracellular localization but also the interaction properties. To avoid these issues, an inducible expression system can be used, which allows to perform the FRET measurements as soon as the protein concentration levels reach the lower detection limit [11, 12].

Here we provide a step-by-step guide to measure FRET *in planta* by FRET-APB and FRET-FLIM.

2 Materials

1. Vectors: Gateway® destination vectors pABindGFP, pABindmCherry and pABindFRET (Fig. 1 and *see* **Note 1**) [12], and the p19 silencing suppressor [13].
2. Enzymes and kits: Invitrogen pENTR™ Directional TOPO® Cloning Kit, Gateway® BP Clonase® II, and Gateway® LR Clonase® II Cloning Kits or compatible.
3. Bacteria and plants: An *Agrobacterium tumefaciens* strain (GV3101 pMP90 is recommended and tested for this protocol) and *Nicotiana benthamiana* plants. This protocol is based on experiments using plants that were grown for 3–4 weeks in a greenhouse under natural daylight conditions. Older plants can be used too, albeit with reduced transformation efficiency.

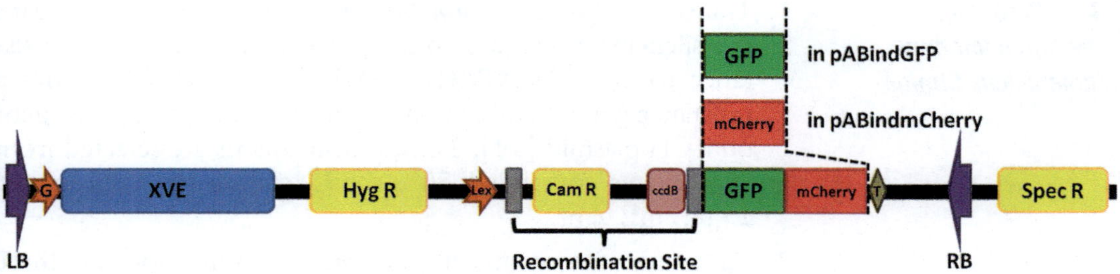

Fig. 1 Vector map of pABindFRET. The backbone of pABindFRET, pABindGFP, and pABindmCherry is identical; the difference is the encoded fluorophore (between the *black dotted lines*). *See* **Note 1** for details

4. Double yeast tryptone medium (DYT): Mix 16 g tryptone, 10 g yeast extract, and 5 g sodium chloride per liter in purified water and autoclave immediately. For solid medium add 18 g agar-agar prior to autoclaving. Medium can be stored at room temperature (RT).
5. Rifampicin (rif) (50 mg/mL in DMSO). Store at −20 °C.
6. Kanamycin (kan) (50 mg/mL in purified water). Store at −20 °C.
7. Gentamicin (gent) (25 mg/mL in purified water). Store at −20 °C.
8. Spectinomycin (spec) (100 mg/mL in purified water). Store at −20 °C.
9. Beta (β)-estradiol: 100 mM in DMSO.
10. Acetosyringone: 450 mM in DMSO.
11. Infiltration medium (IF): 5% sucrose, 0.01% Silwet® L-77, 450 μM acetosyringone, and a spatula tip of glucose in purified water (*see* **Note 2**). Prepare fresh immediately before use and keep on ice or at 4 °C.
12. Induction medium (ID) in a vaporizer: 20 μM β-estradiol and 0.1% polysorbate 20 (i.e., Tween® 20) in purified water. Prepare fresh immediately before use and keep on ice or at 4 °C.

3 Methods

This chapter will focus on using the GFP-mCherry FRET pair as published in Bleckmann et al. (2010) [12]. The principles, however, are applicable to other FRET pairs as well. The corresponding plasmids are pABindGFP, pABindmCherry, and pABindFRET (Fig. 1) [12]. To prepare expression vectors, the gene of interest has to be cloned into a Gateway®-compatible entry vector to use in a Gateway® LR-reaction with the three destination vectors. The fluorescent tag will be attached at the C-terminal end of the protein, and accordingly the gene must be cloned without the stop codon and in frame with the fluorophore.

3.1 Preparing the Agrobacterium tumefaciens Strains

1. Transform *Agrobacterium tumefaciens* GV3101 pMP90 with the silencing suppressor plasmid p19. This will confer resistance to kan. The GV3101 pMP90 strain already carries a resistance gene against rif on its chromosome and against gent on its Ti plasmid [14]. Thus, transformants are selected from DYT plates containing 50 μg/mL of each rif and kan and 25 μg/mL gent.
2. To obtain transgenic *A. tumefaciens* strains for the three pABindGFP, pABindmCherry, and pABindFRET expression vectors, transform one aliquot of the acquired GV3101

pMP90 p19 strain with one of each of the vectors. The expression clones confer resistance to spec, and accordingly transformants are selected from DYT plates containing rif, kan, gent, and 100 μg/mL spec.

3. Plate the transformed *A. tumefaciens* strains on DYT medium containing the appropriate antibiotics, and let them grow for 2 days at 28 °C until distinct colonies are visible (*see* **Note 3**).
4. Inoculate 5 mL liquid DYT medium containing the appropriate antibiotics with one colony, and let it grow overnight at 28 °C on an incubation shaker.
5. Start a fresh liquid culture by adding 1 mL of the overnight culture to 4 mL of fresh liquid DYT medium containing the appropriate antibiotics, and place it back into the 28 °C incubation shaker for 4 h (*see* **Note 4**).
6. Pelletize the bacteria by centrifuging at about 3000 × *g* for 10 min at 4 °C.
7. Decant the liquid medium and carefully take off the remains using a pipette. Then resuspend the bacterial pellet in 5 mL IF medium. Keep on ice.
8. Measure the optical density of the suspension at 600 nm wavelength (OD_{600}) using a spectrophotometer, and adjust it to 0.8 using fresh IF medium.
9. For each infiltration prepare one reaction tube per plant leaf, containing equal amounts of the strains that should be co-infiltrated. The final OD_{600} of the mix should be 0.8. So if two strains are supposed to be co-infiltrated, they will be mixed 1:1, resulting in an OD_{600} of 0.4 for each individual strain and 0.8 overall. Let them rest on ice for roughly 1 h.

3.2 Preparing the Nicotiana benthamiana Plants

1. Optimally, the plants should have leaves with a diameter of roughly 2–3 cm and with strong venation (*see* Fig. 2a). Water them well, and place them on a plant tray under a cover to keep them in a humid environment. This will cause the plants to open their stomata.
2. Label each leaf that will be infiltrated. It is not recommended to infiltrate more than two leaves per plant.
3. When the *A. tumefaciens* strains are ready, infiltrate the mixture into the adaxial side of the leaves using a 1 mL syringe with a flat tip and without a needle (Fig. 2a). Make sure the entire leaf is infiltrated, and use a fresh pair of gloves for every infiltration to avoid cross-transformation through bacteria on the gloves.
4. Place the plants back into the tray and under the cover, and keep them in a continuous light phytochamber at 16 °C (or 21 °C, if 16 °C is not available) for at least 48 h (or up to 7 days).

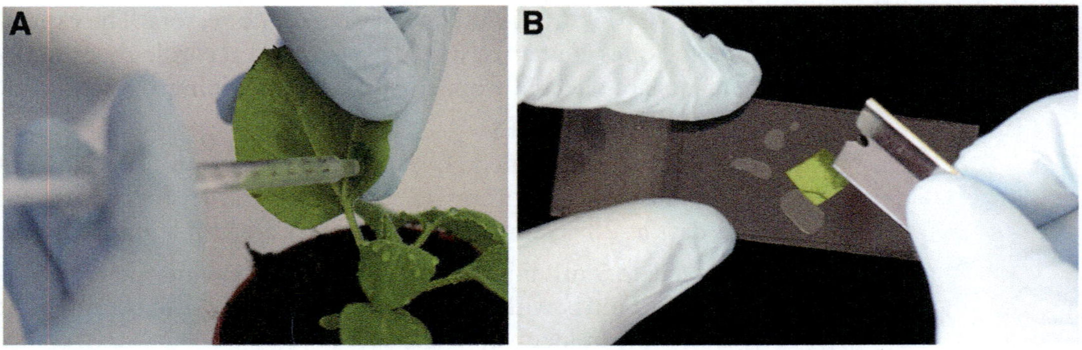

Fig. 2 (**a**) Infiltrating agrobacteria in IF solution into the adaxial side of a *Nicotiana benthamiana* leaf with a flat top syringe. (**b**) Sample preparation on a microscope slide

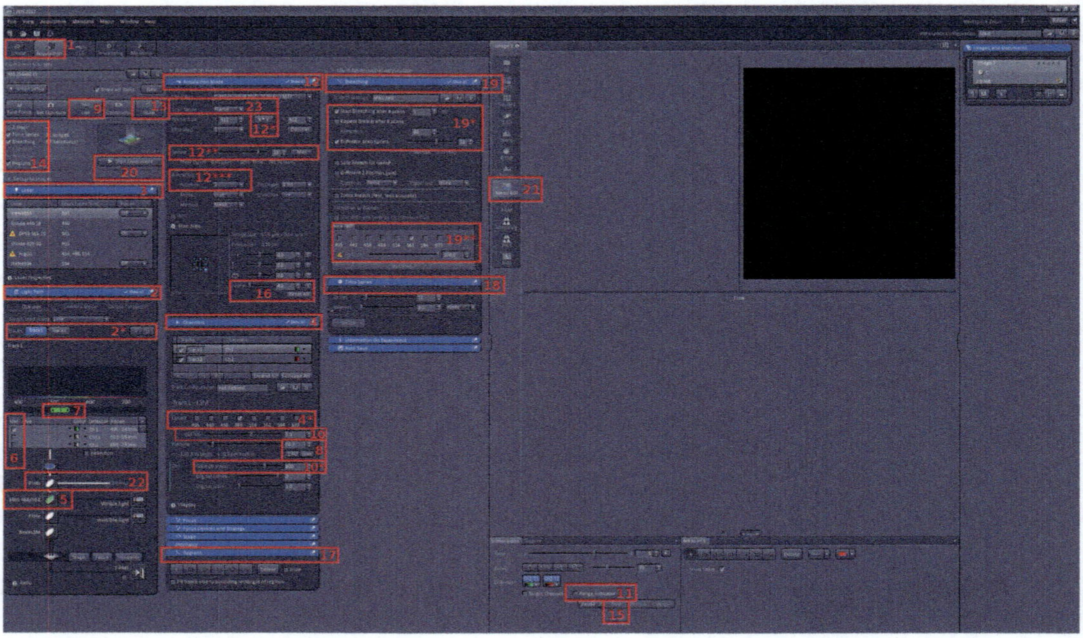

Fig. 3 Interface of the ZEN software in confocal microscope mode with all tool windows on display that will be used in the following descriptions. The different tool windows will be referred to by the names in the *blue* bars on top of the single windows and the *red* number

5. After at least 2 days, spray the leaves that should be induced for gene expression with ID medium using a vaporizer, and place them back into the tray and under the cover in the phytochamber until starting microscopy (*see* **Note 5**).

3.3 Microscopy

The descriptions in this chapter are based on using the ZEISS Efficient Navigation (ZEN) software for ZEISS confocal laser scanning microscopes. However, the principles should be applicable to software from other manufacturers as well. Fig. 3 displays the ZEN interface with all the tool windows that will be mentioned in the following steps. The "Locate" tab is used for scanning of the sample using the light microscope, while the "Acquisition" tab is used for the confocal microscope mode (1 in Fig. 3).

3.3.1 Sample Preparation

1. To prepare your sample, cut a small piece out of the *N. benthamiana* leaf (roughly 0.5 × 0.5 cm in size), and place it in water onto a microscope slide. Cover it with a cover slip, and squeeze the air out of the leaf by carefully stroking over it with a razor blade. You will see the water entering the tissue (Fig. 2b). Then place it into the microscope.

3.3.2 Setting Up the Light Paths

1. Make sure to set up two separate tracks for sequential scanning of the GFP and mCherry channel in the "Light Path" tool (2 and 2* in Fig. 3). Simultaneous scanning of both in one joint track will result in cross talk between the two channels.
2. Activate the correct light source for both channels, i.e., a laser with a wavelength corresponding to the excitation spectrum of the fluorophore, in the "Laser" tool (3 in Fig. 3), and select one of them for each of the two separate tracks you have created in the previous step in the "Channels" tool (4 and 4* in Fig. 3). For GFP (excitation maximum 488 nm), a laser with 488 nm wavelength is suitable, for mCherry (excitation maximum 587 nm) a 561 nm laser.
3. Choose an appropriate major beam splitter (MBS) in the "Light Path" tool (2 and 5 in Fig. 3). Ideally it will match both light sources. For the GFP/mCherry pair, a 488/561 MBS will work well. If there is no MBS available that works for both light sources, each track must have its own specific MBS.
4. Check one of the detector channels for each of the two tracks (6 in Fig. 3).
5. Adjust the detector range to the emission wavelength of the fluorophores by adjusting the sliders position and width (7 in Fig. 3). For GFP that would be around 509 nm, for mCherry around 610 nm (*see* **Note 6**).
6. Adjust the spatial pinhole to one aerial unit (AE) in the "Channels" tool to eliminate out-of-focus signal, and limit detection to the confocal volume in the sample (4 and 8 in Fig. 3).

3.3.3 Record an Image

To do the adjustments described in the following points, use the "Live" mode to scan your sample (9 in Fig. 3). In "Live" mode, the sample is scanned in low resolution and in the highest speed, allowing you to adjust the settings while scanning, but reducing the damage done due to over radiation.

1. To record the ideal image, the laser power and detector gain have to be specifically adjusted for every sample in the "Channels" tool (4 and 10/10* in Fig. 3). The easiest way to

do this is by adjusting the values while doing "Live" scans to immediately see the effect of the changes. For both laser power and detector gain, a good compromise between detected signal intensity, background noise, and fluorophore bleaching must be found. The range indicator in the software is a helpful tool to adjust these values (11 in Fig. 3). *See* **Note 7** for details.

2. Next, adjust frame size, scan speed, and frame averaging in the "Acquisition Mode" tool (12–12*** in Fig. 3). The frame size determines the resolution of the image. A bigger frame size will result in a higher resolution. The scan speed determines how long the signal from a single pixel will be recorded (pixel dwell time). A longer pixel dwell time will result in a stronger signal due to more photons being collected and thereby a higher local resolution. Eventually, averaging the signal from several frames can help to reduce the background noise (*see* **Note 8**). The drawback from these options is always the same: Higher resolution, longer pixel dwell time, and repeated scanning of the same frame can lead to damage due to over-radiation, which can result in photobleaching and destruction of the fluorophores or lead to cell damage. Again, the best compromise has to be found for every individual sample.
3. If all settings are correct, record your image using the "Snap" mode (13 in Fig. 3).

3.3.4 FRET-APB

FRET-APB measurements are based on changes in donor fluorescence (GFP) following the photobleaching of the acceptor fluorophore (mCherry). Accordingly, bleaching prior and following the actual photobleaching event must be prevented, as it would influence the result. To obtain this, low laser power, low-resolution (256 × 256 pixels), and high-speed (maximum) scanning are the keys. Furthermore, all lights in the room should be turned off, and the sample should be covered (e.g., with a little cardboard box) to keep any light from reaching it.

Next to your interaction sample (GFP- and mCherry-tagged proteins co-transformed), you will need the intramolecular FRET-positive control (pABindFRET single transformation) and a negative control (*see* **Note 9**).

In the ZEN software, you will need the "Time series," "Bleaching," and "Regions" tools, so make sure they are activated (14 in Fig. 3).

1. Select and focus on a region of a cell where both fluorophore-tagged proteins are stably co-expressed by quickly scanning the sample using the "Live" mode (9 in Fig. 3) and then using the "Crop" tool to choose a region (15 in Fig. 3).
2. Zoom in on the region of interest in the "Acquisition Mode" tool (*see* **Note 10** and 12 and 16 in Fig. 3). Use the same zoom factor in all measurements of one experiment to ensure comparability of results.

3. Choose the region to bleach using the "Regions" tool (17 in Fig. 3). The scanner of a confocal microscope scans individual lines from left to right, progressing line after line from top to bottom. Accordingly, long and narrow regions along the *x* axis (few long lines) can be scanned quicker than regions spreading along the *y* axis (many short lines) and should therefore be preferred, if possible (see red frame in the upper right image in Fig. 4) (*see* **Note 11**).
4. To account for variation in the fluorescence intensities, the change in donor fluorescence after photobleaching of the acceptor is calculated as change in average fluorescence of five frames before and five frames after the photobleaching event. The "Time Series" tool allows for the recording of several frames (choose 12 frames) (18 in Fig. 3).
5. The bleaching event is controlled in the "Bleaching" tool (19 in Fig. 3). Set the start of bleaching after five frames, and choose around 80 iterations for the bleaching (adjust to more iterations, if 80 are not sufficient to bleach the mCherry to background level.

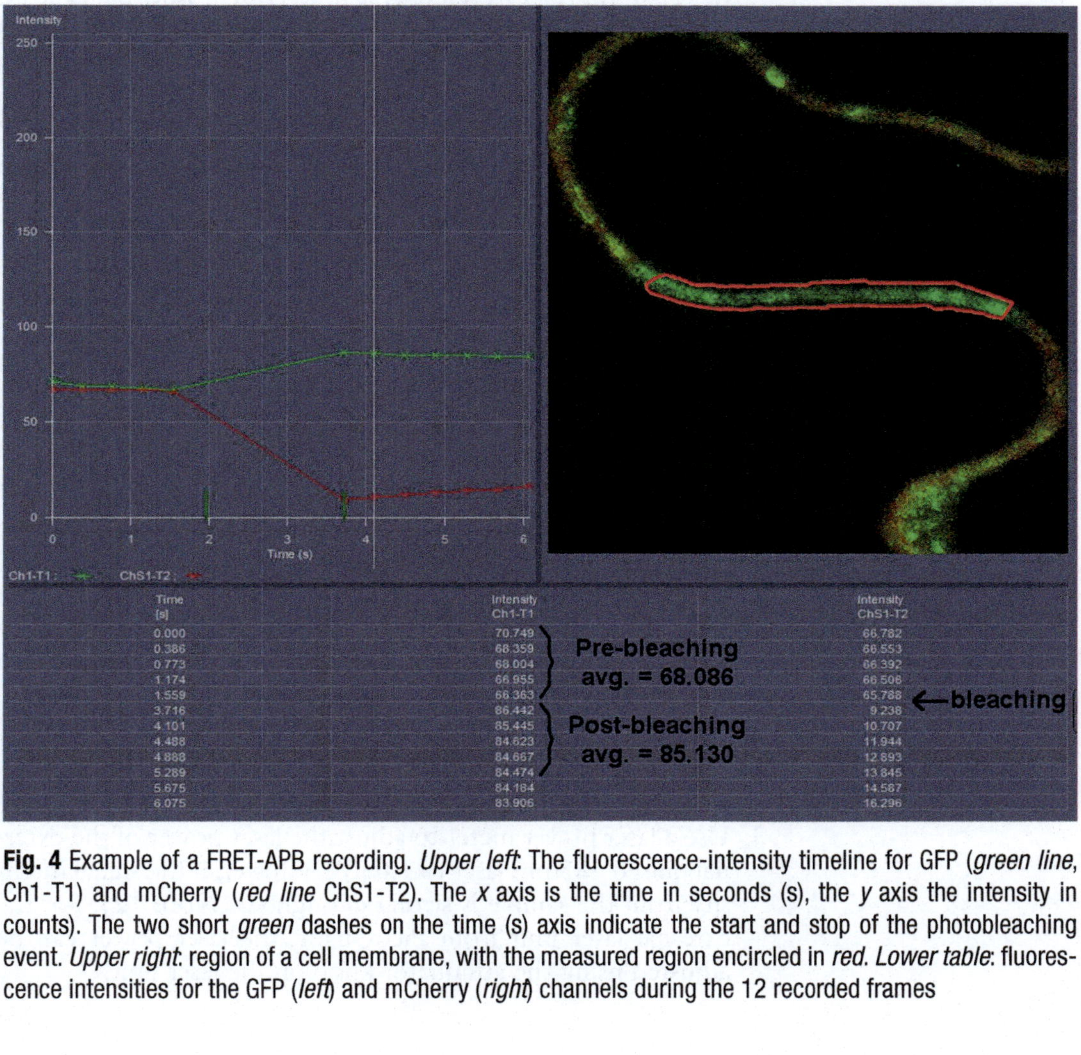

Fig. 4 Example of a FRET-APB recording. *Upper left*: The fluorescence-intensity timeline for GFP (*green line*, Ch1-T1) and mCherry (*red line* ChS1-T2). The *x* axis is the time in seconds (s), the *y* axis the intensity in counts). The two short *green* dashes on the time (s) axis indicate the start and stop of the photobleaching event. *Upper right*: region of a cell membrane, with the measured region encircled in *red*. *Lower table*: fluorescence intensities for the GFP (*left*) and mCherry (*right*) channels during the 12 recorded frames

Less can also suffice, depending on the sample). Make sure the scan speed for the bleaching is set to maximum in both the "Bleaching" and the normal acquisition tool (19* in Fig. 3). Select the acceptor laser (561 nm for mCherry) for the bleaching event and set it to 100% power in the "Bleaching" tool window (19** in Fig. 3).

6. Make sure that frame size is set to 256 × 256 pixels and scan speed is at maximum (12* and 12** in Fig. 3). Then start the measurement by clicking on "Start Experiment" (20 in Fig. 3).
7. Display the "Mean ROI" in the recorded image window, and make sure that the GFP and mCherry intensities are stable (fluorescence fluctuations should be less than 10%) in the five frames prior to the bleaching event, and that they are on a comparable level. Following the bleaching, the mCherry may slowly recover, depending on protein mobility (*see* 21 in Fig. 3 and example in Fig. 4).
8. Copy the table with the time and intensity values (Fig. 4, table at the bottom of the image) to a spreadsheet file for later analysis. Then proceed to the next cell. Record at least ten cells for each interaction. As for all experiments, the results obtained are of higher quality with an increasing number of measurements.

3.3.5 FRET-FLIM

The measurements described here are performed with an external PicoQuant FLIM setup on a Zeiss LSM 780 using a PicoQuant 485 nm pulsed diode laser and PicoQuant Tau-SPAD detectors, as well as the PicoQuant SymPhoTime Software. The basic configuration properties should, however, be comparable in other setups. Fig. 5 displays the SymPhoTime interface with all the tool windows that will be mentioned in the following steps.

1. Create a new workspace for the measurements in the SymPhoTime Software.
2. In the SymPhoTime "Measurement" tab, activate "Point" measurement (1 and 1* in the insert in Fig. 5). In the "Test" tab, at sub-window "Point Recording Parameters," activate "Stop after" 60.0 s (2 and 3 in Fig. 5). In sub-window "TCSPC Settings," set the TCSPC resolution to 1.0 ps (4 in Fig. 5). In the "Laser Settings" sub-window, set the "sync rate" to 32.00 MHz, and activate the 485 nm pulsed laser (5 and 6 in Fig. 5). Make sure that "cw" (continuous wave) is unchecked. Set the relative laser power to a fixed percentage such as 60%. This setting should be constant for all measurements to come. The absolute power will be adjusted in the next step.
3. Use a laser power meter to adjust the laser power of the external pulsed 485 nm laser to 1.0 μW or lower, if the signal intensity from the sample is strong enough (*see* **Note 12**). For the external laser unit from PicoQuant, the laser power can be adjusted using the attenuator knob on the laser unit.

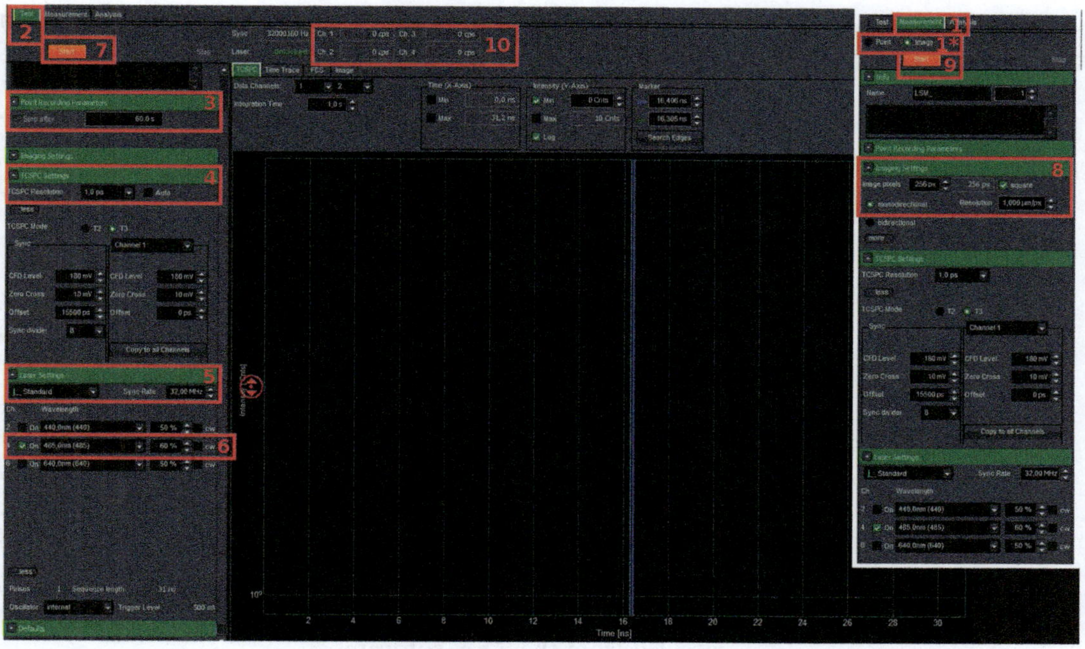

Fig. 5 Interface of the SymPhoTime Software for FLIM recordings. On the left, the control panel for "Test" runs is shown. The corresponding panel for "Measurement" runs is inserted in the *top-right corner*. The different tool windows will be referred to by the names in the *green* bars on top of the single parts

4. Make sure that the detectors are activated, and the correct filters are inserted before them. To split the green and red emission before reaching the detectors, a 560 long pass (LP) filter will be suitable (directing light of all wavelengths longer than 560 to the red and shorter wavelengths to the green detectors), followed by a 520/35 filter for the green channels, and a 607/75 filter for the red channels. A 520/35 filter has its peak passage at 520 nm with 35 nm spread around this value. Accordingly, a 607/75 filter is optimized for 607 nm with a 75 nm spread. Therefore these two filters are compatible to the emission values of GFP (507) and mCherry (610 nm). The exact filters will depend on the available hardware, but must be in accordance with the emission wavelengths of the used fluorophores. The system needs to be calibrated for every session. To do this, the background noise and instrument response function (IRF) have to be determined.
5. Pipette 100 μL of ultrapure water to a cover slip, and place it into the microscope. This will serve to determine the background noise.
6. Use the ocular of the microscope to focus into the middle of the water drop on the cover slip.

7. In the "Light Path" tool window of the ZEN software (2 in Fig. 3), select the laser slot that the external PicoQuant lasers are connected to and move its power slider to 100% (10 in Fig. 3), as the actual power was set externally in one of the previous steps. Chose an appropriate MBS (*see* point 3.3.2.3), and direct the light path away from the internal detectors and toward the external detectors which are connected to the optical fiber output (22 in Fig. 3).
8. In the "Acquisition Mode" tool window of the ZEN software, change the scan mode from "Frame" to "Point" (12 and 23 in Fig. 3).
9. Activate "Time Series" in the ZEN software, and set the number of cycles to the maximum possible number in the "Time Series" tool window to ensure continuous scanning (18 in Fig. 3).
10. In the SymPhoTime Software, start the measurement by clicking on "Start" in the "Measurement" Table (7 in Fig. 5). The actual recording will only start once a trigger signal is received from the ZEN software.
11. In the ZEN software, start the measurement by clicking on "Start Experiment" (20 in Fig. 3). This will also trigger recording in the SymPhoTime Software. Once the 60 s recording in SymPhoTime is done, stop the measurement. If you are correctly focused on the middle of the water drop, you should see an increased count rate of the different detectors (10 in Fig. 5).
12. Repeat this using an erythrosine solution instead of water. Erythrosine has a very short lifetime which will overlap with the IRF. Accordingly, the erythrosine measurement can later be used to subtract the IRF signal from the measurements.
13. Once the calibration measurements are performed, prepare your *N. benthamiana* leaf sample.

3.3.6 FLIM Recordings

1. In the "Measurement" tab of the SymPhoTime Software, activate "Image" measurement (1 and 1* in Fig. 5). In the "Image Settings" sub-window, make sure the "Image pixels" are set to 256 × 256 and that "square" is checked. Make sure "monodirectional" scanning with a resolution of 1000 μm/px is activated (8 in Fig. 5).
2. In the Zen software, select and focus on a region of a cell where both fluorophore-tagged proteins are stably co-expressed by quickly scanning the sample using the "Live" mode with the internal continuous wave (cw) lasers and detectors as described in point 3.3.3. Record an image.
3. Zoom in on the region of interest in the "Acquisition Mode" tool of the ZEN software (*see* **Note 10**), and make sure the image resolution is set to 256 × 256 pixels (12, 12* and 16 in

Fig. 3). Use the same zoom factor and image resolution in all measurements of one experiment to ensure that image and pixel size are the same in all measurements. Make sure that no visible bleaching of the fluorophores occurs.

4. Adjust the scan speed in the ZEN software to ensure that the pixel dwell time is the same in all measurements (12** in Fig. 3). Make sure that no visible bleaching of the fluorophores occurs.
5. Activate time series in the ZEN software. The number of frames scanned in the series is dependent on the sample and has to be determined according to signal strength. While recording, check the counts per second (cps) on the two GFP channels (usually Ch. 1 and Ch. 2). The counts should be in the region of 100,000 to 250,000 (10 in Fig. 5) (*see* **Note 13**).
6. Switch to the pulsed laser and external detectors in the ZEN software, as done previously for the calibration measurements.
7. In the SymPhoTime Software, start the measurement by clicking on "Start" in the "Measurement" Table (9 in Fig. 5).
8. In the ZEN software, start the measurement by clicking on "Start Experiment."
9. In the SymPhoTime Software, stop the measurement by clicking on "Stop" once the time series in the ZEN software is done.
10. Do the same for the mCherry in the SymPhoTime Software, changing the laser source to the internal 561 nm laser and recording only ten frames. This measurement will not be analyzed for the fluorescence lifetime but will inform on the relative amount of acceptor fluorophores in the sample.
11. Proceed to the next cell. Record at least ten cells for each interaction, the negative control (cells expressing the donor-protein only) and the positive control (cells expressing the GFP-mCherry fusion protein or, better, a known interactor of the donor). As for all experiments, the results obtained are of higher quality with an increasing number of measurements.

3.4 Data Analysis

3.4.1 FRET-APB

In APB experiments, FRET is determined by calculating the dequenching of a donor following photobleaching of the acceptor. To account for fluctuations in fluorescence intensity during scanning of a sample, this change is calculated as the change in average intensity of five frames prior and after bleaching.

1. Calculate the average fluorescence intensity of the donor in the five frames prior to bleaching (GFP_{before}) and in the five following frames (GFP_{after}) (*see* Table in Fig. 4).
2. Calculate the apparent donor dequenching in percent (app. $E\%$) using the following formula: app. $E\% = (GFP_{after} - GFP_{before}) \times 100/GFP_{after}$ [15].

3. Determine the average donor dequenching and standard deviation of all measurements performed for one interaction and the negative control to determine if there is FRET between the two fluorophores and thus interaction between the proteins.

3.4.2 FRET-FLIM

Here, we will describe the analysis of FLIM measurements using the PicoQuant SymPhoTime Software.

1. In SymPhoTime, open the workspace where you have saved your measurements. They will be listed on the left side (1 in Fig. 6).
2. In the middle control panel, choose the "Analysis" Table (2 in Fig. 6), and click the "Start" button in the "FLIM" sub-window (2* in the insert in Fig. 6). This will start the FLIM-Analysis interface, displayed in Fig. 6.
3. At the FLIM-Analysis interface, first expand the "Region of Interest" sub-tab and check "Use ROI" (3 in Fig. 6). Then right-click into the measurement image on the right side of the control panel, choose "Free ROI" from the drop-down menu (4 in Fig. 6), and then select your ROI directly in the image, including only the pixels coming from regions you want to include in your analysis.
4. Expand the "FLIM" sub-Table (5 in Fig. 6) and check the data channels for the green emission (mostly 1 and 2). Uncheck the red emission channels. Then check "Threshold" and set the minimum threshold to 300 cnts (5* in Fig. 6 and *see* **Note 13**).

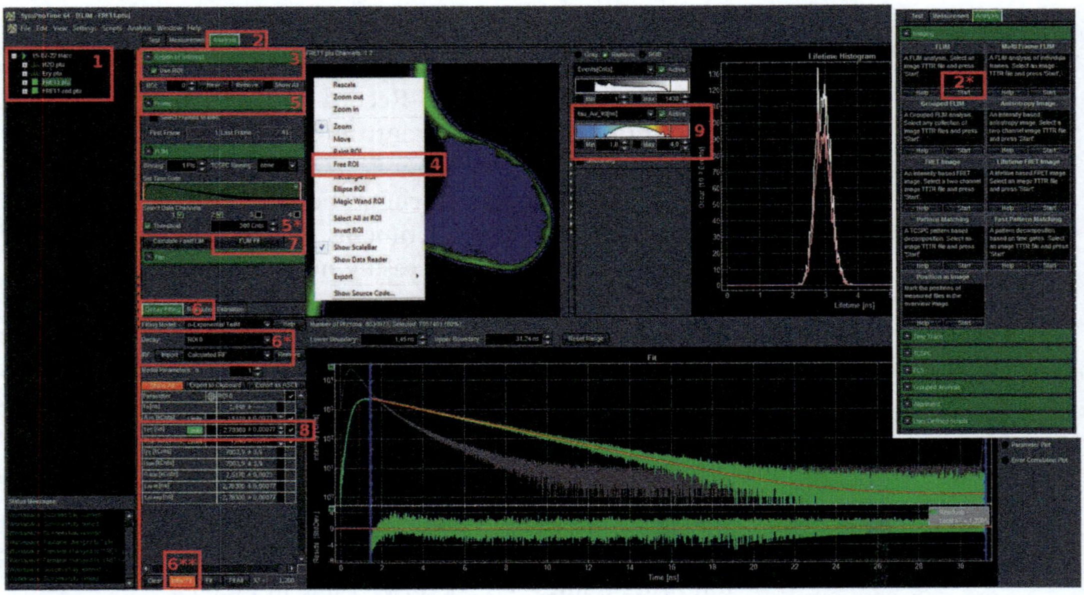

Fig. 6 Interface of the SymPhoTime Software for FLIM analysis. The opened "Analysis" tab with the different analysis techniques is inserted in the *top-right corner*. The different tool windows will be referred to by the names in the *green* bars on top of the single parts

5. At the "Decay Fitting" tab, choose "ROI 0" (the ROI you have selected in the image) for the "Decay," and press "Import" in the "IRF" line to import the IRF you have recorded at point 3.3.5.12. After importing, select it from the drop-down menu next to it (6 and 6* in Fig. 6).
6. Click on the "Initial Fit" button at the bottom of the control panel (6** in Fig. 6), and when the initial fit is done, click on the "FLIM Fit" button in the "FLIM" sub-Table (7 in Fig. 6).
7. The average donor fluorescent lifetime (τ) will be displayed in the table in the "Decay Fitting" Table (6 and 8 in Fig. 6) and also be plotted to the image in color code. The color code can be adjusted in the control panel to the right of the image (9 in Fig. 6).
8. Repeat this for all measurements and then calculate the average fluorescent lifetime of all measurements from the same sample. If FRET between the donor and the acceptor occurred, the average τ of the ten FRET samples will be significantly lower than the average τ of the donor-only samples.

4 Notes

1. pABindGFP, pABindmCherry, and pABindFRET differ only by the fluorophore encoded by them (area between the black dotted lines). The right and left border (RB and LB, lilac arrows) flank the DNA fragment that will be incorporated into the plant cells [16]. The gene of interest is recombined into the vector between the attR1 and attR2 gateway recombination sites (gray boxes) by the LR-clonase reaction and must be in frame with the fluorophore (GFP (green box) and/or mCherry (red box)) coded after the attR2-site [17]. Expression of the fusion protein is controlled by the lexA –46 35S promoter (orange arrow, Lex), which is activated by the XVE (LexA DNA-binding domain (X), VP16 acidic transactivating domain (V), and human estrogen receptor regulatory region (E)) transcription factor (blue box). XVE is encoded on the same vector and is constitutively expressed from the G10–90 promoter (orange arrow, G), but can only bind to the lexA promoter if β-estradiol molecules are present and bound to the XVE [11]. The T3A terminator (brown diamond, T) serves as transcriptional stop. The spectinomycin resistance gene (yellow box, Spec R) will confer spectinomycin resistance in bacteria, the hygromycin resistance gene (yellow box, Hyg R) to hygromycin in plants (and bacteria). Vectors that contain no gene of interest can be selected in *Escherichia coli* through the chloramphenicol resistance gene (yellow box, Cam R) if the cells contain the *ccdA* gene; otherwise the *ccdB* gene (pink box, ccdB) will prevent bacterial growth [18].

2. In order for agrobacteria to infect plant cells, wounding of plant tissue must occur, and the bacterium needs to sense the wounded tissue. In the soil, sugars that are released from wounded plant cells into the ground are used by the bacteria as chemotactic chemicals to guide them to the wound [19]. Subsequently, the plants' phenolic compound acetosyringone, which is associated with wounded plant tissue, is perceived by the bacterium and induces *vir*-gene expression, enabling the bacterium to insert its T-DNA into the plant's genome [20]. Hence, sugars and acetosyringone are used in the IF to enhance transformation efficiency. A spatula tip of magnesium sulfate is sometimes recommended to be added to the medium, e.g., when using the *Agrobacterium tumefaciens* strain LBA4404, to prevent clumping of the bacteria in liquid culture. Scientific evidence for this phenomenon is scarce, however, and magnesium in the IF is therefore missing from some published protocols. We found that in our conditions and working with the strain GV3101 pMP90, magnesium is not needed.
3. While the selection for positive expression clones based on the *ccdB* gene and antibiotics is reliable, it is recommended to additionally check if the colonies carry the desired expression clone by polymerase chain reaction (PCR).
4. To obtain high transformation efficiencies, the bacteria should be in the logarithmic (log) (or exponential) phase of growth, in which the number of living cells doubles per unit of time. After growing overnight, the bacteria will already be in the stationary phase with many of them being dead. Therefore, a fresh culture is started in the morning. Four hours of growth at ideal conditions (fresh medium and 28 °C) will allow for at least two rounds of cell divisions and hence a high number of young, living cells.
5. Plants can be induced 48 h or up to 1 week after infiltration. From 7 days on, efficiency will decrease, but we have successfully used these plants for several more days. The necessary induction time strongly depends on the specific protein expressed and must be determined for each individual case. Small proteins which are quickly folded and matured, and which are localized in the cytoplasm, can reach the lower detection limit within less than 2 h. Larger, more complex protein, such as membrane-integrated receptor-like kinases, may need up to 12 h until they can be stably detected. Generally, measurements should be performed as soon as the intensity is sufficient to record a good image. That way, overexpression artifacts that could influence the proteins intracellular localization and interaction properties can be reduced to a minimum.

6. The exact wavelength range for the detector around those peak emission values has to be determined and adjusted for every single sample. Importantly, it should not overlap with the excitation laser wavelength (i.e., longer than 488 nm for the GFP track and longer than 561 nm for the mCherry track). A narrow range will result in a highly specific signal, but weaker intensity. A broad range will increase the overall detected signal, but also the background noise. Hence, a good compromise needs to be determined. The best way to determine the optimal range for every sample is to start with a broad range, and then narrow it down until you reach the best compromise between signal intensity and background noise.
7. High laser power will produce a strong signal, but will also bleach and eventually destroy the fluorescent proteins. Accordingly, the laser power should be adjusted to a level that allows for repeated scanning of an expressing cell without losing too much of the signal. Low detector gain will result in a highly specific, but weak signal. High gain will increase the signal, but also the overall background noise in the image. The range indicator tool can help to adjust the values. When activated, it will display those pixels in the image with too strong intensity in red and pixels with too low intensity in blue. The pixels in between will be displayed in gray. Ideally, red and blue will be absent from the image.
8. Using frame averaging, a frame will be scanned several times (two, four, or eight times), and eventually the average intensity of every single pixel in all frames is displayed. Since the signal is coming from specific areas in the cell containing the fluorophore-tagged proteins, while the background noise is unspecific to certain pixels and therefore changes between the frames, much of the background will be lost by averaging.
9. The GFP-mCherry fusion protein expressed from the pABindFRET vector serves as an intramolecular FRET-positive control to test the hard- and software setups. Since the donor and acceptor fluorophores are directly fused to each other, they will always be within the FRET distance, and thus, measurements should always result in FRET if hard- and software are set up correctly. A good negative control is essential for the experiment because it allows to judge if the measured app. *E*% from the interaction sample is significant. A simple negative control is to perform FRET-APB measurements on cells expressing only the GFP-fused protein. The app. *E*% value obtained from these measurements represents the system baseline and serves as negative reference [15]. However, the presence of a mCherry-tagged protein can alter this baseline even in the absence of actual protein-protein interaction and FRET between the fluorophores. Accordingly, a preferable

negative control would be the co-expression of a different mCherry-tagged protein that is known to not interact with the donor-tagged protein of interest.

10. Zooming in on a region increases the irradiation in that area, which will lead to stronger photobleaching of the fluorophore. Therefore, the zoom factor has to be chosen carefully.
11. If proteins have a very high mobility (e.g., cytoplasmic proteins), many new proteins will move into the bleached region after the bleaching event, thereby affecting the outcome of the experiment. In such a case, it is useful to select a larger region for the bleaching while choosing a separate, small region in the center of the bleached region for the analysis.
12. It is important that measurements are performed with low laser power. High laser power increases the risk of damaging the cell and the fluorophores (*see also* **Note 6**), which will also influence the lifetime and thus the results. Furthermore, the laser power should be the same for all measurements to ensure comparability.
13. To obtain reliable and high-quality results, a sufficient amount of photons has to be collected for every analyzed pixel. A number of at least 300 photons per pixel (ppp) (or photon counts (cnts)) was proven to work well on our setup. Between 600 and 1000 ppp was determined to be the optimal range. However, the absolute numbers are dependent on the sample and hardware and should be determined experimentally for every setup. Since measurements should be performed with low expression levels to minimize the risk of overexpression artifacts, a single FLIM measurement can require to record around 80 frames to collect the necessary number of photons.

Acknowledgments

M. S. and R. S. thank Petra Žádníková and Maike Breiden for helpful comments and discussion of the manuscript and the CAi of the HHU for technical support. This work was funded by the D. F. G. through the Exc1028 (CEPLAS).

References

1. Gordon GW, Berry G, Liang XH, Levine B, Herman B (1998) Quantitative fluorescence resonance energy transfer measurements using fluorescence microscopy. Biophys J 74(5): 2702–2713
2. Förster T (1948) Zwischenmolekulare Energiewanderung und Fluoreszenz. Ann Phys 2(1–2):55–75
3. Albertazzi L, Arosio D, Marchetti L, Ricci F, Beltram F (2009) Quantitative FRET analysis with the E(0)GFP-mCherry fluorescent protein pair. Photochem Photobiol 85(1):287–297.doi:10.1111/j.1751-1097.2008.00435.x
4. Jares-Erijman EA, Jovin TM (2003) FRET imaging. Nat Biotechnol 21(11):1387–1395. doi:10.1038/Nbt896

5. Jares-Erijman EA, Jovin TM (2006) Imaging molecular interactions in living cells by FRET microscopy. Curr Opin Chem Biol 10(5):409–416. doi:10.1016/j.cbpa.2006.08.021
6. Bastiaens PIH, Jovin TM (1996) Microspectroscopic imaging tracks the intracellular processing of a signal transduction protein: fluorescent-labeled protein kinase C beta I. Proc Natl Acad Sci U S A 93(16):8407–8412. doi:10.1073/pnas.93.16.8407
7. Gadella TWJ, Jovin TM, Clegg RM (1993) Fluorescence lifetime imaging microscopy (FLIM)—spatial-resolution of microstructures on the nanosecond time-scale. Biophys Chem 48(2):221–239. doi:10.1016/0301-4622(93)85012-7
8. Somssich M, Ma Q, Weidtkamp-Peters S, Stahl Y, Felekyan S, Bleckmann A, Seidel CAM, Simon R (2015) Real-time dynamics of peptide ligand-dependent receptor complex formation in planta. Sci Signal 8(388):ra76. doi:10.1126/scisignal.aab0598
9. Bracha-Drori K, Schichrur K, Katz A, Oliva M, Angelovici R, Yalovsky S, Ohad N (2005) Detection of protein-protein interactions in plants using bimolecular fluorescence complementation. Plant J 42(5):781–781. doi:10.1111/j.1365-313X.2005.02444.x
10. Fields S, Song OK (1989) A novel genetic system to detect protein protein interactions. Nature 340(6230):245–246. doi:10.1038/340245a0
11. Zuo JR, Niu QW, Chua NH (2000) An estrogen receptor-based transactivator XVE mediates highly inducible gene expression in transgenic plants. Plant J 24(2):265–273. doi:10.1046/j.1365-313x.2000.00868.x
12. Bleckmann A, Weidtkamp-Peters S, Seidel CAM, Simon R (2010) Stem cell signaling in Arabidopsis requires CRN to localize CLV2 to the plasma membrane. Plant Physiol 152(1):166–176. doi:10.1104/pp.109.149930
13. Qu F, Morris TJ (2002) Efficient infection of *Nicotiana benthamiana* by Tomato bushy stunt virus is facilitated by the coat protein and maintained by p19 through suppression of gene silencing. Mol Plant Microbe Interact 15(3):193–202. doi:10.1094/Mpmi.2002.15.3.193
14. Koncz C, Schell J (1986) The promoter of Tl-DNA gene 5 controls the tissue-specific expression of chimeric genes carried by a novel type of Agrobacterium binary vector. Mol Gen Genet 204(3):383–396. doi:10.1007/Bf00331014
15. Karpova TS, Baumann CT, He L, Wu X, Grammer A, Lipsky P, Hager GL, McNally JG (2003) Fluorescence resonance energy transfer from cyan to yellow fluorescent protein detected by acceptor photobleaching using confocal microscopy and a single laser. J Microsc 209:56–70. doi:10.1046/j.1365-2818.2003.01100.x
16. Zambryski P, Holsters M, Kruger K, Depicker A, Schell J, Vanmontagu M, Goodman HM (1980) Tumor DNA-structure in plant-cells transformed by *A. tumefaciens*. Science 209(4463):1385–1391. doi:10.1126/science.6251546
17. Hartley JL, Temple GF, Brasch MA (2000) DNA cloning using in vitro site-specific recombination. Genome Res 10(11):1788–1795. doi:10.1101/Gr.143000
18. Bahassi EM, O'Dea MH, Allali N, Messens J, Gellert M, Couturier M (1999) Interactions of CcdB with DNA gyrase—inactivation of GyrA, poisoning of the gyrase-DNA complex, and the antidote action of CcdA. J Biol Chem 274(16):10936–10944. doi:10.1074/jbc.274.16.10936
19. Loake GJ, Ashby AM, Shaw CH (1988) Attraction of Agrobacterium tumefaciens C58c1 towards sugars involves a highly sensitive chemotaxis system. J Gen Microbiol 134:1427–1432
20. Schrammeijer B, Beijersbergen A, Idler KB, Melchers LS, Thompson DV, Hooykaas PJJ (2000) Sequence analysis of the vir-region from Agrobacterium tumefaciens octopine Ti plasmid pTi15955. J Exp Bot 51(347):1167–1169. doi:10.1093/jexbot/51.347.1167

Chapter 18

Chemiluminescence-Based Detection of Peptide Activity and Peptide-Receptor Binding in Plants

Mari Wildhagen, Markus Albert, and Melinka A. Butenko

Abstract

In living organisms, physical interaction of ligand molecules with their cognate receptors is an indispensable requirement for the initiation of cellular signaling pathways. To technically prove the biochemical interaction of ligands with their corresponding receptor, a biologically active but labeled peptide is required. Easily scorable bioassays, such as the production of reactive oxygen species, can be used to quantify the activity of a peptide. By using chemiluminescent probes, such as acridinium esters, as conjugates to label peptide ligands of interest, quantitative measurements of ligand-receptor binding can be performed in standard luminometers. Here we describe how this binding approach can be used to reveal peptide ligand-receptor binding in plant material.

Key words Chemiluminescence detection, Oxidative burst, Ligand-receptor binding, Acridinium ester label

1 Introduction

Small molecules that are ligands for receptor kinases can be identified by genetic approaches. In *Arabidopsis* several peptide ligand-receptor systems have been discovered by studying mutant phenotypes and investigating genetic interactions. Based on the genetic approach, putative peptide ligand-receptor pairs can be identified by similarities of their mutant phenotypes and by genetic studies where phenotypes of enhanced expression of the peptide encoding gene are lost in a receptor mutant background. Although several peptide ligand-receptor pairs have been identified this way [1], biochemical evidence for ligands interacting with specific receptors is needed to supplement the genetic results.

Studying the biochemical interaction of ligands with their corresponding receptors requires a highly bioactive peptide, ideally in sub-nanomolar ranges, that retains its activity in the presence of a label and a highly sensitive system for detecting and monitoring the bound ligand. Testing of synthesized versions of the peptide

Wolfgang Busch (ed.), *Plant Genomics: Methods and Protocols*, Methods in Molecular Biology, vol. 1610,
DOI 10.1007/978-1-4939-7003-2_18, © Springer Science+Business Media LLC 2017

and eventual modifications is facilitated by bioassays allowing quantitative readout of peptide activity. Once the active peptide is characterized, labeled versions can be tested for activity by the same bioassay.

In this protocol we first lay out a standardized version of an assay that monitors the production of reactive oxygen species (ROS), also termed oxidative burst, that can be used to quantify the activity of a peptide and then describe how a chemiluminescent labeled version of the peptide can be used to measure peptide ligand-receptor interaction.

The oxidative burst is a typical cellular response of animals and plants to diverse biotic and abiotic stresses. Here, we outline the detection of the oxidative burst in plant leaf pieces using a luminol-based bioassay which allows for the detection of chemiluminescence. The assay was originally developed by Keppler et al. 1989 [2] and subsequently adapted for other plant cells and tissues [3]. We then describe a peptide-binding assay using an acridinium ester-labeled peptide. Chemical oxidation of acridinium esters triggers chemiluminescence, allowing quantitation of this compound to at least femtomol concentrations in standard luminometers. This sensitive binding approach can be used to reveal peptide ligand-receptor binding in plant material or in material with receptors immobilized to affinity beads after immunoprecipitations.

The oxidative burst assay has successfully been used to monitor the activity of a highly active plant peptide involved in regulating floral abscission in *Arabidopsis*, and an acridinium-labeled version of this inflorescence deficient in abscission (IDA) peptide was used to study interaction with the corresponding receptor haesa-like 2 (HSL) [4, 5].

2 Materials

2.1 Luminol-Dependent Oxidative Burst Components

Prepare all solutions and experiments using demineralized (dd) H_2O.

1. Plant leaf pieces (*see* **Note 1**).
2. Peptides of interest; labeled and unlabeled.
3. Peptide for positive control (e.g., flg22).
4. 10× luminol master mix: 200 μM luminol L-012 (Wako Chemicals USA), 10 μg/mL peroxidase (horseradish peroxidase). Prepare 1 mL luminol master mix in an Eppendorf tube: 880 μL H_2O, 20 μL from a 10 mM luminol stock solution, 2 μL from a 5 mg/mL peroxidase stock solution (*see* **Note 2**).
5. Scissors, small (*see* **Note 3**).
6. Petri dishes, normal size; 9 cm diameter.
7. Small spatula, the size should be small enough for it to fit in a well or tube used in the experiment.

8. 96-well plates, white, flat-bottomed (e.g., LIA-plate, white, 96-well, flat-bottom, Greiner Bio-One GmbH) or suitable cuvettes for single-cell instruments (*see* **Note 4**).
9. Luminometer, either a 96-well plate reader (e.g., Centro LB 960, Microplate Luminometer, Berthold Technologies) or a single-cell instrument.

2.2 Acridinium Ester Label Binding Assay Components

1. Plant material harboring the receptor of interest (*see* **Note 5**).
2. Appropriate binding buffer. Example of binding buffer used in [4]: 25 mM MES (2-ethanesulfonic acid, pH 6.0), 150 mM NaCl, 0.49 g MES, 0.88 g NaCl, 100 mL H_2O. Stir until dissolved (*see* **Note 6**).
3. DTT (*see* **Note 7**).
4. Proteinase inhibitor mix.
5. 5 mM citric acid. Weigh 10.5 mg citric acid, and transfer to a small beaker containing 10 mL H_2O.
6. 0.03% H_2O_2 in 100 mM NaOH (prepare fresh right before use). Measure 0.1 mL of a 10 M NaOH stock solution, and place in a 15 mL tube containing 10 mL H_2O. Add 10 μL H_2O_2 from a 30% H_2O_2 solution.
7. Table top centrifuge, with cooling.
8. Luminometer (e.g., single tube machines like FB 12, Berthold Technologies, that allows injection).
9. Acridinium-labeled peptide (*see* **Note 8**).
10. Unlabeled peptide (synthetic unlabeled peptides can be ordered by, e.g., Biomatik).

3 Methods

3.1 Oxidative Burst Measurements

1. Fill petri dishes half full with H_2O.
2. Cut leaves of plants (*see* **Note 9**) into equally sized small pieces, size: 3 mm × 3 mm (*see* **Note 10**). You will get the best result if you use the area of the leaf marked with boxes in Fig. 1.
3. Let leaf pieces float on water in the half full petri dishes. Orientation does not matter, but avoid submergence of the leaf piece.
4. Incubate overnight or a minimum of 8 h at room temperature (*see* **Note 11**), avoid shaking.
5. The next day: 90 μL H_2O and 10 μL luminol master mix (10×) are pipetted into each well of a 96-well plate (*see* **Note 2**). For single-cell luminometers, this might be scaled up, depending on the size of the cuvette used for measurements.

Fig. 1 *Nicotiana benthamiana* leaf. Cutting leaf pieces from the parts of the leaf in the *boxed areas* will give you the best response in an oxidative burst assay. Cut away the edge of the leaf to leave only the *boxed areas*. Cut the *boxed areas* into small pieces, size 3 mm × 3 mm, leave the middle vein

6. Carefully add one leaf piece per well using the spatula, avoid mechanical damage and wounding. The leaf pieces should float in the middle of the well.
7. Per planned treatment, prepare three to four replicates.
8. Measure the background level to ensure that a constant value of emitted light is detected over time (*see* **Note 12**). If the background levels are fluctuating, it might indicate that the plant material is stressed. Usually a measurement over a 5 min period is sufficient to control if the baseline is stable. If the baseline is not constant within this time, the background measurement might be increased up to 60 min.
9. Interrupt the background measurements and add your peptides (usually 1 μL) at a desired concentration, slightly shake the plate horizontally on the table, and continue or restart the measurement. The peptides can be added by using an injector integrated in the machine or by pipetting directly into the wells. If the latter is chosen, it is crucial that the addition of peptide is done quickly in order to avoid delay between the samples for a better comparison.
10. Record the oxidative burst over time to obtain a kinetics curve with different peptide concentrations (*see* **Note 12**).
11. Plot the maximum values of emitted light against peptide concentrations to obtain a classical dose-response curve where the half-maximum effective concentration (EC_{50}) can be determined as explained in [5].

When the peptide has been tested for activity, it can be used for direct binding experiments using an acridinium ester-labeled peptide as described in the next section.

3.2 Acridinium Ester Label Binding Assay

Here, the procedure is described for the use of frozen and ground leaf tissue. However, the method should be easily adaptable for other materials, e.g., affinity beads with immunoadsorbed receptor proteins.

1. Make sure your material is expressing the receptor you are interested in investigating (*see* **Note 13**).
2. Shock-freeze leaf tissue or any material of choice (cell suspension culture, e.g., *Nicotina benthamiana* (*N. benthamiana*) leaves (see **Note 14**)) in liquid nitrogen.
3. Grind tissue in liquid nitrogen to a fine powder using a mortar and pestle.
4. Using a precooled spatula, place the right amount of tissue in a precooled 1.5 mL reaction tube (*see* **Note 15**).
5. Add 1 mL binding buffer, DTT, and proteinase inhibitor mix per 500 mg tissue (*see* **Note 16**).
6. Centrifuge suspension to create a pellet (13,000 × *g*, 1 min, 4 °C).
7. Resuspend the pellet in binding buffer (500 mg tissue/mL binding buffer).
8. Aliquot 80 μL in 1.5 mL Eppendorf tubes ($n \geq 3$ per treatment) (*see* **Note 17**).
9. Add 20 μL of a solution containing the desired combination of acridinium-labeled peptide and unlabeled peptide (peptides should be diluted in binding buffer) to reach 100 μL total reaction volume (*see* **Note 18**).
10. Twenty minutes incubation, 4 °C. The incubation time might vary (*see* **Note 19**).
11. Sediment the sample by centrifugation, 13,000 × *g*, 1 min, 4 °C, pipette gently, and discard the supernatant.
12. To remove unbound, labeled ligand, wash the pellet 2–3 times by resuspension in 1 mL binding buffer and centrifugation under the same conditions as step 11 (*see* **Note 20**).
13. Resuspend the final pellet in 100 μL 5 mM citric acid, and place in a tube suitable for the single-tube luminometer used for these measurements.
14. Place tube in the luminometer.
15. Start the software and the program so it is ready to start collecting light immediately after induction with the NaOH/H_2O_2 solution (*see* **Note 21**).
16. Inject 100 μL of 100 mM NaOH containing 0.03% H_2O_2 (induction of luminescence) into the tube/cuvette containing your sample.

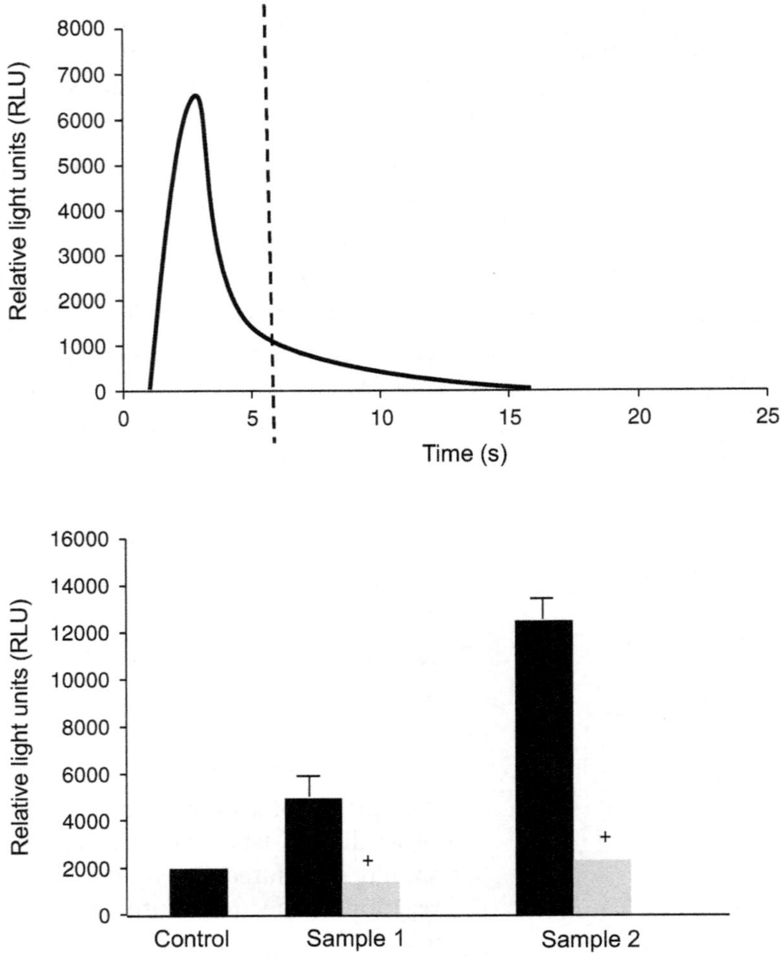

Fig. 2 *Upper panel*: By integrating the values as relative light units (RLU) to the left of the *dotted line* (flash peak), the data can be transferred to a diagram for comparison (*lower panel*). *Lower panel*: An example derived from integrating data points of emitted light from experiments of different samples. A control sample should always be included; this is a sample of plant material that is not expressing a receptor that is incubated with the same amount of acridinium-labeled peptide and washed the same way as all other samples. In this example sample 1 and 2 represent plant tissue expressing different levels of the receptor. Both samples are grinded plant material that are incubated with acridinium-labeled peptide alone (*black columns*) or with unlabeled (1000× excess) as competitor (*gray columns*, competitor indicted with a +). The differences between the *black* and *gray* columns indicate the specific binding of the labeled peptide to the corresponding receptor within each sample

17. Collect data for 10–30 s. To compare samples, light collected during the flash peaks should be integrated (*see* **Note 22**) and (Fig. 2).

4 Notes

1. Any plant of interest, e.g., *Arabidopsis* or *N. benthamiana*, expressing a receptor of interest. If the receptor is expressed by the introduction of a construct, either Western blot or imaging should be used to detect expression level of the protein.
2. Luminol is light sensitive; keep the stock solution in a tube for light-sensitive solutions. The luminol master mix should be prepared fresh right before use.
3. It is important that the scissors are sharp to create clean cuts, in order to avoid too much damage to the plant tissue.
4. It is important that the plates or tubes are white flat-bottomed, to avoid background signals in the luminometer.
5. In order to allow efficient washing to remove unbound ligand, it is important that the receptor is immobilized. Suggested plant materials could be intact cells from suspension cultures, ground plant tissue, or isolated receptors immobilized on immunoprecipitation beads. Some receptors have a low expression level. Therefore, overexpression, e.g., by transient expression in *N. benthamiana* leaves [6], may be required to obtain sufficient expression levels. Use the same material lacking a functional receptor as a negative control when possible.
6. The binding buffer might be adapted or modified by adjusting to different pH (e.g., 25 mM MES pH 5.5 or 25 mM TRIS pH 7.5) or different salt concentrations (reduce nonspecific ionic interactions, e.g., 10, 50, or 100 mM NaCl) in order to optimize binding while keeping nonspecific binding at a minimum.
7. Do not add DTT if receptors or ligands are sensitive to reducing agents.
8. Peptides can be ordered from a company of choice or synthesized using solid-phase technology with Fmoc-protected amino acids. Acridinium esters can be conjugated to the N-terminal amino groups of the peptides on resin by coupling with N-hydroxysuccinimide-activated acridinium esters (Cayman) before deprotection and purification of the peptides via HPLC.
9. The plants you use should be 4–6 weeks old, and they should not flower.
10. These pieces should easily fit into the wells of a 96-well plate. It is recommended that you cut about two to three times more leaf pieces than you need.
11. It is important that the leaf pieces are left in room temperature around 22 °C. If the temperatures are too high, the plant material will be stressed and the assay might not work.

12. Depending on the requirements of the experiment or the available machine, the luminometer settings might be set up accordingly. Example of luminometer settings to measure kinetics in a 96-well reader: 1 s measurement duration per well (can be shorter if necessary; this is dependent on the sensitivity of the machine and needs to be tested). Some software or programs calculate the cycle time depending on the measurement duration and number of samples. Repetitions of measurement cycles or total time should be set in order to monitor the response for a minimum of 20 min and if necessary up to 60 min. It is recommended that maximum half of the plate is occupied with samples. If more than half of the plate is measured in one experiment, the time difference between the measurement of the first and the last sample is too long. One round of measurement should be less than 60 s, in order for each sample to be monitored approximately once per minute. If you use a single-cell luminometer, you might use a continuous mode or use a comparable program such as described above. The emission maximum wavelength for luminol is 425 nm [7].
13. If using transient expression by agroinfiltration, or other transient expression methods, the time until your protein is expressed at high levels might vary, depending on the protein, promoter, etc. Make sure to use a tag in order for expression level to be detected.
14. If leaf tissue is used, it is important that the tissue is healthy and viable; no senescence symptoms should be visible.
15. We recommend using 500 mg frozen tissue for 10–12 individual measurements when ~ 40–50 mg tissue powder is used per sample.
16. Per mL of binding buffer, add 1 μL DTT (from a 1 M DTT stock solution) and 15 μL proteinase inhibitor mix (or less if you use less than 500 mg plant material).
17. Cutting off the tip of the pipette helps to avoid clogging of tissue at the tip and to get an even distribution of tissue in the samples.
18. Use a final concentration of 0.1–10 nM acridinium-labeled peptide in your 100 μL sample. Run replicates ($n \geq 3$) and samples containing an excess (>1000-fold) of unlabeled peptide to determine nonspecific binding (e.g., 1 nM acridinium-peptide plus 1 μM unlabeled peptide as competitor).
19. Reaching binding saturation may differ, and experiments to test this for individual peptides might be necessary. If no specific binding is detected, parameters such as incubation time or temperature could be increased.

20. It is crucial that the total time required for the washing steps are kept at a minimum. Depending on their dissociation rates, prolonged washing may lead to considerable loss of ligand binding.
21. Depending on your luminometer/software, you should monitor emitted light in a continuous mode. If not possible, monitor in shortest possible intervals (each 0.2 or 0.5 s) over time.
22. The oxidation and light emission proceeds rapidly and should lead to a short flash of light (flash peak) [4, 5, 8]. The light emission should be integrated over time to collect the light during the flash peak (Fig. 2). In most cases an integration time of 10–30 s should be sufficient to collect >90% of the light emitted. A sensitive photometer allows detection of a functional acridinium label down to the low femtomol or attomol range.

Acknowledgments

This work was supported from grants 13785/F20 and 230849/F20 from the Research Council of Norway.

References

1. Murphy E, Smith S, De Smet I (2012) Small signaling peptides in arabidopsis development: how cells communicate over a short distance. Plant Cell 24:3198–3217
2. Keppler LD, Baker CJ, Atkinson MM (1989) Active oxygen production during a bacteria-induced hypersensitive reaction in tobacco suspension cells. Phytopathology 79: 974–978
3. Felix G, Duran JD, Volko S et al (1999) Plants have a sensitive perception system for the most conserved domain of bacterial flagellin. Plant J 18:265–276
4. Butenko MA, Wildhagen M, Albert M et al (2014) Tools and strategies to match peptide-ligand receptor pairs. Plant Cell 26:1838–1847
5. Wildhagen M, Butenko MA, Aalen RB et al (2015) A chemiluminescence based receptor-ligand binding assay using peptide ligands with an acridinium ester label. Bio-Protocol 5(6):e1422
6. Li X (2011) Infiltration of Nicotiana benthamiana protocol for transient expression via Agrobacterium. Bio-Protocol Bio101: e95.
7. White H, Roswell DF, Burr JG (eds) (1985) Chemi- and bioluminescence. Marcel Dekker, New York, p 215
8. Joss UR, Towbin H (1994) Acridinium ester labelled cytokines: receptor binding studies with human interleukin-1 alpha, interleukin-1 beta and interferon-gamma. J Biolumin Chemilumin 9: 21–28

Chapter 19

Application of Chemical Genomics to Plant–Bacteria Communication: A High-Throughput System to Identify Novel Molecules Modulating the Induction of Bacterial Virulence Genes by Plant Signals

Elodie Vandelle, Maria Rita Puttilli, Andrea Chini, Giulia Devescovi, Vittorio Venturi, and Annalisa Polverari

Abstract

The life cycle of bacterial phytopathogens consists of a benign epiphytic phase, during which the bacteria grow in the soil or on the plant surface, and a virulent endophytic phase involving the penetration of host defenses and the colonization of plant tissues. Innovative strategies are urgently required to integrate copper treatments that control the epiphytic phase with complementary tools that control the virulent endophytic phase, thus reducing the quantity of chemicals applied to economically and ecologically acceptable levels. Such strategies include targeted treatments that weaken bacterial pathogens, particularly those inhibiting early infection steps rather than tackling established infections. This chapter describes a reporter gene-based chemical genomic high-throughput screen for the induction of bacterial virulence by plant molecules. Specifically, we describe a chemical genomic screening method to identify agonist and antagonist molecules for the induction of targeted bacterial virulence genes by plant extracts, focusing on the experimental controls required to avoid false positives and thus ensuring the results are reliable and reproducible.

Key words Phytopathogenic bacterial virulence, Plant extract, Reporter gene, Virulence gene promoter activity, Chemical library, High-throughput screening

1 Introduction

Unlike vertebrates, plants lack an adaptive immune system and instead rely on the innate immunity of individual cells and systemic signals originating from infection sites. Plants have an innate immune system with two branches to overcome disease and employ a battery of immunity mechanisms [1]. In the first defense layer, plants perceive the so-called microbe-associated or pathogen-associated molecular patterns (MAMPs or PAMPs) through pattern recognition receptors (PRRs), leading to the induction of PAMP/pattern-triggered immunity (PTI) [2], which reduces

Wolfgang Busch (ed.), *Plant Genomics: Methods and Protocols*, Methods in Molecular Biology, vol. 1610,
DOI 10.1007/978-1-4939-7003-2_19,

microbial colonization of the apoplast. However, successful bacterial pathogens counter this defense by delivering PTI-suppressing effectors, resulting in effector-triggered susceptibility (ETS) and ultimately disease [1, 3–5]. Plants have evolved additional receptors that specifically recognize such effectors, resulting in the initiation of a further layer of defense, named effector-triggered immunity (ETI) [1], which is characterized by the induction of the hypersensitive cell death at the infection site. The working model of the plant innate immune system, introduced by Jones and Dangl [1], is therefore named the "zigzag model" and reflects the adaptations of both plants and microbes during the course of their coevolution through a molecular dialog based on the recognition of particular molecules from both sides.

Plants release chemicals potentially involved in plant–microbe communication, such as sugars, amino acids, and phenolics, which are thought to play a role in chemical signaling and have the ability to induce bacterial virulence. However, the nature of the plant signals and corresponding bacterial signaling pathways is still largely unknown. A remarkable example of the induction of bacterial virulence by plant-derived molecules is the regulation of the phytopathogenic bacteria type III secretion system (T3SS), composed of long pili that allow T3SS-dependent protein delivery into host cells. The production and assembly of the T3SS pilus are dependent on activities carried out by the many T3SS regulon members (known collectively as *hrp* genes), which are induced at the start of infection. It is well known that the expression of *hrp/hrc* genes is first induced by environmental conditions: they are repressed in nutrient-rich media but induced in artificial *hrp*-inducing minimal medium that mimics the apoplast, particularly in terms of the pH, carbon source, and osmolarity [6–11]. Interestingly, environmental conditions are not the only regulators of *hrp* gene expression. The cocultivation of *Pseudomonas syringae* and plant cells induces a much higher level of *hrpA1* gene expression than cultivation in *hrp*-inducing minimal medium [11]. Furthermore, *hrpA1* protein accumulation and secretion are also strongly induced in the presence of plant-derived compounds in vitro [9]. Finally, the mutation of *prhA*, which encodes a membrane protein related to siderophore receptors, disrupts the plant contact-dependent induction of T3SS but not the induction by *hrp*-inducing minimal medium, suggesting that distinct signals are perceived from the host cells and apoplast [12]. However, the chemical nature of these plant signals and the bacterial signaling components they regulate to induce *hrp* gene expression are unknown.

Signaling pathways have classically been studied by the forward genetic screening of randomly mutagenized bacteria [13]. However, this conventional method is unsuitable when (1) multiple genes are responsible for a single trait, (2) a gene product is necessary for survival, or (3) a single gene is responsible for multiple phenotypes. In this context, chemical genetics is a suitable alternative [14]. Chemical genetics refers to the use of small molecules to induce phenotypes and modulate target protein functions, allowing the dissection of complex biological systems [15, 16]. In high-throughput mode, bacterial cells can be grown in 96-well plates and tested systematically with unique compounds from chemical libraries of molecules with known activity or with novel molecules or extracts. Recently, chemical genomics has increasingly been used to study bacterial functions, aiming to uncover new therapeutic candidates and to probe virulence mechanisms [14]. However, these works have focused almost exclusively on human pathogens and have not considered the response of pathogens to the host [14].

Given the complexity of plant–pathogen interaction, chemical genetics is a promising strategy to decrypt the role of plant signaling in the induction of virulence in phytopathogenic bacteria. We therefore developed a method based on a reporter gene encoding green fluorescent protein (GFP) controlled by a bacterial virulence gene promoter, to follow in vivo bacterial virulence gene expression induced by plant extracts in a 96-well plate system and to apply chemical genetics to decipher the signaling pathways involved in such responses. The method described herein has been implemented to study the effect of kiwifruit molecules on the virulence of *Pseudomonas syringae* pv. *actinidiae* (Psa), an emerging plant pathogen responsible for wood cankers and leaf spots in the genus *Actinidia* [17], which has been spreading rapidly [18]. Several studies have already been published concerning Psa whole genome sequencing, the definition of diagnostic tools, and insights into the diffusion and evolution of the different biovars [19–21], but the molecular mechanisms underlying kiwifruit/Psa interaction are still unknown.

Here we describe the different steps, from the production of transformed Psa strains carrying the GFP gene under the control of Psa *hrpA1* promoter to the validation of the method for studying *hrp* gene expression induced by signaling molecules in plant extracts, as well as the application of the chemical library with successive controls in vitro for hit confirmation. We focus particularly on the identification and elimination of false-positive hits at the early stage of the validation pipeline, such as molecules with fluorescence-modulating activity and bactericidal or bacteriostatic effects. The workflow described herein (Fig. 1) can reveal not only bacterial virulence requirements but also novel bioactive small molecules that selectively modulate virulence gene induction by plant signals, thus providing promising candidates for eco-sustainable plant protection. In this context, we recommend additional confirmation steps to validate the candidate molecules *in planta* (e.g., biocompatibility, phenotype confirmation), which are fundamental prior to field applications. Similarly, a deeper study of the structure–activity relationship would help to define the active moiety and to identify any modifications that increase efficacy.

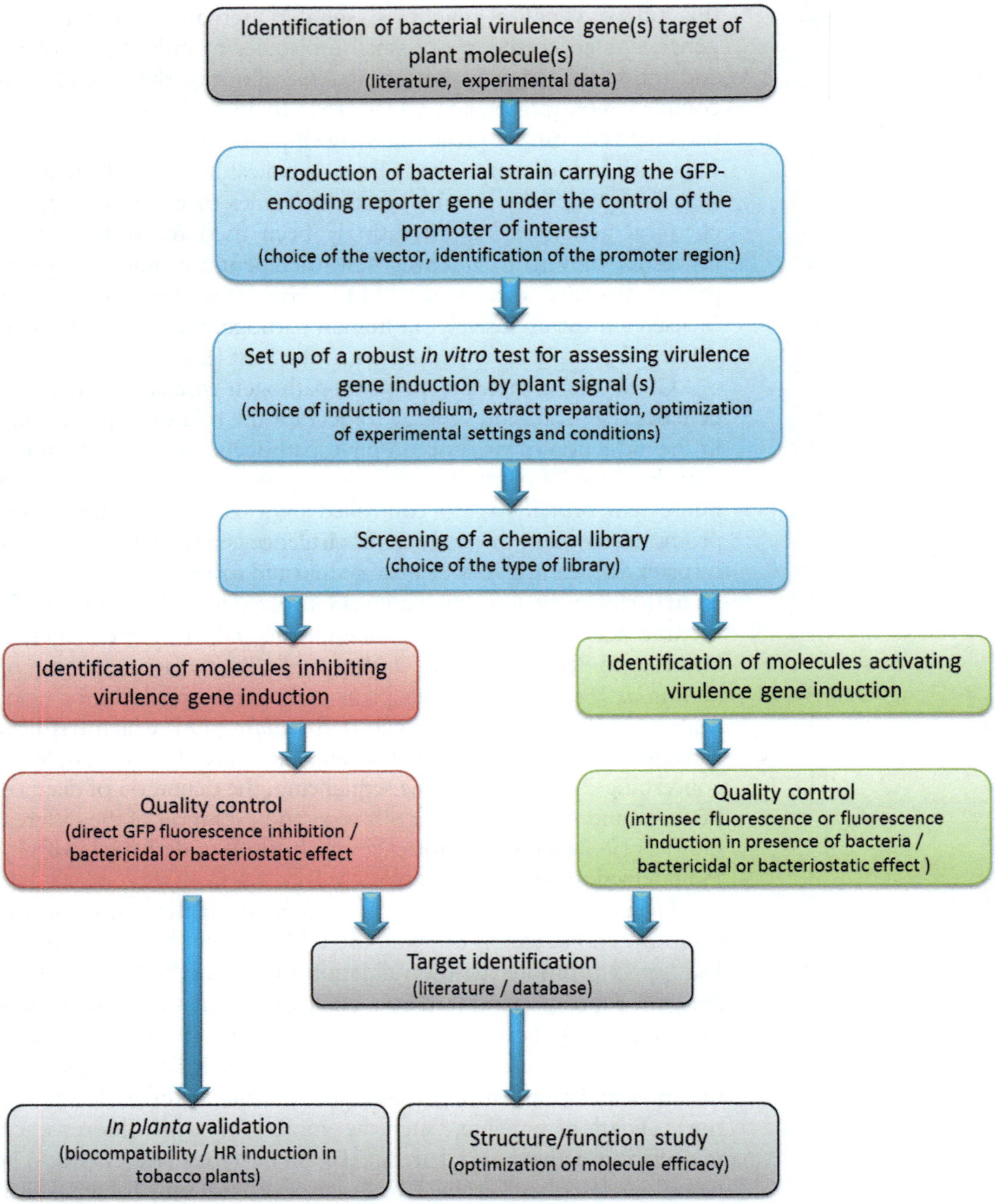

Fig. 1 Chemical genomics workflow for the analysis of compounds that interfere with plant signal-induced bacterial virulence. The different steps in *blue*, *pink*, and *green boxes* are described in the text

2 Materials

2.1 Bacterial Strains and Cultivation Equipment

1. *Pseudomonas syringae* pv. *actinidiae* (Psa) strain CRAFRU8.43.
2. Ampicillin (100 mg/mL), dissolved in water, filter sterilized. Stored in 1-mL aliquots at −20 °C.
3. Gentamicin (40 mg/mL), dissolved in water, filter sterilized. Stored in 1-mL aliquots at −20 °C.
4. Kanamycin (50 mg/mL), dissolved in water, filter sterilized. Stored in 1-mL aliquots at –20 °C.
5. Nitrofurantoin (50 mg/mL), dissolved in N,N-dimethylformamide, filter sterilized. Stored in 1-mL aliquots at −20 °C.
6. King's B broth: 20-g peptone, 1.5-g $MgSO_4.7H_2O$, 1.5-g K_2HPO_4, 10-mL glycerol, topped up to 1 L with deionized water.
7. King's B agar: King's B broth with 15-g/L agar.
8. Lysogeny broth (LB): 10-g tryptone, 5-g yeast extract, 10-g NaCl, topped up to 1 L with deionized water.
9. Lysogeny broth (LB) agar: LB with 15-g/L agar.
10. 50-mL Falcon tubes.
11. Bench centrifuge for Falcon tubes.
12. Incubators set at 28 °C or 37 °C with shaker.
13. Spectrophotometer set at a fixed wavelength of 600 nm.

2.2 Plant and Cultivation Equipment

1. Kiwifruit (*Actinidia deliciosa* cv. Hayward) plants.
2. Plastic pots (11-cm diameter).
3. Soil.
4. Greenhouse.

2.3 Reagents and Equipment for Bacterial Transformation

1. 0.5 M EDTA (ethylenediaminetetraacetic acid) stock solution: 93.05 g of EDTA powder in 400 mL of deionized water. Adjust the pH with sodium hydroxide (NaOH) (*see* **Note 1**). Top up the solution to a final volume of 500 mL.
2. 50× TAE (Tris-acetic acid-EDTA) stock solution: 242-g Tris base in approximately 750 mL of deionized water, 57.1-mL glacial acetic acid, 100 mL of 0.5 M EDTA (pH 8.0). Adjust solution to final volume of 1 L (*see* **Note 2**).
3. X-Galstocksolution:20-mg/mLX-GalinN,N-dimethylformamide.
4. Expand High Fidelity PCR System (Roche).
5. 2.5 mM each of dATP, dCTP, dGTP, and dTTP.
6. GoTaq® Flexi DNA Polymerase (Promega).
7. pGEM®-T Easy vector (Promega).

8. pBBR1MCS-5 (Gm^R) vector [22].
9. 1–1.5% agarose gels: 1–1.5-g agarose powder in 100 mL of TAE buffer 1× (*see* **Note 3**).
10. EUROGOLD Gel Extraction Kit (Euroclone).
11. EUROGOLD Plasmid Miniprep Kit (Euroclone).
12. Restriction enzymes (New England Biolabs).
13. T4 DNA ligase (Promega).
14. Shakers.
15. Thermocycler.
16. Microfuge.
17. Bench centrifuge.

2.4 Reagents and Equipment for GFP Fluorescence Measurement, Chemical Library Screening, and In Vitro Validation

1. *hrp*-inducing medium (HIM): 5.5-g KH_2PO_4, 1.5-g K_2HPO_4, 1-g $(NH_4)_2SO_4$, 0.344-g $MgCl_2$, 0.1-g NaCl, 2-g glycerol. Top up to 1 L using deionized water. Adjust pH to 5.5 with HCl. Autoclave before use.
2. 15-mL Falcon tubes.
3. Small chemical molecule library (96-well plate format).
4. Commercial recombinant green fluorescent protein.
5. Flat-bottomed black 96-well plates.
6. Fluorescence microplate reader.

3 Methods

The protocol can be adapted for other bacteria depending on specific requirements.

3.1 Preparation of Transformed Bacterial Strains Carrying the *GFP* Reporter Gene

3.1.1 Preparation of the Reporter Vector pBGFP Carrying the GFP Sequence

1. To construct a GFP-based reporter plasmid, the *gfp*mut3 gene, deprived of its promoter, is amplified from the plasmid pBBR2-GFP (*see* **Note 4**) [23] using the primers GFPEF (5′-GGA ATT CAA GAG GAG AAA TTA AGC ATG-3′) and GFPSR (5′-AGT CGA CTC AGC TAA TTA AGC TTA TT-3′). These primers provide EcoRI and SalI sites respectively at their 5′ termini, allowing successive correct insertions in the pBBR1MCS-5 plasmid.
2. Prepare two PCR mixes in sterile microfuge tubes as follows:

MIX1	Volume
Sterile ddH_2O	19 μL
dNTP mix 2.5 mM each	4 μL
Primer GFPEF 50 μM	0.5 μL

MIX1	Volume
Primer GFPSR 50 μM	0.5 μL
Template pBBR2-GFP plasmid	1 μL = 100 ng
Final volume	25 μL

MIX2	Volume
Sterile ddH_2O	19.25 μL
High fidelity buffer 10× with 15 mM $MgCl_2$	5 μL
High fidelity enzyme mix	0.75 μL
Final volume	25 μL

3. Combine MIX1 and MIX2 in a thin-walled PCR tube, and immediately place it in the thermocycler block.
4. Set the thermocycler using the following thermal profile:

	Temperature	Time	Number of cycles
Initial denaturation	93 °C	5 min	1
Denaturation	95 °C	30 s	8
Annealing	42 °C	30 s	
Elongation	72 °C	1 min	
Denaturation	95 °C	30 s	18
Annealing	50 °C	30 s	
Elongation	72 °C	1 min	
Final elongation	72 °C	7 min	1
Cooling	4 °C	Unlimited	

5. Once completed, load the total volume of the PCR product on a 1% agarose gel to ensure the correct size of the amplicon.
6. Cut the band corresponding to the amplified GFPmut3 product from the gel, and purify it using the EUROGOLD Gel Extraction Kit. The DNA is eluted in 30-μL sterile ddH_2O.
7. The purified fragment should be ligated into the pGEM-T Easy vector according to the reaction below:

Reagent	Volume
Sterile ddH_2O	1.5 μL
T4 DNA ligase buffer 10×	1.5 μL
pGEM-T Easy vector	1 μL
Purified fragment	10 μL
T4 DNA ligase	1 μL
Final volume	15 μL

8. Incubate the ligation reaction at room temperature for 3 h.

3.1.2 Construction of the pBGFP Plasmid

1. Transform 200 μL of competent *E. coli* DH5α cells with half of the volume of ligation reaction in a sterile Eppendorf tube.
2. Incubate the cells on ice for 30 min.
3. Heat shock at 42 °C for 90 s.
4. Incubate on ice for another 5 min and then dilute with 1 mL of LB medium.
5. Incubate for 1 h of growth at 37 °C.
6. Transfer one tenth of the culture onto the selective LB medium plate containing 100-μg/mL ampicillin and 20-μg/mL X-Gal to differentiate between white (positive) and blue (negative) colonies.
7. Incubate plates overnight at 37 °C.
8. Choose four white colonies, and transfer them into 4 mL of liquid LB medium containing 100-μg/mL ampicillin.
9. Incubate overnight at 37 °C.
10. Purify the plasmids from each culture using the EUROGOLD Plasmid Miniprep Kit.
11. Send the plasmids for sequencing to verify the insert is correct.
12. Excise the *gfp*mut3 gene from the pGEM-T Easy vector using the restriction enzymes EcoRI and SalI, and transfer the isolated fragment into the corresponding sites in the pBBR1MCS-5 plasmid (*see* **Note 5**), resulting in the final reporter vector pBGFP (*see* **Note 6**).

3.1.3 Insertion of the Promoter of Interest into the pBGFP Reporter Vector

1. Extract Psa genomic DNA to use as a template for promoter sequence amplification.
2. Amplify the sequence of the promoter of interest (here *hrpA1*) using PCR with specific primers (*see* **Note 7**) containing the proper restriction sites for the successive correct orientation of the promoter in relation to the *gfp* gene. In the example discussed here, the *hrpA1* promoter was amplified using forward (5′-AGG ATC CTT TTT TGC AAA GAC GCT GG-3′) and reverse (5′-

GGA ATT CTC CTG CAA ATG CGA CCA T-3′) primers carrying BamHI and EcoRI restriction sites, respectively.

3. Load the PCR product onto a 1–1.5% agarose gel to verify the correct size of the amplicon.
4. Purify the PCR product and insert it into the pGEM-T Easy vector (*see* **steps 6–8**, Subheading 3.1.1).
5. Repeat section **steps 1–12** (Subheading 3.1.2) to ligate the promoter sequence between the corresponding restriction sites (here BamHI and EcoRI) in the pBGFP plasmid.

3.1.4 Mobilization

Here the reporter construct (here pBGFP-*hrpA1*) is brought into Psa strain CRAFRU8.43 by triparental mating conjugation (*see* **Note 8**). The "donor" is the *E. coli* strain carrying the plasmid to be transferred; the "helper" is the *E. coli* DH5α (pRK2013) strain, which contains the helper plasmid pRK2013 (*see* **Note 9**); and the "recipient" is the Psa strain CRAFRU8.43.

1. Inoculate the donor strain onto an LB agar plate containing 10-μg/mL gentamicin, the helper *E. coli* DH5a (pRK2013) onto an LB agar plate containing 50-μg/mL kanamycin, and the recipient Psa strain CRAFRU8.43 on a KB agar plate (*see* **Note 10**).
2. Incubate the plates for 1 day.
3. Take a scraping of each bacterial strain with a wire loop and transfer to a fresh KB agar plate, mixing the strains together. As a negative control, mix together the donor and the recipient without the helper.
4. Incubate at 28 °C for 6 h to overnight.
5. At the end of the incubation period, streak the bacterial conjugation mix on KB agar plates supplemented with 50-μg/mL gentamicin and 150-μg/mL nitrofurantoin (*see* **Note 11**).
6. Incubate at 28 °C. Positive colonies will be visible after 48 h.

3.2 Cultivation of Kiwifruit Plants and Extract Preparation

Most standard protocols for kiwifruit cultivation will be suitable, and the following is provided as an example. Commercial in vitro-cultivated kiwifruit plantlets should be used as the starting material.

1. Transfer the multiplied shoots onto a root initiation medium comprising half-strength MS medium containing 0.02-mg/L indole-3-butyric acid (IBA).
2. Transfer rooted plantlets into 5-cm pots filled with soil, and move to the greenhouse for acclimatization. Water the plants regularly by irrigating the base of the pot.

3. After the acclimation period, transfer the plants into 11-cm pots filled with soil, and water the pots regularly by irrigating the base of the plant.
4. Cut the leaves and petioles from the plants at the four to five leaf stage.
5. Place the cut plant material in a kitchen juice extractor, and squeeze until no more water is released. Collect the whole extract in a 50-mL Falcon tube, and keep the extract on ice for all subsequent steps.
6. Centrifuge the crude extract (5000 × *g*, 4 °C, 10 min) and discard the pellet.
7. Repeat **step 6** until no pellet is visible at the bottom of the tube.
8. Sterilize the clarified extract by passing through a 20-μm filter.
9. Store aliquots of the sterile clarified kiwifruit extract at −20 °C.

3.3 Preparation of the Bacterial Suspension

1. Inoculate a single colony of Psa carrying pBGFP-*hrpA1* or pBGFP (negative control) from a fresh KB plate into 20 mL of KB liquid medium supplemented with 50-μg/mL gentamicin (*see* **Note 12**).
2. Incubate overnight (at least 16–20 h) at 28 °C with agitation at 200 rpm (*see* **Note 13**).
3. Pellet the bacteria by centrifugation (4500 × *g*, room temperature, 15 min).
4. Decant the supernatant.
5. Resuspend the pelleted cells in an equal volume of fresh HIM or HIM supplemented with plant extract, here from kiwifruit leaves (*see* **Note 14**).
6. Measure the OD at 600 nm (OD_{600}) using a spectrophotometer (*see* **Note 15**), and dilute further with autoclaved fresh minimal medium if necessary to reach a final OD_{600} of 1.0 (*see* **Note 16**), corresponding to 10^9 colony forming units (cfu)/mL.

3.4 Induction of hrpA1 Expression in the Presence of Plant Extract: Determination of Optimal Screening Conditions

1. Aliquot 200 μL of the bacterial cultures carrying *hrpA1*::GFP or the promoterless GFP reporter gene (PROM0), resuspended in HIM or HIM supplemented with 1% kiwifruit extract, in the wells of a black microtiter plate (*see* **Note 17**).
2. Place the plate in the microplate reader, and set the instrument parameters as follows: λexc = 485 nm, λem = 535 nm, large emission aperture, count time = 1 s, and medium shaking (*see* **Note 18**).
3. Measure the fluorescence emission every 15 min for 20–24 h.
4. Calculate the GFP fluorescence associated with the induction of the *hrpA1* promoter by subtracting the fluorescence values from the PROM0 Psa strain from those of the *hrpA1*::GFP

strain (*see* **Note 19**) for each time point. Figure 2 shows an example of such measurements.

5. Define the optimum time point for chemical screening according to the time course of fluorescence emission (*see* **Notes 20** and **21**).

3.5 Screening a Chemical Library in 96-Well Plates

In the example discussed here, *hrpA1* promoter activity is analyzed in minimal medium supplemented with kiwifruit extract (to identify natural molecules that inhibit the pathway induced by plant signals) and in minimal medium alone (to identify natural molecules that activate *hrpA1* expression, mimicking plant signal induction). In both cases, this allows the investigation of bacterial signaling pathways modulated by plant signals.

1. Aliquot 200 μL of the bacterial culture into each well of a microtiter plate. The first and last columns of the plate should be used for positive and negative controls, respectively, i.e., conditions that always induce or never induce the promoter of interest, as comparison for the library screening results to ensure technical reliability (*see* **Note 22**) [24, 25].

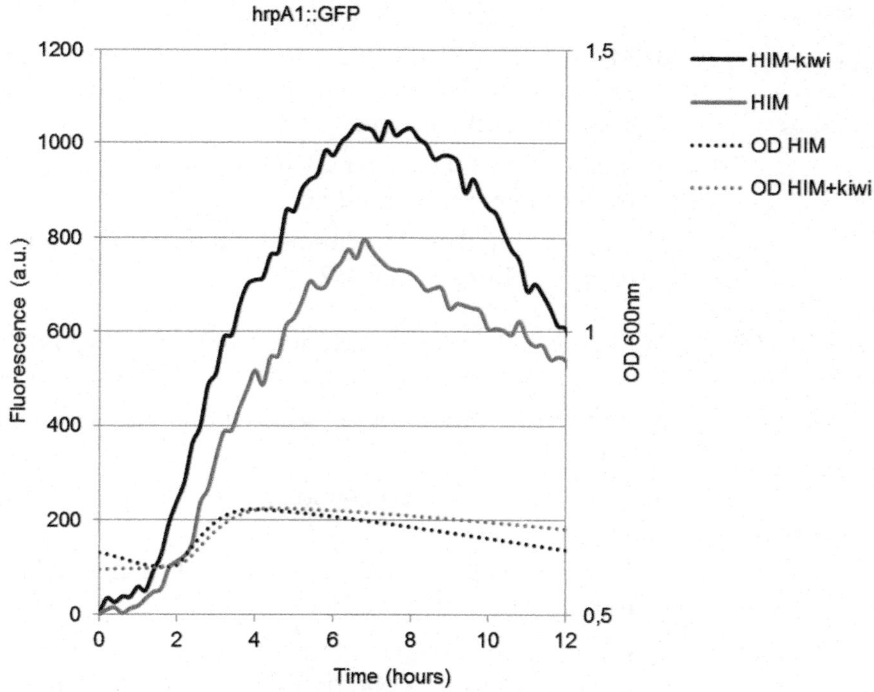

Fig. 2 Exemplary time course of fluorescence emission by a *Pseudomonas syringae* pv. *actinidiae* strain carrying *hrpA1*::GFP. The fluorescence signal was recorded with bacteria incubated in *hrp*-inducing medium (HIM) supplemented (*filled black line*) or not (*filled gray line*) with kiwifruit extract (HIM+kiwi). The bacterial OD_{600} is shown on the *right* axis

2. Expose the bacteria to the chemical library using a multichannel pipette, with each chemical at a final concentration of 10 μM (*see* **Notes 23** and **24**).
3. Incubate the microplate at room temperature (~24 °C) for 4 h with gentle shaking (*see* **Notes 25** and **26**).
4. Measure fluorescence after 0, 2, and 4 h, and calculate the promoter-dependent fluorescence as described above (*see* Subheading 3.4).
5. Perform three biological replicates with the entire library (*see* **Note 27**), and calculate the average of the triplicates for each molecule for successive hit selection.
6. Select for further characterization active molecules that cause the GFP fluorescence to significantly decrease or increase relative to the positive control (here *hrpA1*::GFP fluorescence in HIM supplemented with kiwifruit extract); these are considered as positive hits (*see* **Notes 28** and **29**).

3.6 In Vitro Validation of Positive Hits: Controls and Dose–Response Testing

3.6.1 Assessment of Bactericidal or Bacteriostatic Effects

1. Prepare a serial dilution of the selected positive hits in a 96-well microplate.
2. Aliquot 200 μL of the *hrpA1*::GFP Psa culture into the wells as described above (*see* Subheading 3.5).
3. Mix the bacteria with the serially diluted candidate molecule, and incubate the microplate at room temperature (~24 °C) for 4 h with gentle shaking.
4. Measure the OD_{600} after 24 h using a microplate reader spectrophotometer (*see* **Note 30**).
5. Discard candidates that inhibit bacterial growth or kill the bacteria (*see* **Notes 31** and **32**).

3.6.2 Assessment of GFP Fluorescence Quenching

1. Aliquot 250 ng of commercial recombinant GFP into the wells of a microtiter plate with three technical replicates per condition.
2. Add serial dilutions of the selected candidates to the recombinant protein.
3. Incubate at room temperature (~24 °C) for 4 h with gentle shaking.
4. Measure GFP fluorescence as described in Subheading 3.4 using the microplate reader.
5. Calculate the average of the fluorescence emission using the triplicates for each condition (non-treated protein and protein treated with selected molecules) as well as the corresponding standard deviations.
6. If the fluorescence values of the treated protein are significantly lower than the fluorescence values obtained with the non-treated protein, discard the candidates that quench GFP fluorescence in this way (*see* **Note 33**).

3.6.3 Assessment of Nonspecific Fluorescence Emission

1. Aliquot 200 μL of the PROM0-GFP Psa culture into the wells of a microtiter plate as described above (*see* Subheading 3.5) with three technical replicates per condition.
2. Add serial dilutions of the selected candidate molecule to the bacterial cells.
3. Incubate at room temperature (~24 °C) for 4 h with gentle shaking.
4. Measure GFP fluorescence as described in Subheading 3.4 using the fluorescence microplate reader.
5. Calculate GFP fluorescence values, using the untreated PROM0-GFP bacterial cells as a reference.
6. Discard candidates that produce a fluorescent signal unrelated to GFP (*see* **Note 34**).

3.6.4 Dose–Response Assay to Define Optimal Concentration of Active Molecules

1. Aliquot 200 μL of the *hrpA1*::GFP and PROM0-GFP Psa cultures prepared in HIM or HIM supplemented with kiwifruit extract into the wells of a microtiter plate as described in Subheading 3.5, with three technical replicates per condition.
2. Add serial dilutions of the putative inhibitory candidate molecule (HIM supplemented with kiwifruit extract) or putative stimulatory candidate molecule (HIM).
3. Incubate the microplate at room temperature (~24 °C) with gentle shaking.
4. Measure the fluorescence measurements after 0, 2, and 4 h, and calculate the fluorescence value as described in section (*see* Subheading 3.4). The optimal concentration will correspond to the concentration giving the maximum of gene expression inhibition/activation without any bactericidal/bacteriostatic effect.

3.6.5 Analysis of hrpA1::GFP Fluorescence in Liquid Culture in Flasks

1. Aliquot 20 mL of the *hrp*::GFP Psa culture prepared in HIM or HIM supplemented with kiwifruit extract into 100-mL flasks (*see* **Note 35**).
2. Mix the bacterial cells with the selected inhibitory or stimulatory candidates at the optimal concentrations defined above (*see* **Note 36**).
3. Incubate at room temperature (~24 °C) for 4 h shaking at 200 rpm.
4. Aliquot 200 μL of bacterial cells into the wells of a 96-well microplate, and measure the fluorescence using the microplate reader (Subheading 3.5).
5. Calculate the *hrpA1*::GFP fluorescence by subtracting the fluorescence values obtained by treating PROM0-GFP Psa cultures with the same candidate molecule to get the GFP-related fluorescence emission. If the activity of the

chemical molecule (inhibitory or stimulatory) is confirmed in this second assay, it can be considered as a robust candidate for further characterization.

4 Notes

1. EDTA will not dissolve completely until the pH is adjusted to ~8.0.
2. This stock solution can be stored at room temperature. The pH of this buffer is not adjusted and should be ~8.5.
3. The solution should be heated until the powder dissolves completely and then cooled before use.
4. Any other GFP gene sequence can be used if preferred.
5. The pBBR1MCS-5 (Gm^R) vector was chosen because it is a broad-host-range, medium-copy-number plasmid that contains many unique restriction sites.
6. In this way, any of the unique restriction sites upstream of the EcoRI site can be used to insert the promoter of interest, which will control the *gfp*mut3 gene.
7. Many different primer design software platforms are available on the Internet. We recommend PRIMER3 (http://primer3.ut.ee/).
8. Triparental mating conjugation is a simple and effective method for the introduction of foreign DNA into bacteria that are not amenable to standard methods [26].
9. The helper plasmid pRK2013 provides the *tra* and *mob* genes required to transfer the plasmid from the donor strain to the recipient strain.
10. It is important to start the conjugation process using fresh cultures.
11. The selection plate should kill both the helper and donor *E. coli* but not the recipient.
12. Antibiotics should be chosen according to the resistance phenotype of the transformed strains.
13. Growth conditions can be adapted to any other bacteria selected for analysis.
14. Medium selection depends on the aim of the experiment. In the example reported here, *hrp*-inducing medium (HIM) is used to induce the *hrpA1* promoter [10]. Moreover the kiwifruit leaf extract is used to mimic *in planta* conditions, not only restricted nutrient availability and low pH but also the presence of tissue components and molecules derived from the plant.

15. Dilute the bacterial cultures 1:10 to ensure that the OD_{600} is within the range of the instrument. Take the dilution factor into account when calculating the final OD_{600}.
16. A large initial OD_{600} (0.5–1) is recommended in case the bacteria grow slowly in the medium or produce limited fluorescence, to ensure that the signal can be detected.
17. Both white and black microtiter plates are suitable for the assay. A preliminary experiment is recommended to determine which plates minimize the fluorescence background most effectively while ensuring the fluorescence signal is detected.
18. The parameters should be adjusted according to the selected reporter gene. Ensure the correct excitation and emission wavelengths are used [27].
19. The subtraction of the PROM0 fluorescence values removes the background signal produced by the bacterial cells and plant extract.
20. At least three technical replicates are recommended when selecting the conditions in order to determine the robustness of the signal.
21. We recommend that both the fluorescence intensity and signal duration should be taken into account. Screening is therefore performed at both 2 h (earlier induction) and 4 h (stronger induction). This should guarantee the robustness of the effect observed with candidate molecules.
22. We recommend that several technical replicates are prepared for the controls to define precisely the range of fluorescence values anticipated when candidate molecules in the chemical library produce positive and negative effects.
23. The most suitable concentration of the candidate molecules should be determined empirically. The optimum concentration is assessed in more detail during the second phase of the experiment (*see* Subheading 3.6.3), when each candidate is validated.
24. Prepare a working plate at a concentration corresponding to the reliable range of the pipette. The DMSO concentration should not exceed 0.5–1% to avoid unwanted effects on the bacteria.
25. The plates should ideally be prepared and treated under sterile conditions, but sterility is not necessary for the brief treatment periods (4 h) described herein.
26. After reading the fluorescence at the appropriate time points, the cultures should be incubated until the 24 h postexposure, and the OD_{600} should then be measured to identify any potential bactericidal or bacteriostatic effects.

27. Biological replicates are recommended to ensure the robustness of the screening process. Although this is time-consuming, the best candidates are selected based on a robust statistical analysis thus avoiding the need for further validation.
28. Both increases and decreases in fluorescence should be followed up. A candidate that inhibits the bacterial signaling cascade leading to promoter activation would reduce the fluorescence, whereas a candidate that activates the cascade, possibly mimicking the induction conditions, would increase the fluorescence.
29. For positive hit selection, it is possible to define a percent inhibition cutoff using normalized percent inhibition (NPI) values or to define a threshold according to the median+k MAD [28].
30. The time point selected for OD_{600} measurement should be defined based on the experimental conditions.
31. Bacteriostatic candidates could be suitable for other applications, such as the suppression of bacterial infections without killing, thus preventing the emergence of resistant strains.
32. Bactericidal candidates can be evaluated in more detail using vital stains, e.g., 2-(4-iodophenyl)-3-(4-nitrophenyl)-5-phenyl tetrazolium, fluorescein diacetate, or propidium iodide.
33. As an alternative to recombinant GFP, the wells could be inoculated with bacteria producing GFP constitutively, and these cells could be tested for fluorescence quenching.
34. Although autofluorescence can be detected by measuring the signal from the selected candidates directly in culture medium without bacterial cells, the PROM0-GFP strain can also identify nonspecific signals that are produced indirectly, e.g., if the candidate lacks intrinsic fluorescence but can induce the production of fluorescent compounds by the bacteria.
35. For the cultivation of Psa, the 1:5 ratio between the culture volume and flask capacity is necessary for the adequate oxygenation of the cells, but this ratio may need to be modified for other bacterial strains.
36. The concentration of the candidate molecules should be determined according to the results of the different validation assays. The optimum concentration would achieve the most potent inhibitory or stimulatory effect in the absence of autofluorescence and fluorescence quenching, with no effect on bacterial growth or viability.

Acknowledgments

This work was supported by a grant from the Italian Veneto region "Progetto di innovazione per la difesa della pianta del kiwi e per la valorizzazione dei suoi frutti" (DGR n. 2587–23/12/2014). Work in AC's laboratory was funded by the Spanish Ministry for Science and Innovation grant BIO2013–44407-R. We thank Prof. Scortichini (CREA-FRC, Caserta, Italy) for providing the strain of *Pseudomonas syringae* pv. *actinidiae* CRAFRU8.43.

References

1. Jones JDG, Dangl JL (2006) The plant immune system. Nature 444:323–329
2. Dodds PN, Rathjen JP (2010) Plant immunity: towards an integrated view of plant-pathogen interactions. Nat Rev Genet 11:539–548
3. Block A, Li G, Fu ZQ, Alfano JR (2008) Phytopathogen type III effector weaponry and their plant targets. Curr Opin Plant Biol 11:396–403
4. Espinosa A, Alfano JR (2004) Disabling surveillance: bacterial type III secretion system effectors that suppress innate immunity. Cell Microbiol 6:1027–1040
5. Chisholm ST, Coaker G, Day B, Staskawicz BJ (2006) Host-microbe interactions: shaping the evolution of the plant immune response. Cell 124:803–814
6. Rahme LG, Mindrinos MN, Panopoulos NJ (1992) Plant and environmental sensory signals control the expression of hrp genes in Pseudomonas syringae pv. phaseolicola. J Bacteriol 174:3499–3507
7. Xiao Y, Lu Y, Heu S, Hutcheson SW (1992) Organization and environmental regulation of the Pseudomonas syringae pv. syringae 61 hrp cluster. J Bacteriol 174:1734–1741
8. Aldon D, Brito B, Boucher C, Genin S (2000) A bacterial sensor of plant cell contact controls the transcriptional induction of Ralstonia solanacearum pathogenicity genes. EMBO J 19:2304–2314
9. Haapalainen M, van Gestel K, Pirhonen M, Taira S (2009) Soluble plant cell signals induce the expression of the Type III secretion system of pseudomonas syringae and upregulate the production of pilus protein HrpA. Mol Plant Microbe Interact 22:282–290
10. Bonas U (1994) hrp genes of phytopathogenic bacteria. In: Dangl JL (ed) Bacterial pathogenesis of plants and animals: molecular and cellular mechanisms. Springer, Berlin, Heidelberg, pp 79–96
11. Lee J-S, Cha J-Y, Baik H-S (2011) Plant cell contact-dependent virulence regulation of hrp genes in pseudomonas syringae pv. tabaci 11,528. J Life Sci 21:227–234
12. Marenda M, Brito B, Callard D, Genin S, Barberis P et al (1998) PrhA controls a novel regulatory pathway required for the specific induction of Ralstonia solanacearum hrp genes in the presence of plant cells. Mol Microbiol 27:437–453
13. Hayes F (2003) Transposon-based strategies for microbial functional genomics and proteomics. Annu Rev Genet 37:3–29
14. Anthouard R, DiRita VJ (2015) Chemical biology applied to the study of bacterial pathogens. Infect Immun 83:456–469
15. Hicks GR, Raikhel NV (2014) Plant chemical biology: are we meeting the promise? Front Plant Sci 5:455
16. Fonseca S, Rosado A, Vaughan-Hirsch J, Bishopp A, Chini A (2014) Molecular locks and keys: the role of small molecules in phytohormone research. Front Plant Sci 5:709
17. Serizawa S, Ichikawa T, Takikawa Y, Tsuyumu S, Goto M (1989) Occurrence of bacterial canker of kiwifruit in Japan: description of symptoms, isolation of the pathogen and screening of bactericides. Jpn J Phytopath 55:427–436
18. Scortichini M, Marcelletti S, Ferrante P, Petriccione M, Firrao G (2012) Pseudomonas syringae pv. actinidiae: a re-emerging, multifaceted, pandemic pathogen. Mol Plant Pathol 13:631–640
19. Marcelletti S, Ferrante P, Petriccione M, Firrao G, Scortichini M (2011) Pseudomonas syringae pv. actinidiae draft genomes comparison reveal strain-specific features involved in adap-

tation and virulence to actinidia species. PLoS One 6:e27297

20. Vanneste JL, Poliakoff F, Audusseau C, Cornish DA, Paillard S et al (2011) First report of pseudomonas syringae pv. actinidiae, the causal agent of bacterial canker of kiwifruit in France. Plant Dis 95:1311–1311
21. McCann HC, Rikkerink EHA, Bertels F, Fiers M, Lu A et al (2013) Genomic analysis of the Kiwifruit pathogen *Pseudomonas syringae* pv. *actinidiae* provides insight into the origins of an emergent plant disease. PLoS Pathog 9:e1003503
22. Kovach ME, Elzer PH, Steven Hill D, Robertson GT, Farris MA et al (1995) Four new derivatives of the broad-host-range cloning vector pBBR1MCS, carrying different antibiotic-resistance cassettes. Gene 166:175–176
23. Passos da Silva D, Castañeda-Ojeda MP, Moretti C, Buonaurio R, Ramos C et al (2014) Bacterial multispecies studies and microbiome analysis of a plant disease. Microbiology 160:556–566
24. Chini A (2014) Application of yeast-two hybrid assay to chemical genomic screens: a high-throughput system to identify novel molecules modulating plant hormone receptor complexes. In: Hicks RG, Robert S (eds) Plant chemical genomics: methods and protocols. Humana Press, Totowa, NJ, pp 35–43
25. Serrano M, Kombrink E, Meesters C (2015) Considerations for designing chemical screening strategies in plant biology. Front Plant Sci 6:131
26. Wise AA, Liu Z, Binns AN (2006) Three methods for the introduction of foreign DNA into agrobacterium. In: Wang K (ed) Agrobacterium protocols. Humana Press, Totowa, NJ, pp 43–54
27. Shaner NC, Steinbach PA, Tsien RY (2005) A guide to choosing fluorescent proteins. Nat Methods 2:905–909
28. Goktug AN, Chai SC, Chen T (2013) Data analysis approaches in high throughput screening. In: PHA E-S (ed) Drug discovery. InTech, Rijeka

Index

Wolfgang Busch (ed.), *Plant Genomics: Methods and Protocols*, Methods in Molecular Biology, vol. 1610, DOI 10.1007/978-1-4939-7003-2, © Springer Science+Business Media LLC 2017

V

W

Y

Printed by Printforce, the Netherlands